KT-480-471

DE MONTFORT
UNIVERSITY
Riseholme Campus
Library
Reference only
PC899

Fish Nutrition in Aquaculture

CHAPMAN & HALL AQUACULTURE SERIES

Available or forthcoming titles

1 Fish Nutrition in Aquaculture
Sena S. De Silva and Trevor A. Anderson

Fish Larval Nutrition
Edited by J. Verreth

Water Quality for Fish Culture
M.G. Poxton

Aquaculture Effluent Control
A. Midlen and T.A. Redding

Fish Pharmacology and Therapeutics
C.A. Brown

Fish Nutrition in Aquaculture

Sena S. De Silva
School of Aquatic Science and Natural Resources Management
Deakin University
Victoria
Australia

and

Trevor A. Anderson
Department of Zoology
James Cook University
Queensland
Australia

CHAPMAN & HALL
London · Glasgow · Weinheim · New York · Tokyo · Melbourne · Madras

Published by Chapman & Hall, 2–6 Boundary Row, London SE1 8HN

Chapman & Hall, 2–6 Boundary Row, London SE1 8HN, UK

Blackie Academic & Professional, Wester Cleddens Road, Bishopbriggs, Glasgow G64 2NZ, UK

Chapman & Hall GmbH, Pappelallee 3, 69469 Weinheim, Germany

Chapman & Hall USA, One Penn Plaza, 41st Floor, New York NY 10119, USA

Chapman & Hall Japan, ITP-Japan, Kyowa Building, 3F, 2-2-1 Hirakawacho, Chiyoda-ku, Tokyo 102, Japan

Chapman & Hall Australia, Thomas Nelson Australia, 102 Dodds Street, South Melbourne, Victoria 3205, Australia

Chapman & Hall India, R. Seshadri, 32 Second Main Road, CIT East, Madras 600 035, India

First edition 1995

Typeset in Palatino 10/12 by EXPO Holdings, Malaysia

Printed in Great Britain by St Edmundsbury Press, Bury St Edmunds, Suffolk

ISBN 0 412 55030 X

A catalogue record for this book is available from the British Library

Library of Congress Catalog Card Number: 94–72448

Printed on permanent acid-free text paper, manufactured in accordance with ANSI/NISO Z39.48-1992 and ANSI/NISO Z39.48-1984 (permanence of Paper).

Contents

Acknowledgements

The authors are indebted to a large number of people for their support, encouragement and assistance in the preparation of this book.

The views and ideas expressed in this book owe much to the stimulation provided by our colleagues, students and to the farmers with whom we have worked. The initial impetus for the production of the manuscript came from the development of a postgraduate aquaculture course at Deakin University, which was generously funded by the Victorian Education Foundation. We wish to acknowledge the efforts of Mr Ian Kennedy of the Foundation for his role in providing this funding.

We are particularly grateful to Mr Peter Appleford, who critically reviewed the manuscript, Dr John Carragher for his useful comments and support, Dr Brad Mitchell and the Head of the School of Aquatic Science and Natural Resources Management, Professor Bob Loch, for their encouragement, Mrs Kathleen Hope and Mrs Vicki Hicks, who patiently typed many drafts, and Ms Alison Bridle for drawing the diagrams. Our thanks are also due to Mr Stuart Hosking and Mr Bill Potter for their editorial help and encouragement.

We are grateful to the International Development Research Centre (Canada), and in particular to Dr Brian Davy, for initiating and funding the Asian Fish Nutrition Network which provided the opportunity for one of us (S.S.DeS.) to interact with a large number of fish culturists and scientists in the region and to understand associated nutritional problems of a wide variety of aquaculture systems. Our gratitude is also due to the enormous patience and support of our wives, Celine and Margaret, and our families both during the preparation of this manuscript and during those periods when fish nutrition took us for lengthy periods far away from them.

Preface

The global aquaculture industry was worth over US$26 billion in 1990. It is an industry which has grown at a significant rate over the last two decades. During this period it has begun to be transformed from an art to a science; however, even to this date this transition remains far from complete. However, in the midst of an era of increasing competition for primary resources, such as land and water, and equally increasing environmental concerns, aquaculture can no longer be considered as an industry which utilizes marginal land. Aquaculture was previously considered an environmentally sound practice because some of the traditional practices were based on optimum utilization of farm resources. This has changed, and aquaculture is now considered to be a potential polluter. Increasing environmental restrictions are being imposed on aquacultural activities all over the world.

As in other forms of animal husbandry, feeds and feeding are crucial elements in the culture of aquatic animals. However, in aquaculture, the problems associated and encountered in feeds and feeding differ from those found in other forms of husbandry. These arise because of the large number of aquatic species cultured, which differ drastically in anatomy, behaviour and physiology and because a multitude of culture practices are utilized to raise a single species. Other problems arise because of the nature of the aquatic medium in which the animals need to spend some or all of their life.

Feed cost is considered to be the highest recurrent cost in aquaculture, often ranging from 30% to 60%, depending on the intensity of the operation. Any reduction in feed costs either through diet development, improved husbandry or other direct or indirect means is therefore crucial to the development and well-being of the industry.

Generally, most texts discussing finfish nutrition have tended to concentrate on the nutrient requirements of cultured species, and in particular those species which are cultured intensively. In contrast, little consideration is given to feeds and feeding of semi-intensive finfish culture even though such practices account for nearly 80% of global finfish culture production.

In this book we have attempted to strike a balance and provide a more holistic view to finfish nutrition in aquaculture. We have therefore approached the subject with due consideration of basic energetics and metabolism of finfish, and ontogenetic and physiological features of importance to fish nutrition in aquaculture. We have played down the nutrient requirements of individual finfish species not because they are unimportant, but because a number of specialized texts and manuals are available on this aspect. Prominence has been given to simple feed formulations, utilization of non-conventional feedstuffs and feed processing. In this approach to the treatment of the subject we have also dealt with aspects related to feeds and feeding which are becoming increasingly important to the aquaculture industry, in particular feeds and feeding in relation to the environment and the aquafeed industry.

A concerted effort has been made to draw examples from existing practices, covering a wide range of species, from the simplest to the more complex type of aquaculture practices, ranging from tropical to temperate zones and marine, estuarine and freshwater species.

Finally, although the book is entitled *Fish Nutrition in Aquaculture*, the contents are confined to finfish only.

Sena S. De Silva
Trevor A. Anderson

Series foreword

During the last decade aquaculture production has greatly increased. Over 26% of shrimp and 20% of salmon consumed are now farm-grown, both far higher than the 1% of ten years ago. Indeed, this dynamic worldwide industry is now growing at nearly 10% annually.

Although extensive culture in Asia is still extremely important in terms of overall production, it is the culture of expensive commodities such as salmon and shrimp in intensive and semi-intensive systems that has aroused most recent interest. Cultures of this sort have focused attention on the need for new diets, new equipment, new management practices and on environmental impacts. A tremendous impetus has been given to the industry by the flood of research papers in the last two decades, yet it remains the case that there are still relatively few universities offering well-designed courses in aquaculture and relatively few aquaculture textbooks, manuals and comprehensive but succinct and affordable treatises on which such courses can draw support.

The purpose of this new aquaculture book series is to fill this gap by encouraging authors to synthesize the bulging literature and to distil the information to provide common sense, practical advice and accessibility. The level of the series will be suitable for undergraduate students, farm management staff and administrators to understand. By being both topical and up-to-date the series will be of interest to researchers who require to keep abreast of developments in what is after all the ultimate multidisciplinary subject.

The aim of the Chapman & Hall Aquaculture Series is to present timely volumes reviewing important areas of aquaculture. Books in the series should also be of interest to students and researchers in biology, bioengineering, biotechnology, physiology and zoology. The applied nature of the topics to be covered will appeal to a wide audience of fish farmers, fisheries scientists and consultants.

It is entirely appropriate that the first book in the Chapman & Hall Aquaculture Series should be concerned with fish nutrition as feeds and feeding are among the most important aspects of day-to-day management of cultured fish. A thorough understanding of this area is vital for the health of the fish being cultured and the profitability of any commercial, intensive fish culture operation.

Sena De Silva and Trevor Anderson are to be congratulated for producing a comprehensive text which I found both informative and very readable due to their excellent style. *Fish Nutrition in Aquaculture* addresses both the basic concepts of the nutrition of fin fish and the practical aspects of nutrition pertaining to the aquaculture industry. The book discusses the present status of the aquafeed industry and deals with new ideas on feed-cost saving, particularly in semi-intensive situations. Potential problems relating to environmental pollution associated with waste feeds and with feeding are also discussed.

I thoroughly recommend this book to all those involved in the aquafeed industry, including fish feed technologists and manufacturers, aquaculture consultants, fish farmers, advanced aquaculture students and all those involved in teaching and research in this area.

Michael G. Poxton
Editor, *Chapman & Hall Aquaculture Series*

Abbreviations

ACP	acyl-carrier protein
ADG	average daily growth
ADP	adenosine diphosphate
AIA	acid-insoluble ash
APD	apparent protein digestibility
ATP	adenosine triphosphate
BHA	butylated hydroxyanisole
BHT	butylated hydroxytoluene
BOD	biological oxygen demand
BSN	brine shrimp nauplii
cd	chemical determination
CF	crude fibre
CMC	carboxymethylcellulose
CP	crude protein
CSC	critical standing crop
DE	digestible energy
DHA	docosahexaenoic acid
E	energy
EAA	essential amino acid
EE	ether extract
EFA	essential fatty acid
EPA	eicosapentaenoic acid
ERV	energy retention value
FAD	flavin–adenine dinucleotide
FCR	food conversion ratio
FE	faecal energy
FM	fish meal
FME	fish meal equivalent

GET	gastric emptying time
GFM	global fish meal
GH	growth hormone
GnRH	gonadotropin-releasing hormone
GR	growth rate
GTP	guanosine triphosphate
HE	heat production
HND	high nutrient density
HRA	hydrolysis-resistant ash
HROM	hydrolysis-resistant organic matter
IE	intake energy
IGF	insulin-like growth factor
LTS	low-temperature shock
ME	metabolizable energy
NAD	nicotinamide adenine dinucleotide
NADP	nicotinamide adenine dinucleotide phosphate
NAG	N-acetyl-D-glucosamine
NFE	nitrogen-free extract
NPU	net protein utilization
PEP	phosphenolpyruvate
PEPCK	PEP carboxykinase
PER	protein efficiency ratio
PRV	protein retention value
PUFA	polyunsaturated fatty acid
RD	reference diet
RE	retained energy
RGL	relative gut length
s.d.	standard deviation
s.e.	standard error
SDA	specific dynamic action
SE	surface energy
SGR	specific growth rate
TC	temperature constant
TD	test diet
TFI	total feed intake

TRH	thyroid releasing hormone
TSH	thyroid-stimulating hormone
UDP	uridine diphosphate
UE	urinary energy
VLDL	very low-density lipoprotein
ZE	gill excretion energy

CHAPTER 1

Aquaculture

1.1 INTRODUCTION

Aquaculture, or the farming of aquatic organisms, is known to have been practised in various forms and to varying degrees of intensity for over 2000 years. It is thought to have originated in mainland China, and most certainly the first documentation on aquaculture was in Chinese. Farming in the aquatic medium, like most other traditional forms of agriculture and animal husbandry, was practised much as a craft, with knowledge of the methods being transferred from generation to generation, and improvements being achieved through trial and error procedures. The science of aquaculture, on the other hand, is relatively new, with perhaps the first breakthrough being the development of induced reproduction by a hypophysation technique in Brazil in 1935 (Van Ihering, 1937). Despite a long history aquaculture has only had an impact as a major source of animal protein within the last three decades, when it was realized that the landings from many wild fisheries were in excess or close to sustainable levels. As a consequence of this realization a concerted effort was initiated to meet the shortfall in production through aquaculture.

Aquaculture is defined variously. The most recent and internationally accepted definition has been arrived at by the Food and Agriculture Organization (FAO) of the United Nations Organization (UNO) (FAO, 1990):

> ... the farming of aquatic organisms including fish, molluscs, crustaceans and aquatic plants. Farming implies some form of intervention in the rearing process to enhance production, such as regular stocking, feeding, protection from predators etc. Farming also implies individual or corporate ownership of the stock being cultivated.

Even though, for all intents and purposes, most culture-based fisheries fall strictly within the realm of aquaculture, in general those practices in which target species are grown in specialized containers such as ponds, cages, pens or raceways, or in bags, on ropes and in various forms of nets, as in the case of molluscs, are considered as true forms of aquaculture.

It is not intended in this chapter to go into details of the multitude of culture practices that is witnessed globally, nor will it be a general treatise on aquaculture, for it is impossible to do that in a single chapter; in any case a number of authors, including Hickling (1962), Iversen (1968), Bardach *et al.* (1972) and Pillay (1990), have done this admirably. Instead, the primary objective of this chapter is to give the reader a brief overview of the present aquaculture industry and to focus upon the relationship and importance of feeds and feeding in aquaculture. This approach will hopefully enable the reader to appreciate the diverse topics that are dealt with in later chapters.

1.2 FORMS OF AQUACULTURE

It is evident from the definition of aquaculture given earlier that aquaculture can take various forms. It ranges in scope from releasing artificially reared young into a lake or a reservoir where they have to fend for themselves until they are captured, to capturing young and growing them to market size in specialized containers, to caring and nurturing from the egg, through the various phases of the life-cycle, until harvesting (Figure 1.1). All forms of fish culture will necessarily involve either the provision

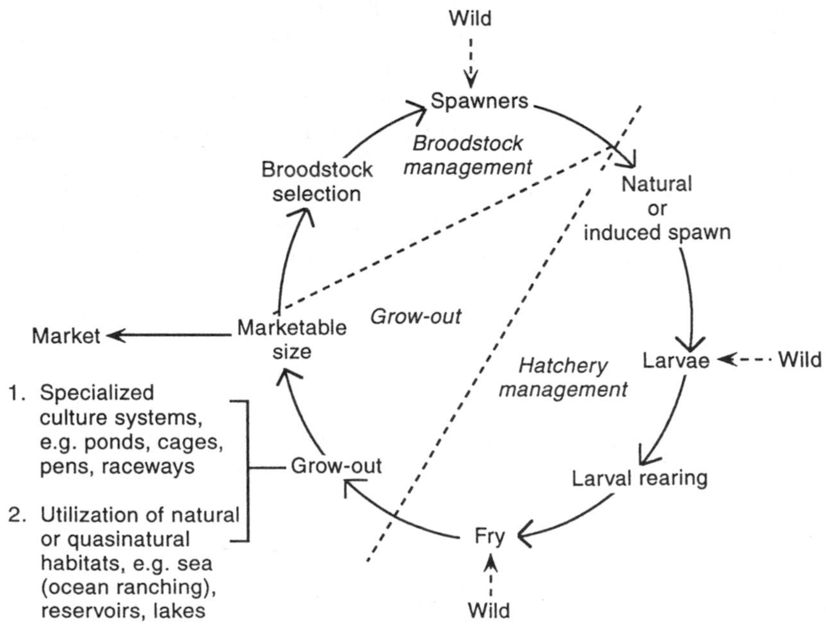

Figure 1.1 The different stages in aquaculture operations. A single aquaculture enterprise may encompass all or any of these stages.

of an artificial feed or the stimulation of natural food production, or both, in order to make a luxuriant food supply available for the target species.

Aquaculture can be broadly categorized as mariculture (seawater), freshwater culture or brackish water (estuarine) culture depending on the salinity of the culture medium. The culture of some species may require access to more than one of these media, depending on their life-cycle. This classification, however, is very broad and does not give any further insight into the culture practices. Another method is to define the type of culture according to the container used to grow the species, such as pond, cage, raceway or tank culture, etc. This classification also does not reveal details about the culture practices.

All types of finfish culture, irrespective of the nature of the medium or the container, can be classified according to the degree of intensity of culture. They can belong to extensive, semi-intensive or intensive culture practices (Figure 1.2). One key to this classification method is the type of nutrition provided for the cultured species. In intensive culture, all the nutrition of the target species is provided by the aquaculturist, in the form of a nutritionally balanced, easily utilizable, wholesome diet, usually but not always in the form of a pellet. In contrast, in extensive culture practice the target species is completely dependent on the natural food (endogenous) supply from the culture medium. Falling within this continuum are various degrees of semi-intensive practices. Although the distinctions between extensive and intensive fish culture practices are clear and somewhat obvious, the continuum between is rather ill defined. Apart from feeding, each of these culture practices is characterized by a number of other features, foremost among which is the degree of management required for the operation to function effectively (Figure 1.2).

It is important to note that the great bulk of fish culture practices in the world falls within the realm of semi-intensive culture. This means that the nutrition of the target species is a mix of endogenous and exogenous food supplies.

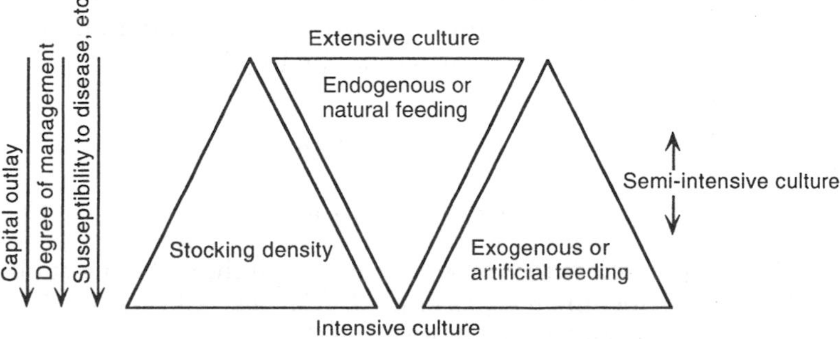

Figure 1.2 A conceptual representation of extensive, semi-intensive and intensive culture practices. (Modified after Tacon, 1988.)

Contrary to the common belief, aquaculture is not restricted to the culture of organisms for direct or indirect human consumption. Species may be cultured for a variety of other reasons, among which are:

- conservation, such as the production of native fish species to augment the declining wild stocks, e.g. the Australian species trout cod (*Maccullochella macquariensis*), Murray cod (*Maccullochella peelii*);
- the maintenance of wild capture fisheries, such as the Japanese abalone fishery and the salmonid fisheries of the Pacific and the Atlantic;
- recreation, such as the production of 'sport' fish for stocking into lakes, streams and reservoirs for recreational angling, e.g. rainbow trout (*Oncorhynchus mykiss*);
- the production of ornamental fish for the aquarium industry, e.g. goldfish (*Carassius auratus*);
- the production of industrial products such as pharmaceuticals;
- the production of specialist foods often sold as health products, e.g. the algal species *Spirulina*; and
- environmental manipulation, such as for removing aquatic vegetation (e.g. use of grass carp, *Ctenopharyngodon idella*, in the USA and New Zealand).

1.3 THE STATUS OF THE AQUACULTURE INDUSTRY

The status and growth of aquaculture, as of any other industry, may be evaluated using raw statistics. These enable an evaluation of an industry and/or a sector both nationally and globally. Reliable statistics or estimates of aquaculture production and economics are required by the public and private sectors to establish priorities, and to plan and invest wisely and profitably. Equally, such estimates are useful, and indeed essential, in making long-term policy decisions which influence the surrounding environment in a multitude of ways. The collation of global aquaculture-related statistics is carried out by the FAO.

1.3.1 Present status

(a) General aspects

The analysis presented here is based primarily on the statistics published by the FAO in its document *Aquaculture Production* 1985–89 (FAO, 1990), 1986–90 (FAO, 1991) and 1984–90 (FAO, 1992).

The total aquaculture production in 1990 was just over 15.0×10^6 tonnes, valued at about US$26.5 billion (Josupeit, 1992), thus confirming that worldwide aquaculture is a significant industry. The industry grew from 6 102 289 t in 1975 to 14 369 291 t in 1989. However, it needs to be pointed out that the production statistics for 1975 were far from complete when only those from a limited number of countries were available. The collation of aquaculture production has considerably improved over the years, particularly from 1985 onwards.

The present status of the aquaculture sector, in regard to both the species groups and monetary returns, is given in Table 1.1. The most salient points that are evident from Table 1.1 are as follows:

- The aquaculture industry is divided into four major species groups: finfish, crustaceans, molluscs and aquatic plants.
- The contribution of each of the major commodities to the industry as a whole has remained essentially unchanged over the last 5 years, except perhaps that crustacean production has more than doubled.
- In the longer term (1975–89), there have been significant changes in the quantity produced and the importance of aquatic plants, molluscs and crustaceans has increased.
- The value of the produce has increased at a faster rate than the quantity of commodities produced (value has shown a 15.8% increase per annum as opposed to a 5.9% for the quantity produced from 1985 to 1989).

The trends in aquaculture production were reviewed in some detail by Hempel (1993) when he concluded that it is not impossible for the aquaculture industry to increase production from 15×10^6 t to over 60×10^6 t in three decades.

(b) *Distribution of aquaculture activities in the world*

The next point to consider is the 'spread' of the industry in different parts of the globe.

Historically, Asia is considered to be the cradle of aquaculture, where it has been practised at a subsistence level for thousands of years. Generally, agricultural and agriculture-related commodities are initially grown/produced in response to a domestic demand. There is some parallel to this principle with respect to aquaculture, although admittedly the trends are not very clear and, in recent years, there have been major deviations from this which have resulted in random and ill-conceived introduction of species between countries and continents.

With aquaculture becoming a high-return enterprise in the last two decades or so, it has spread across those countries which have suitable land and water resources, even if fish is not a staple component of the traditional diet. One of the best examples is the recent rapid development of shrimp culture in Ecuador, which was encouraged by the government as a source of employment and foreign exchange.

Table 1.1 World aquaculture production by different species groups (in tonnes), the percentage contribution of each to the total production and the value of the total aquaculture produce in US$1000

Species group	*1975*[a]	*1985*	*1986*	*1987*	*1988*	*1989*	*1990*
Finfish	3 980 042	5 017 065	5 744 308	6 546 959	7 114 507	7 678 603	8 410 734
(%)	(65.0)	(45.3)	(45.5)	(50.4)	(49.2)	(53.4)	(54.9)
Freshwater	–	4 131 633	4 789 015	5 463 276	5 292 293	6 061 612	6 968 053
(%)		(37.3)	(35.6)	(42.1)	(40.9)	(42.2)	(45.5)
Diadromous	–	604 303	721 975	882 460	907 816	994 001	1 135 118
(%)		(5.5)	(5.9)	(6.8)	(6.3)	(6.9)	(7.4)
Marine	–	221 129	233 319	261 223	281 488	273 029	307 563
(%)		(2.0)	(1.9)	(2.0)	(1.9)	(1.9)	(2.0)
Crustaceans	15 663	276 506	389 090	589 067	604 438	611 869	715 586
(%)	(0.26)	(2.5)	(3.2)	(4.5)	(4.2)	(4.2)	(4.7)
Molluscs	1 051 341	2 229 953	2 441 298	2 688 413	3 076 967	3 032 788	2 965 040
(%)	(17.2)	(20.2)	(20.2)	(20.7)	(21.3)	(21.1)	(19.4)
Aquatic plants	1 054 793	3 484 188	3 507 051	3 155 657	3 632 281	2 997 471	3 187 691
(%)	(17.3)	(31.5)	(28.9)	(24.3)	(25.1)	(20.9)	(20.8)
Others	–	28 224	30 416	27 702	34 203	48 560	43 652
(%)		(0.3)	(0.2)	(0.2)	(0.2)	(0.3)	(0.3)
Total	6 102 289	11 065 936	12 105 163	12 987 798	14 466 486	14 369 291	15 322 703
Value	–	428 281	16 666 893	20 513 130	23 934 000	24 236 147	26 455 328

Sources: [a]Pillay (1979); others FAO (1990, 1991, 1992).

Aquaculture is spread unevenly in the world (Table 1.2). The epicentre of aquaculture activity is in Asia, and is likely to remain so in the foreseeable future. Asia contributes approximately 85% of the total production, followed by Europe (about 7–10%). Three continents (as defined by the FAO), Africa, South America and Oceania, each contributes less than 1% to the total production.

Approximately 150 countries in the world are known to have some sort of aquaculture activity. Needless to say, the magnitude of this activity varies widely, some countries producing less than 1 tonne per annum. There are also differences between countries in the range and balance of the organisms cultured. China is the exception. It is the leading producer of all aquaculture commodities. This is not a recent development; China has a long history and tradition of utilizing its water resources and aquatic species.

1.4 ORGANISMS CULTURED

The aquatic organisms cultured for food in the world can be categorized into five main groups (Table 1.1): finfish (freshwater, marine and diadromous species), crustaceans, molluscs, aquatic plants and others [this category includes tunicates (sea-squirts), frogs, turtles, etc.].

By far the most important cultured group in terms of total production is finfish, followed by molluscs and aquatic plants. Amongst finfish, freshwater finfish is the major category cultured in the world.

Table 1.2 Aquaculture production by continents[a] (weight in tonnes)

Continent	*1985*	*1986*	*1987*	*1988*	*1989*
Africa	58 749	59 952	62 433	69 950	93 960
(%)	(0.57)	(0.53)	(0.51)	(0.51)	(0.76)
America (North)	438 444	487 523	537 915	512 698	552 015
(%)	(4.3)	(4.3)	(4.4)	(3.8)	(4.4)
America (South)	59 388	58 716	110 179	133 679	151 381
(%)	(0.57)	(0.52)	(0.91)	(0.98)	(1.2)
Asia	8 596 129	9 614 642	10 209 531	11 630 547	11 040 680
(%)	(83.7)	(84.3)	(83.9)	(85.1)	(82.3)
Europe	801 270	815 359	860 352	911 734	1 174 062
(%)	(7.8)	(7.2)	(7.1)	(6.7)	(8.8)
Oceania	23 025	29 253	30 671	40 019	42 154
(%)	(0.22)	(0.26)	(0.25)	(0.29)	(0.35)
USSR (former)	293 292	323 501	348 824	364 783	354 165
(%)	(2.9)	(2.9)	(2.9)	(2.67)	(2.6)

[a]As defined by the FAO, Australia is included under Oceania.
Source: FAO, FIDI/C815 Revision 2 (1991).

The degree to which production of the major commodities occurs in the three media, i.e. freshwater, brackish water and seawater, reflects the habitats of the important species being cultured. For example, the bulk of finfish culture operations are in freshwater, while for crustaceans and aquatic plants they are in brackish water and marine environments respectively. The bulk of molluscan culture is also marine.

The most diverse group of cultured organisms, in terms of both the number of species and their feeding habits, is finfish. Among finfishes the non-carnivores constitute by far the major component of culture. Carnivores are cultured primarily in Japan, Europe and North America, and essentially consist of marine species and the diadromous and freshwater salmonids; all are high-value species. A few freshwater carnivorous fish are cultured, catfish, snakeheads and freshwater eels being the main ones. The non-carnivores are dominated by the cyprinids and the tilapiine fish cultured in Asia, Africa and South America. The non-carnivorous species have a wide range of food and feeding habits; they include plankton filter-feeders, aquatic macrophyte feeders and detritivores.

1.4.1 Species cultured

The number of species for which culture is developed to date is 195, which includes 92 species of finfish (22 species groups belonging to 12 orders), 23 species of crustaceans, 36 species of molluscs and 44 species of aquatic plants. Although many species are cultured, relatively few are important commercially. Of the finfish species currently cultured, based on the volume produced, the most important ones belong to the orders Cypriniformes, Salmoniformes, Siluriformes, Perciformes and Anguilliformes.

Summaries of the major species cultured in different environments are given in Tables 1.3, 1.4 and 1.5. The data presented indicate that, when all forms of finfish culture are considered, there are only 11 finfish species which yield more than 100 000 t per annum. Of these seven are truly freshwater species. Two species yield more than 1 million tonnes per annum, and another two species over 500 000 t. Moreover, both of the last two species feed low in the food chain and they are mostly cultured semi-intensively.

1.5 FINFISH CULTURE – FOOD AND FEEDING

Clearly finfish culture involves a number of species. Each of these species has its own nutritional requirements, particular feeding behaviour and food preferences. The species cultured range from strict carnivores to

Table 1.3 The important freshwater fishes cultured globally. For each species up to three leading producing countries are also given

Group/species	*Cultured in continents*	*Countries*	*Type of culture*	*Production 1990 (t)*
Cyprinids (carps/barbs)	All			**4 980 559**
Hypophthalamichthys molitrix (silver carp)	NAm, As, USSR, Eur	China, USSR, Hungary	P	1 515 262
Cyprinus carpio (common carp)	All	China, USSR, Indonesia	P, C	1 112 726
Aristichthys nobilis (bighead carp)	NAm, As, Eur	China, Hungary, Hong Kong	P	671 460
Ctenopharyngodon idella (grass carp)	NAm, As	China, Malaysia, USSR	P	1 047 050
Parabramis pekinensis (bream)	As	China	P, C	161 615
Cichlids	All			**390 825**
Oreochromis niloticus	Af, SAm, As, Oc	China, Thailand, Philippines	P, C	201 933
Oreochromis mossambicus[a]	Af, SAm, As	Indonesia, Malaysia, Thailand	P, C	37 386
Oreochromis aureus	Af, SAm, As	Cuba, Nicaragua, Panama	P	18 683
Catfishes	Af, NAm, SAm, As, Eur			**240 084**
Ictalurus punctatus	NAm	Mexico, USA	P	164 948
Pangasius pangasius	As	Thailand, Malaysia	P, C	14 000
Pangasius sutchi	As	Cambodia	P	971

Table 1.3 (Continued)

Group/species	*Cultured in continents*	*Countries*	*Type of culture*	*Production 1990 (t)*
Anabantids	As			**67 610**
Anabas testudineus (climbing perch)	As	India, Thailand	P	40 260
Trichogaster pectoralis (snakeskin gourami)	As	Indonesia, Malaysia, Thailand	P, F	15 701
Osphronemus goramy	As	Indonesia, Malaysia, Philippines	P, F	5656
Helostoma temminicki (kissing gourami)	As	Indonesia	C	5500

[a]Also cultured in brackish water.
Continents: Af, Africa; NAm, North America; SAm, South America; As, Asia; Eur, Europe; Oc, Oceania. Type of culture: P, ponds; C, cages; R, raceways; Pe, pens; F, paddy fields.
Source: FAO statistics.

Table 1.4 Important diadromous groups and species cultured globally. For each species up to three leading producing countries are given

Group/species	Cultured in continent	Countries	Predominant habit	Type of culture	Production (t)		
					1988	1989	1990
Salmonids	All	–	FW, M	–	422 751	501 590	593 428
Oncorhynchus mykiss (rainbow trout)	All	Japan, Denmark, France	FW, M[a]	P, C, R	241 608	249 976	271 478
Salmo salar (Atlantic salmon)	SAm, NAm, Eur, Oc	Norway, UK, Ireland	M	C	112 363	171 286	231 543
O. gorbuscha (pink salmon)	NAm	USA	M	C	26 486	57 765	–
O. kisutch (coho salmon)	NAm, SAm, As, Eur	Japan, Chile, Canada	M	C	25 791	30 309	39 402
Eels	As, Eur, Oc	–	–	–	99 061	89 960	104 068
Anguilla japonica	As	Japan, S. Korea	FW	P	91 737	83 597	95 817
A. anguilla	Eu	Italy, France, Portugal	FW	P	7293	6312	7961
Other species							
Lates calcarifer (sea perch)	As, Oc	Thailand, Indonesia, Malaysia, Australia	BW	C, P	3998	8323	7394
Chanos chanos (milkfish)	As, Oc	Philippines, Indonesia, Thailand	FW, BW	P, C, Pe	346 006	333 660	429 870

[a]Also cultured in brackish water.
P, ponds; C, cages; R, raceways; Pe, pens; FW, freshwater; M, Marine; BW, brackish water.

Table 1.5 Important marine fish species cultured globally; most of these species are cultured in marine cages, with a few exceptions, which are cultured in ponds

Group/species	*Cultured in*		*Type of culture*	*Production (t)*		
	Continent	*Country*		*1988*	*1989*	*1990*
Carangids	As	–	–	174 522	161 650	192 944
Seriola quinqueradiata (yellow tail)	As	Japan, Korea	C	167 186	154 733	161 568
Trachurus japonicus	As	Japan, Korea	C	6455	6000	6200
Sea breams, basses	As, Eu	–	–	54 147	58 255	70 610
Sparus auratus[a] (gilthead bream)	Eu, As	Italy, France, Tunisia	C, P	1326	2621	3096
Sparus major	As	Japan	C	45 220	46 000	50 921
Rhabdosargus sarba	As	Hong Kong	C	1529	950	830
Dicentrarchus labrax (sea bass)	Eu	France, Italy, Spain	C, P	955	1551	1631
Flatfishes	As	–	C	3280	3550	4823
Paralichthys olivaceus (bastard halibut)	As	Japan, Korea	C	3113	3249	4137

[a]Also cultured in brackish water.

those feeding very low in the food chain. When the global finfish aquaculture industry is considered (Table 1.6) the dominance of non-carnivorous finfish is obvious: they currently constitute about 85% of the global and nearly 90% of Asian finfish production.

Carnivorous fish are almost always cultured intensively and all the nutrition the fish needs is supplied through an artificial diet. Therefore, the process of developing feeds and feeding practices for carnivorous finfish is relatively straightforward; the nutritional requirements are evaluated and suitable diets and feeding regimes that give an optimal economic return are determined. Generally, it is considered that any endogenous food supply does not contribute to the nutrition of the species being cultured. As will be appreciated by now, this approach only applies to a small fraction of the global finfish aquaculture industry: the great bulk of production depends on non-carnivorous species cultured extensively and semi-intensively. Moreover, almost all of this production is confined to tropical and subtropical climates, where the endogenous food supply in the culture system can be maintained at significant levels throughout the year. This means that in non-carnivore culture artificial feeds have a secondary role to play and thus any exogenous feeds are referred to as supplemental feeds. The matter is further complicated because supplemental feeds can range from a single ingredient to a mix of ingredients, presented in various forms, to a properly formulated practical diet (New *et al.*, 1993). The role of supplementary feeding in pond fish culture of non-carnivores was aptly dealt with by Hepher (1988a), when the dearth of knowledge on this subject was highlighted.

Table 1.6 Trends in finfish aquaculture production 1985 to 1989 (quantities are expressed in millions of tonnes)

	1985	*1986*	*1987*	*1988*	*1989*
Total finfish production	5.047	5.743	6.550	7.150	7.327
Non-carnivorous fish					
World	4.471	4.918	5.600	6.097	6.166
Asia/Pacific	3.694	4.312	4.964	5.414	5.484
Asia (%)[a]	82.62	87.7	88.6	88.8	88.9
Asia (%)[b]	73.2	75.1	75.8	75.7	74.9
Carnivorous fish					
World	0.576	0.825	0.950	1.053	1.161
Asia	0.347	0.352	0.394	0.439	0.427
Asia (%)[a]	65	42.8	41.6	41.8	36.9
Asia (%)[b]	6.9	6.1	6.0	6.1	5.8

[a]Asian contribution to the particular group.
[b]Asian contribution of the commodity to total world finfish production.
Source: Based on data from FAO (1990, 1991).

The majority of finfish nutrition research and documentation has tended to concentrate on the feeds and nutrition of carnivorous fish. This has obviously left tremendous gaps in our understanding of the rest of the industry and consequently limits its development in the long term.

In the ensuing chapters an attempt is made to address the nutrition of finfish species in a way which we consider to be a more equitable representation of the global aquaculture industry. However, before dealing with the industry-related aspects of finfish nutrition it is imperative to have a basic understanding of the general principles of nutrition in relation to energetics, metabolism and digestion, and aspects of growth. These aspects are presented in Chapters 2, 3 and 4 respectively. More practical aspects of fish nutrition and feeding are covered in subsequent chapters.

CHAPTER 2

Energetics

2.1 INTRODUCTION

Energetics is the study of energy requirements and flow of energy within a system, in this case a biological system.

Energy, defined as the capacity to do work, is required by all organisms to sustain life. Work done in a biological system drives the chemical reactions required to build new tissues, maintains salt and water balance, moves food through the digestive tract, respires, reproduces and moves the muscles to provide locomotion. Animals obtain the energy they require from their food or, in periods when they are deprived of food, from body stores.

This chapter discusses bioenergetics in relation to the nutritional requirements of fish, relating the requirement to satisfy energy needs to the requirement for other nutrients when developing artificial feeds for fish. Growth, which is the end-point of net energy increase, and its measurement is considered further in Chapter 12.

Energy is not a nutrient itself, but is present in the chemical bonds that hold the molecules in the nutrients together. There are many different types of bonds, each containing a different amount of energy. Accordingly, the amount of energy in the various nutrients that make up a feed is of great importance. In addition, the capacity of different species to utilize the energy contained in different nutrients varies considerably. For example, some species are able to use carbohydrates as a major energy source, whilst others use carbohydrates poorly and rely to a greater extent on protein for energy.

Therefore, it is apparent that in any discussion of nutrition the energetic requirements of the animal for maintenance and growth, and its capacity to utilize the energy contained in the food, must be considered.

2.2 DEFINITION OF ENERGETICS

Energy can only be measured as it is converted from one form to another. In determining the energy contained in a food or a nutrient, the energy must first be converted to heat. This is done because energy is quantitatively converted to heat, which is then relatively easily measured. Heat energy is usually expressed as one of two units. A calorie is the amount of heat energy needed to raise the temperature of one gramme of water by one degree Celsius (°C). The alternative and more commonly used unit is the joule (J). One joule is the energy required to accelerate a mass of 1 kg at 1 $m s^{-1}$ over a distance of 1 m. One calorie is equivalent to 4.184 J. Joules and calories are extremely small units. In practice, energy content of food is usually expressed as kilojoules (kJ $=10^3$ J) or megajoules (MJ $=10^6$ J).

A number of terms are commonly used in discussions of energetics, and it is necessary to define these. The abbreviations used in this text are those of Smith (1989) but reflect general conventions.

Gross energy (E) is the energy that is released as heat when a substance is completely oxidized to carbon dioxide, nitrous oxide or water. Determination of gross energy is usually undertaken using a bomb calorimeter. A known amount of a dried sample is placed in a crucible inside a sealed pressure vessel, pure oxygen is introduced into the pressure vessel and the sample is combusted. The energy contained within the sample is measured as heat generated from its oxidation.

Intake energy (IE) is the gross energy consumed by an animal in its food. The majority of intake energy is present in the form of carbohydrate, protein or lipid. Other food components make up relatively small amounts of the intake energy.

Faecal energy (FE) is the gross energy of the faeces. Faeces consist of undigested food and metabolic products, which may include sloughed gut epithelial cells, digestive enzymes and excretory products. Accordingly, FE is composed of the energy of undigested food (F_iE) and energy of compounds of metabolic origin (F_mE). In practice, F_iE and F_mE are not differentiated.

The digested energy within a food is called the apparently digested energy (DE) and is determined as the energy in food minus the energy in faeces, i.e.

$$DE = IE - FE \tag{2.1}$$

It is generally assumed that all DE is absorbed by the fish.

Urinary energy (UE) is the gross total energy in urinary products. It includes energy of compounds absorbed from the food but not utilized (U_iE) and the energy of products of metabolic processes such as ammonia (U_mE). Again, differentiation between U_iE and U_mE is generally not made.

Gill excretion energy (ZE) is the gross energy of the compounds excreted through the gills of aquatic animals. Its equivalent in mammalian energetics is the energy of the compounds excreted through the lungs. However, in mammals this energy is extremely low and generally not considered. In fish ZE can be quite high and may constitute a major component of the energy balance of an aquatic organism.

Surface energy (SE) is the energy lost from the surface of an organism. In aquatic organisms, this may take the form of mucus or scales sloughed from the skin.

Metabolizable energy (ME) is the energy in the food less the energy lost in faeces, urine and through excretion from the gills. It is the energy available for the conduct of the metabolic processes of an animal. It may be described by the equation:

$$ME = IE - (FE + UE + ZE) \tag{2.2}$$

Total heat production (HE) is the energy lost from the animal in the form of heat. The heat is produced as a result of metabolism and so HE is a measure of the metabolic rate of an animal. Heat energy can be determined either by measuring an animal's temperature change (calorimetry) or by determining the metabolic rate of the animal by measuring oxygen consumption. Heat energy can be divided into a number of constituents.

Basal metabolic rate or basal metabolism (H_eE) is the heat energy produced by the conduct of those activities that are necessary to maintain the life of the animal (i.e. cellular activity, respiration and blood circulation). Basal metabolic rate is determined when the animal is in a fasting and resting state and in a thermoneutral environment. In fish basal metabolic rate can only be determined in that part of the year when the animal is not developing gonads.

The heat of activity (H_jE) is the heat produced by muscular activity associated with locomotion and maintaining position in the water column.

The heat of thermal regulation (H_cE) is the heat produced as a result of an animal's efforts to maintain body temperature when environmental temperature goes above the zone of thermal neutrality. Maintenance of body temperature when environmental temperature is below the zone of thermal neutrality is usually achieved by locomotive activity.

The heat of waste formation and excretion (H_wE) is the heat production associated with the synthesis and excretion of waste products, for instance urea. In teleosts, conversion of ammonia to urea is unusual and so H_wE is generally negligible.

Heat increment (H_iE) or specific dynamic action (SDA) is the increase in heat production following consumption of food by an animal in a thermoneutral environment. This heat is produced as a result of the metabolic processes associated with digestion of the food and the subsequent molecular transformation of the digested nutrients in the liver and elsewhere. H_iE is appreciable when expressed in terms of the energy content of the food ingested, being about 6% for sucrose, 13% for fats and 30% for proteins (Gordon *et al.*, 1977). This is particularly relevant in the case of fish which rely heavily on protein as an energy source. Experimental determination of H_iE is likely to include some part of H_wE.

Heat production may be summarized by the equation:

$$HE = H_eE + H_jE + H_cE + H_wE + H_iE \quad (2.3)$$

Retained energy (RE) is that portion of the energy contained in the food that is retained as part of the body or voided as a useful product such as gametes. The aim of nutrition in aquaculture is to maximize RE and minimize all other energy losses in a cost-effective manner, whilst maintaining an acceptable body composition of the end product.

The energy balance of an organism therefore can be described by the equation:

$$IE = FE + UE + ZE + SE + HE + RE \quad (2.4)$$

This equation can be rewritten using an alternative nomenclature adopted by some authors as:

$$C = F + U + R + P \quad (2.5)$$

In this form C is the energy of the food consumed (IE), F the energy lost in the faeces (FE), U the energy lost in excreta (UE + ZE + SE), R the energy of metabolism (HE) and P the energy deposited in the body as growth (RE). The use of this form of the expression reflects the parameters that can be readily and accurately measured by experimentation. The values of C and F are obtained through direct determination by bomb calorimetry. The value of U is obtained either by measuring ammonia and urea excreted and converting to equivalent energy values using 23.05 J/mg ammonia and 24.85 J/mg urea (Elliott, 1976) or by using the model of Brett and Groves (1979) relating energy lost in excretion to energy consumed as:

$$U = 0.07\,C \quad (2.6)$$

The value of R is determined using a respirometer and converting to an equivalent energy value using 13.56 J/mg oxygen (Elliott and Davison, 1975).

2.2.1 Gross conversion efficiency

Gross conversion efficiency (K) is often used as an indicator of the bioenergetic physiology of a fish under different experimental conditions instead of determining a complete energy budget. This parameter measures growth rate relative to feed intake of the fish. Both growth rate and feed intake are related to body size. Gross conversion efficiency is described by the equation

$$\text{Growth conversion efficiency } (K) = \frac{\text{SGR}}{\text{Relative food intake}} \times 100 \quad (2.7)$$

where

$$\text{Specific growth rate (SGR)} = \frac{\ln(Wt_2) - \ln(Wt_1)}{t_2 - t_1} \times 100 \tag{2.8}$$

$$\text{Relative food intake} = \frac{F}{0.5(Wt_2 - Wt_1 \times (t_2 - t_1)} \times 100 \tag{2.9}$$

ln (Wt_1) = natural log of the weight of animals at time 1 (t_1) and ln (Wt_2) = natural log of the weight of animals at time 2 (t_2).

2.3 FACTORS AFFECTING ENERGY PARTITIONING

The partitioning of energy between the various parameters described by the terms on the right-hand side of the equations above is affected by a variety of environmental and metabolic processes.

2.3.1 Factors affecting the basal metabolic rate

A number of factors affect basal metabolic rate, most of which are related to the animal's environment. Maintaining these parameters at the optimal level for the species of interest is the best way of ensuring that the minimal energy expenditure occurs, maintaining the animal

(a) Body size

It may seem reasonable to assume that oxygen consumption, and hence metabolic rate, should increase directly with increasing body size, each additional unit of mass providing for an increase of equal magnitude in the metabolic rate. However, this is not the case, and the energy demand of a piece of tissue depends on the size of the animal of which it is a part. Thus, the energy demand of 1 g of a mouse is 20 times that of 1 g of an elephant. This effect is called scaling and is a well-known physiological phenomenon. It can be described mathematically by the formula:

$$Y = aX^b$$

where Y is any physiological variable, in this case metabolic rate, a is a proportionality constant, X is body mass, and b describes the effect of size on Y.

Considerable controversy has arisen over the value of b in all vertebrates, and fish are no exception. Figure 2.1 shows oxygen consumption at 25°C versus body mass in four species of freshwater fish. Analysis of these data produced a value for b of 0.7. In a similar study of salmon held at 15°C, b was determined to be 0.846 (Brett and Glass, 1973).

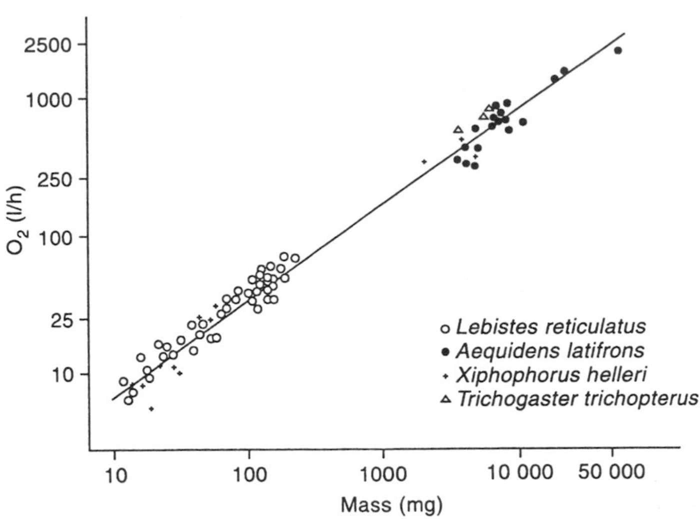

Figure 2.1 The relation of oxygen consumption at 25°C to body mass in four species of freshwater fish. The slope of the regression line is 0.70 (s.d. = 0.02). (From Gordon *et al.*, 1977.)

However, despite the controversy, the value of *b* is generally considered to be between 0.7 and 0.8. Thus it can be seen that the larger the fish, the less energy per unit body weight is required to maintain basal metabolic rate.

(b) Oxygen availability

Aquatic animals consume oxygen either at a rate directly dependent upon ambient oxygen tensions (conformers) or at a rate independent of ambient oxygen tensions (non-conformers or regulators). The latter constitute the majority of the vertebrate aquatic species.

In the case of conformers, the metabolic rate is greater at higher oxygen tensions and so the energetic cost of maintenance is also increased, although this is countered by the fact that growth rates are greater at higher oxygen tensions. With non-conformers, oxygen consumption remains constant until the oxygen tension in the ambient water reaches a critically low level, the critical point, whereupon oxygen consumption begins to decline. Eventually the oxygen tension of the water will reach a point at which the animal becomes asphyxiated. In non-conformers the energy diverted to maintenance remains constant until the critical point. Often, however, both growth and feed intake are reduced at oxygen levels slightly higher than the critical point. It is therefore of advantage to operate the culture system with oxygen tensions in excess of the critical point, ideally as near as possible to saturation (Jobling, 1988).

(c) Temperature

Since fish are poikilotherms (i.e. their body temperature reflects the temperature of their surroundings), water temperature plays an extremely important role in energy partitioning. Two effects of temperature can be observed in aquatic animals.

When an animal is acclimated to a certain temperature and then introduced to an environment where the temperature is greater, its metabolic rate will increase. If the animal is soon returned to the original temperature, its metabolic rate will return to the original rate. However, if the animal is kept at the higher temperature for a period of time, then its metabolic rate will decrease until it becomes acclimatized. A diagrammatic representation of this is shown in Figure 2.2. Since temperature fluctuations generally occur slowly in the wild, the usual temperature compensation effect on metabolic rate is one of acclimatization. In nature this means that the effect of increasing temperature on energy consumed by basal metabolism is minimal. In an aquaculture situation, however, it is possible that the temperature of an animal's environment may vary considerably and quite quickly. As a result, temperature effects on metabolic rate can become quite significant.

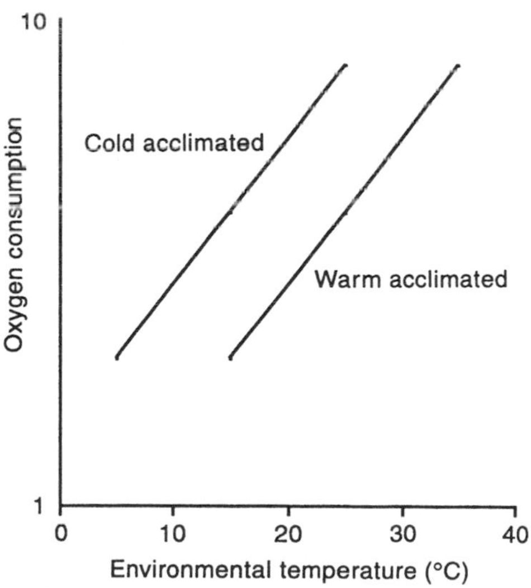

Figure 2.2 An example of temperature compensation in the oxygen consumption of a fish. Units on the ordinate axis are arbitrary.

A more detailed discussion of mechanisms involved in temperature acclimation can be found in a review by Hazel and Prosser (1974).

(d) Osmoregulation

The salinity of a fish's environment plays an important part in the energetic cost of osmoregulation, which is a major energy-consuming process in aquatic vertebrates. Freshwater fish live in an environment that is hyposaline to their body tissues, and hence they must continually excrete water and sequester ions. Saltwater fish, on the other hand, have exactly the opposite problem and expend energy avoiding the loss of water and in excreting ions.

Most species have evolved mechanisms to deal efficiently with osmoregulation in the particular environment in which they are normally found, and therefore the salinity of the water in which they are to be cultured is usually determined by reference to their natural habitat. It is important to be aware, however, that any variation from the optimal salinity for a given species is likely to result in increased energetic costs of osmoregulation and therefore in reduced growth rates.

(e) Stress

Stress results in increased basal metabolic rate and can be induced by a variety of factors, including accumulation of waste products in water, low oxygen, crowding, handling, external disturbances, water pollution, poor-quality feed or aggression. The energetic cost involved in dealing with these stressors will reduce growth rates. Stress also invokes a process of tissue breakdown in the animal, independently of a need for increased nutrients or energy substrates. This is most clearly manifest in the hypoglycaemia exhibited by stressed fish (Braley and Anderson, 1992). The energy required for degrading and subsequently resynthesizing body tissue is energy not otherwise available for growth. Attempts should therefore be made at all times to reduce stress in cultured fish in order to maximize growth rates.

(f) Cycles

All animals display cycling of their physiological processes. Some of these cycles are clear, such as the seasonality of reproduction or the variation in growth rate with season. On the other hand, many cycles are much more subtle. Standard metabolic rate has been shown to cycle with season in the perch *Perca fluviatilis* (Karas, 1990), while lunar cycles have been demonstrated in coho salmon (*Oncorhynchus kisutch*) (Farbridge and Leatherland, 1987).

Interactions between the effect of environmental variables are also observed. Zanuy and Carrillo (1985) used gross conversion efficiency (K) as an indicator of the bioenergetic physiology of sea bass, *Dicentrarchus labrax*, in a study of the effect of salinity on the annual cycle of growth. It was found that, despite a slightly higher temperature of the higher salinity water, K values for fish held at high salinity were lower than those of fish held at lower salinity during June and July. However, the reverse was true during October and November when the fish were developing their gonads. Zanuy and Carrillo proposed increasing the efficiency of production of sea bass by varying the salinity of the culture medium according to season.

2.3.2 Factors affecting non-basal metabolic rate

A number of factors also influence the metabolic energy consumption of an animal other than its basal metabolic rate.

(a) Gonadal growth

Gonadal growth results in the diversion of large amounts of energy away from growth of muscle and other activities. At the peak of the reproductive season, gonads may account for a large proportion of a fish's body weight – even up to 30–40%. As such it is generally of advantage to the aquaculturist to limit the growth of gonads in animals being produced for sale. This, of course, is not the case in animals in which the gonad constitutes a major portion of the saleable product (e.g. caviar), or in broodstock, in which gamete production is the objective.

Diversion of energy to gonad development may be limited by a number of techniques. By far the most common method used is to grow the animal out only while it is immature, and so limit gonadal growth by the natural biological control mechanisms.

In many species there is a specialist market for larger animals. The premium price obtained usually compensates for the additional cost (much of it feed cost) involved in growing out the fish through one or more reproductive cycles.

Techniques are available for producing sterile fish by inducing triploidy. This is generally achieved by subjecting fertilized ova to temperature shocks and pressure. The production of sterile juveniles is widespread in the salmonid culture industry.

(b) Locomotion

The energy cost of locomotion is a major part of the total energy consumption of an animal and varies between species depending upon body shape and behavioural patterns. The energy cost of locomotion is also dependent upon size. Animals that live in an aquatic medium have a much lower relative cost of locomotion than animals that live on land. Figure 2.3 shows a comparison of the energetic cost of locomotion for a number of different groups of animals. The energetic advantage of swimming is derived in a number of areas. Swimmers are passively supported by the medium they move through, and therefore all of the energy used during locomotion can be directed to moving the body forward rather than providing lift. The greater viscosity of water relative to air also allows more efficient transfer of energy from the muscles to assist locomotion in aquatic versus terrestrial animals. The relatively low cost of locomotion results in aquatic animals generally being very efficient in converting feed to body tissue.

The effects of locomotion are not all negative. Sustained exercise has been shown to be related to increased growth rates in fry of Arctic charr (*Salvelinus alpinus*), although food intake was probably increased in exercised versus non-exercised animals (Christiansen *et al.*, 1989).

The significance of energy expenditure on gonadal growth and locomotion is apparent from a study of energy partitioning in roach, *Rutilus rutilus* (Koch and Weiser, 1983). Using video-monitoring of activity and respirometry, it was found that activity was closely

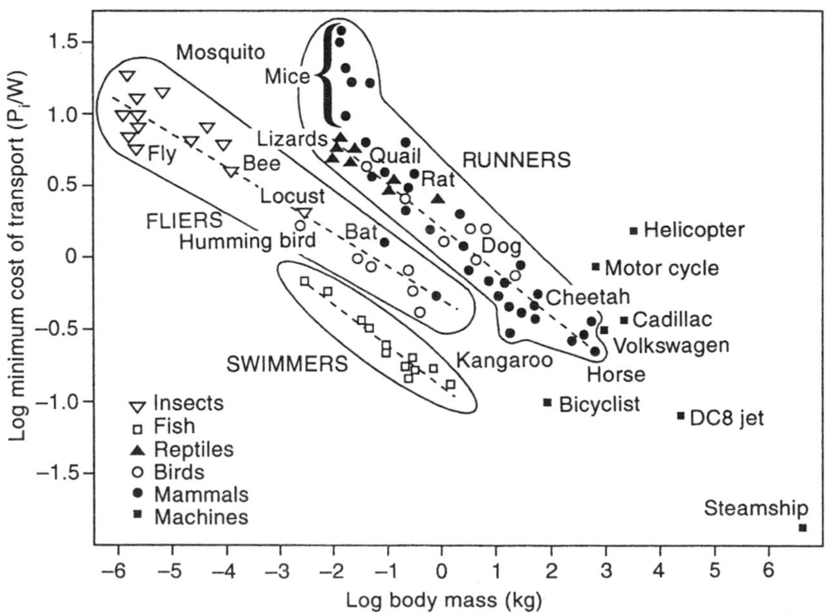

Figure 2.3 The minimum cost of transport in relation to size. The geometric figures enclose the values of animals using the same general types of locomotion. The dotted lines are the linear regressions of the log-transformed values of body mass and minimum cost of transport. (From Gordon *et al.*, 1977.)

correlated with oxygen consumption in the first half of the year (December to June). In the second half of the year (July to December), following spawning when the gonads were recuperating, activity declined whilst oxygen consumption remained elevated. Apparently the energy expenditure on either gonadal development or locomotion is so great that it is advantageous for roach to compensate for one by reducing the other.

2.3.3 Heat increment of specific dynamic action

Determination of H_iE is achieved by monitoring the increase in oxygen consumption of either fish fed different ration levels or types or fish following feeding. H_iE varies with diet depending upon the composition, energy content, ration size and the temperature of the water (Jobling, 1981a). Depending on these factors the magnitude of H_iE as a proportion of intake energy is reportedly as high as 30% (Pierce and Wissing, 1974). The duration of post-prandially elevated oxygen consumption is apparently related to the rate at which food passes through the digestive tract (Jobling, 1981a); it also increases following feeding after a period of food deprivation (Soofiani and Hawkins, 1982). Since the heat increment due to feeding constitutes a significant portion of the energetic expenditure of aquatic animals, it seems appropriate to attempt to minimize it in order that the energy can be redirected into growth. This may be achieved by limiting ingredients that are associated with a high H_iE and to consider H_iE in determining the temperature of culture and the feeding regime. An example of the latter would be to avoid the practice of feeding five working days with no feed provided on weekends.

2.3.4 General hypotheses of partitioning

On the basis of the analysis of different sets of data, it was concluded that fish spend about 60% of their energy intake on metabolism and 40% on growth when fed *ad libitum* (Brett and Groves, 1979; Cui and Liu, 1990). Whilst this is a useful generalization, the pattern of energy allocation by a species will depend on a wide variety of factors as described above. Closer analysis of the data presented by Cui and Liu (1990) shows that the proportion of energy devoted to growth in the six species investigated ranged from 21.3% to 63.4%. Similarly, Xie and Sun (1993) found that *Silurus meridionalis*, a Chinese catfish, expended approximately 60% of the energy it consumed on growth. Since the behaviour of the species is likely to play an extremely important role in energy partitioning, it is unlikely that general hypotheses of fish energy budgets will prove to be of great use to the aquaculturist.

2.4 ENERGETICS AND FEEDING

Fish feed to provide sufficient energy for their requirements (Nose and Halver, 1981), and hence dietary energy is important in regulating food intake. Feeding adequate amounts of energy is necessary for the economical production of fish, while feeding excess energy will result in obesity and deterioration of flesh quality. A number of factors associated with feeds and feeding must be considered when determining feed formulations and the level of feeding for aquaculture.

2.4.1 Diet composition

The majority of the gross energy of feed is contained in one of three types of molecules, namely carbohydrates, lipids and proteins. The total amount of energy contained in a diet is usually determined by bomb calorimetry. The proportion of the relative components (i.e. carbohydrates, lipids or proteins) in the diet can be determined by chemical analysis. The energetic contribution of each component to the total energy of the diet is then calculated by multiplying the proportion of that component in the diet by an appropriate factor known as the physiological fuel value. Standard values for these factors are 19 $kJ\,g^{-1}$ for protein, 36 $kJ\,g^{-1}$ for lipid and 15 $kJ\,g^{-1}$ for carbohydrate (adapted from Smith, 1981). However, the numerical values of these factors are controversial (for example bovine albumin, a protein, contains 22.7 $kJ\,g^{-1}$) and will vary according to the species, digestibility, the actual composition of the protein, lipid or carbohydrate being consumed and the nature of waste products produced by their degradation. As such, it is important that in the formulation of feeds the energy content of each component and the total energy content of the complete diet are determined.

2.4.2 Determining the quality of a diet

The quality of a feed is a function of how well that feed meets the nutrient requirements of an animal. Not only must the feed contain the correct proportions of nutrients, but the nutrients must be able to be digested and absorbed in a form that makes them available for providing energy and substrates for growth to the animal. This is termed bioavailability. The digestibility of the food is currently the primary determinant of bioavailability. The major problem with using digestibility is that it varies with species, source of nutrient, the temperature at which it is evaluated and often between two samples of exactly the same feedstuff that are treated in different ways (e.g. different heats of drying) (Pfeffer *et al.*, 1991). These factors make it difficult to relate the data obtained by separate groups of workers, or even by one group of workers at one time with those obtained by the same group at different times. Nevertheless,

Table 2.1 Digestibility of different components of feed by the African catfish, *Clarius gariepinus*

Energy in diet (kJ/g)	*Apparent protein digestibility (%)*	*Apparent fat digestibility (%)*	*Apparent energy digestibility (%)*
8.4	94.0	90.5	50.0
12.4	95.5	94.0	65.0
16.8	86.5	73.0	60.0

Source: Adapted from Machiels and Henken (1985).

digestibility remains the most widely used method of determining how much of a given food component is bioavailable.

It is important to note that the digestibility of carbohydrates, lipids and proteins contained within a single food source, for instance soya meal, also varies from one another. Examples of this variability in digestibility are given in Table 2.1. This fact influences the proportion of the energy requirements of a fish that are met by the different components of a feed.

An alternative term, metabolizable energy, more closely estimates the energy available to the fish for growth and metabolism since it allows for energy lost in urine and gill excretions, in faeces and from the body surface.

A third term, retained energy, is easily measured simply by determining the energy in the whole carcass of the fish. This term takes account of the digestibility of the diet and is equal to metabolizable energy minus loss of energy lost during metabolism and activity. Retained energy is a measure of the energy used to increase body mass. Figure 2.4 shows the relationship between DE, ME and RE.

There is some debate about the relative value of DE, ME and RE for the evaluation of the quality of fish feeds. In practice, RE is rarely used despite its relatively simple determination. RE values give little information about why one feed may be better than another. ME is also rarely used because determination of the energy contained in urine and gill excretions involves a number of technical difficulties and assumptions, and so DE serves as the main determinant of the value of feeds. However, problems are also inherent in the determination of DE; these are discussed in detail in Chapter 4.

Other parameters are used to determine the value of feeds for providing the necessary energy for growth. These include food conversion ratios (FCRs) also known as utilization efficiency and protein efficiency ratios (PERs). Food conversion ratio is described by the equation:

$$\text{FCR} = \frac{\text{Mass of food consumed (dry)}}{\text{Increase in mass of animal produced (wet)}} \tag{2.10}$$

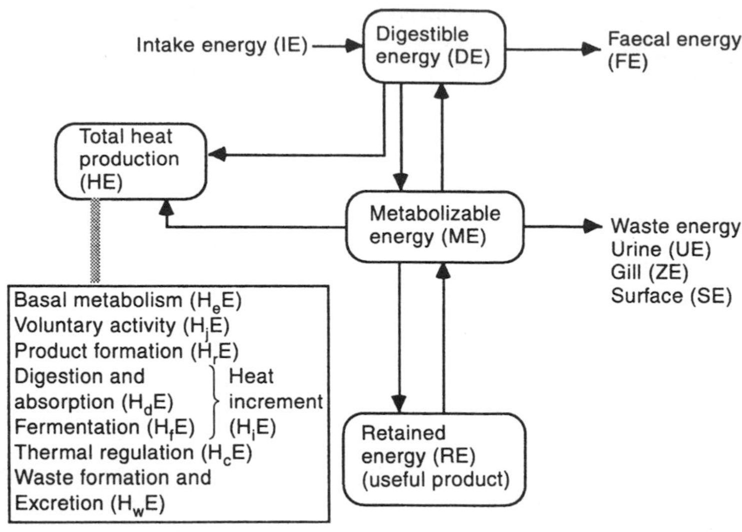

Figure 2.4 Energy flow in an aquatic organism.

Food conversion ratios (for which a lower value indicates an improved outcome) as low as 1 have been reported in fish, although generally they range between 1.2 and 1.5 for animals fed carefully prepared diets. Table 2.2 shows a comparison between the food conversion ratios of rainbow trout, catfish, chicken, pigs and cattle.

Protein efficiency ratio is described by the equation

$$\text{PER} = \frac{\text{Increase in the mass of animal produced (wet wt)}}{\text{Mass of protein in fed (dry wt)}} \tag{2.11}$$

This parameter gives a measure of how well the protein source in the diet provides for the essential amino acid requirement of the animal (see Chapter 3) as well as how well the diet is balanced for energy and protein.

PER measures the deposition of fat as well as protein, which means that diets producing fatty fish can be associated with high PER values.

Table 2.2 Food conversion ratio of a range of species

Species	*Air-dry feed/wet gain ratio*
Rainbow trout	1.5
Catfish	1.8
Chicken (broiler)	2.5
Swine	4.0
Cattle (beef)	8.0

Source: Smith (1989).

Because of this constraint, it has been recognized that net protein utilization (NPU), also known as protein retention (%PPV), is a better measure of the feed quality than protein efficiency ratios (Lie *et al.*, 1988). It is given by the formula

$$\mathrm{NPU}\ (\%) = \frac{\text{Protein gain in fish (g)}}{\text{Protein intake in food (g)}} \times 100 \qquad (2.12)$$

2.4.3 Dietary sources of energy

Fish growth involves the laying down of muscle, fat, epithelial and connective tissue. The proportion of protein or fat laid down in these tissues is highly dependent upon the diet. In order for protein synthesis to occur, the correct ratio of essential amino acids must be provided. Essential amino acids are those amino acids that an animal is incapable of synthesizing and that therefore have to be provided in the diet. The requirement for essential amino acids of aquatic animals varies between species and is discussed in detail in Chapter 3. For any given protein source, one essential amino acid will be limiting. Protein synthesis will continue until the limiting amino acid is completely consumed, although this process is not 100% efficient, that is some of the limiting amino acid will be used to provide energy. The amino acids will be deaminated and used for energy. The process of protein synthesis has an energy demand on top of the basal metabolic rate. Unless a form of non-protein energy is provided in the diet, an animal has to direct amino acids into energy-liberating pathways in order to provide the energy needed for basal metabolism and growth. This will limit the scope for growth. The alternative energy sources that can be included in the diet to meet these needs are carbohydrate and lipid.

The ability or capacity to utilize carbohydrate and lipid for energy varies between species. Inclusion of non-protein energy sources in a diet formulation allows the formulator to reduce the protein content of the diet. This capacity is called the protein-sparing effect, and it is of great interest to those involved in formulating fish diets. The relative prices of ingredients containing high levels of protein, high levels of fat and high levels of carbohydrates are different enough to make the protein-sparing effect of lipids and carbohydrates cost-effective.

Lipids contain more energy per unit weight than any other dietary component, and they are used efficiently by fish as energy sources. Lipids increase the palatability of feeds (up to a point), and assist in reducing dust and stabilizing the pellets during manufacture, transportation and storage. As a result, lipid is an excellent source of non-protein energy.

Carbohydrate is utilized less efficiently by fish as an energy source, but is much less expensive than alternative sources of energy for fish diets. There is considerable controversy about the optimal level of carbohydrate that may be included in an aquatic animal feed. This level varies according to species, from less than 12% in rainbow trout (Phillips *et al.*, 1948) to as high as 33% for channel catfish (Wilson, 1991).

2.4.4 Protein as an energy source

As can be seen from the preceding discussion, there is considerable interaction between protein, carbohydrate and lipid as energy sources. However, each will be dealt with separately as there are some important differences between them. Detailed protein requirements are presented in Chapter 3, so the discussion here will centre on protein as an energy source. For the purposes of this discussion, it will be assumed that the composition of the protein provides a balanced mix of the essential amino acids. This assumption allows us to ignore any growth-limiting effects associated with a restricted supply of essential amino acids. Not all studies apply this assumption validly, and when interpreting the results presented by various investigators about the protein requirement of cultured species careful attention should be paid to the amino acid composition of the tested diet. For instance, in a study of the protein requirement of hatchery-produced juvenile white sturgeon (*Acipenser transmontanus*) the researchers used graded amounts of a fixed protein mixture (casein – wheat gluten – egg white; 62 : 30 : 8). In this way the amino acid composition of the protein was maintained at fixed ratios (Moore *et al.*, 1988). However, in a study of the dietary protein requirement of fingerlings of an herbivorous carp tawes (*Puntius gonionotus*) the researchers used fish meal, casein and blood meal as protein sources, but varied only the amount of fish meal throughout the study (Wee and Ngamsnae 1987). As a result the amino acid ratios would have varied. Studies using designs of the latter type generally overestimate the protein requirement of the species under investigation.

There has been a traditional reliance upon fish meal as a protein source in diets for farmed species. Fish meal is expensive relative to other protein sources, and a number of investigations have addressed the effectiveness of replacing the fish meal with some other protein source, for instance some types of grain meal (e.g. soybean meal, cottonseed meal). It has generally been found that most alternative protein sources are able to replace fish meal to some extent. A number of factors affect the proportion of fish meal that can be replaced, and these depend upon the nature of the protein source. In some grain meals, the essential amino acid composition is not adequate, while many others contain anti-nutritional factors discussed in Chapter 9. Despite these limitations there can often be considerable cost advantage in replacing some of the fish meal with alternative protein sources.

Since dietary protein is relatively expensive, the nutritionist aims to formulate the diet in such a manner that the energy required by the animal is provided by non-protein sources. This will allow the majority of the protein in the diet to be directed toward protein synthesis by the fish. An imbalance in the ratio will lead to either wasted protein or the production of lower value fatty animals.

The optimal dietary protein–energy ratios that have been reported for finfish have an average of about 22 mg of protein per kJ; values range from 17 $mg\,kJ^{-1}$ for tilapia (De Silva *et al.*, 1989) and 17.9 $mg\,kJ^{-1}$ for cod, *Gadus morhua* (Jobling *et al.*, 1981a), to 22.5 $mg\,kJ^{-1}$ for Chinese major carps (De Silva and Gunasekera, 1991) and 28.7 $mg\,kJ^{-1}$ for channel catfish, *Ictalurus punctatus* (Reis *et al.*, 1989). Clearly the optimum dietary protein–energy ratio varies significantly between species, and will also vary within a species according to the digestibility and amino acid composition of the protein source. Optimal protein–energy ratios are known to be affected by water temperature (Hidalgo and Alliot, 1988) and are also likely to be affected by other environmental parameters that affect the partitioning of energy.

As implied, within any one species, protein–energy ratios of diets are correlated with the body composition of the product. Body composition is the amount of moisture, protein, carbohydrate and fat contained in a fish carcass. Optimizing body composition to yield the maximal dressing percentage (the proportion of saleable product expressed as a percentage of the whole fish) and sensory quality (taste) is important for an aquaculturist since it is directly related to profitability (Reis *et al.*, 1989; Parazo, 1990).

2.4.5 Carbohydrate as an energy source

For a variety of reasons that will be discussed in detail in Chapter 3, carbohydrate has limited use as an agent providing energy and so for sparing protein in the diets of finfish. This is unfortunate, since carbohydrate can be a particularly cheap source of dietary energy. This is not to say that carbohydrate has no role in fish diets. For example, a study by Degani and Viola (1987) showed that European eel (*Anguilla anguilla*) had an increased specific growth rate (SGR; equation 2.8, section 2.2) when fed a 40% protein, 38% wheat (carbohydrate) diet relative to both lower carbohydrate (50% protein, 20% wheat) and higher carbohydrate (30% protein, 56% wheat) diets. All diets were isocaloric (having the same energy value). In addition, the FCR, NPU and percentage energy retained in the carcass were all increased in the 40% protein, 38% wheat diet. Notably, the PER was greatest in fish fed the highest carbohydrate diet, indicating that there was a large amount of fat being deposited in these fish (Table 2.3).

Table 2.3 Feed utilization by the European eel (*Anguilla anguilla*) fed differing proportions of protein and energy. The specific growth rate (SGR), percentage of dietary protein retained in the carcass (%PV), percentage dietary energy retained in the carcass (%ERV) are all greatest and food conversion ratio (FCR) lowest in group 2. The protein efficiency ratio (PER) is greatest in group 3, which indicates that these fish are depositing body fat rather than protein.

Group no.	*Diet*	*SGR*	*FCR*	*%PV*	*PER*	*%ERV*
1	50% protein (20% wheat)	1.63	1.84	18.52	108.30	49.74
2	40% protein (38% wheat)	1.80	1.61	29.11	154.86	102.93
3	30% protein (56% wheat)	1.66	1.95	26.32	169.81	67.37

ERV, energy retention value; PV, protein retention value.
Source: Degani and Viola (1987).

2.4.6 Lipid as an energy source

Lipid is digested and metabolized with greater relative ease and so serves as a much better source of energy for protein sparing than carbohydrate. Again, too much lipid can be included in the diet, which results in production of fatty fish.

The protein-sparing effect of lipid varies between species but appears to be optimal at about 15–18% of the diet (Lie *et al.*, 1988; De Silva *et al.*, 1991). The effect is most clearly observed when the amount of dietary protein consumed is low, whether this is the result of a lower proportion of protein in the diet (De Silva *et al.*, 1991) or diminished rations (Beamish and Thomas, 1984).

Rainbow trout had reduced excretion of nitrogen (a measure of amino acid metabolism: section 3.6) when fed at 0.5% wet body weight per day a diet which contained higher amounts of lipid, but no effect of lipid level was observed when they were fed 0.1% wet body weight per day (Beamish and Thomas, 1984). De Silva *et al.* (1991) found improved NPU in red tilapia fed to satiety with diets containing increasing amounts of lipid to 18% dry weight of diet, but thereafter increasing dietary lipid caused a reduction in NPU.

2.5 THE EFFECT OF RATION ON GROWTH

The ideal relationships between ration (feeding rate) and two parameters commonly used to determine the nutritional value of a diet – specific growth rate (SGR) and food conversion ratio (FCR) – is shown in Figure 2.5. Specific growth rate is defined in section 2.2.1. Food conversion ratio is defined in section 2.4.2.

The relationship between ration and growth is curvilinear (Figure 2.5). There is a rapid increase in growth rate with increasing feeding rate until SGR = 0 (the animal no longer loses weight). This point is termed the maintenance feeding rate or maintenance ration and occurs when the energy consumed equals the energy required by the animal to maintain itself without mobilizing endogenous energy reserves. From this point, SGR increases with ration size at a decreasing rate to a point which is the maximum rate of feeding. This level varies according to the species, water temperature and size of the fish. The FCR decreases from the maintenance ration to a point termed the optimum feeding rate and then increases as the feeding rate increases towards the maximum rate. It is clear then that there is an economic advantage in identifying and feeding fish at the optimal rate. However, it is sometimes of benefit to increase the feeding rate above the optimum to shorten the grow-out period and thereby reduce other costs associated with culture.

In order to understand why these relationships between feeding rate and specific growth rate and between feeding rate and food conversion ratio occur, it is useful to refer to equation 2.4, which described the energy balance of aquatic animals.

Figure 2.5 shows that as IE (feeding rate) increases to optimum ration, RE (specific growth rate) increases rapidly. After this point the increase in RE per unit IE decreases. Consideration of each component of the right-hand side of the equation will give us some information about the relationship between ration and growth. Figure 2.6 shows the results of a classic study by Elliott (1976) of the energetic losses of brown trout (*Salmo trutta*). It can be seen from this figure that the energy lost in the

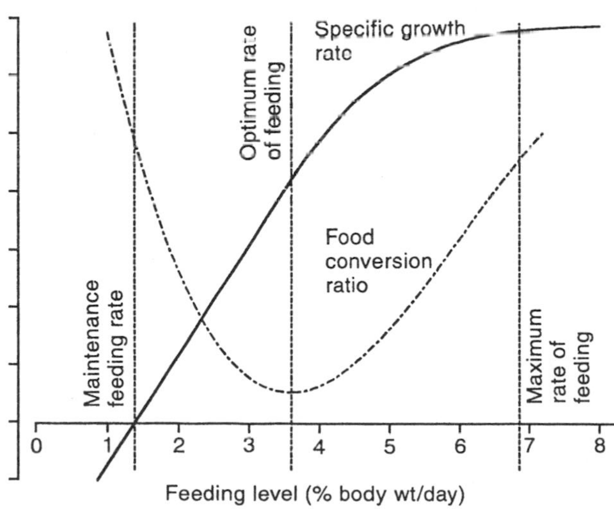

Figure 2.5 Theoretical relationship between specific growth rate, food conversion ratio and feeding rate in fish.

faeces (FE) as a proportion of the energy intake (IE) increases as the animal's ration size increases, that is the efficiency of digestion and absorption decreases with increasing ration size. Therefore, a proportion of the observed decrease in growth rate with increasing ration size is due to increased loss of energy in the faeces (increased FE). This relationship applies at a range of temperatures (Figure 2.6) and has been confirmed in a variety of species since Elliott's original finding.

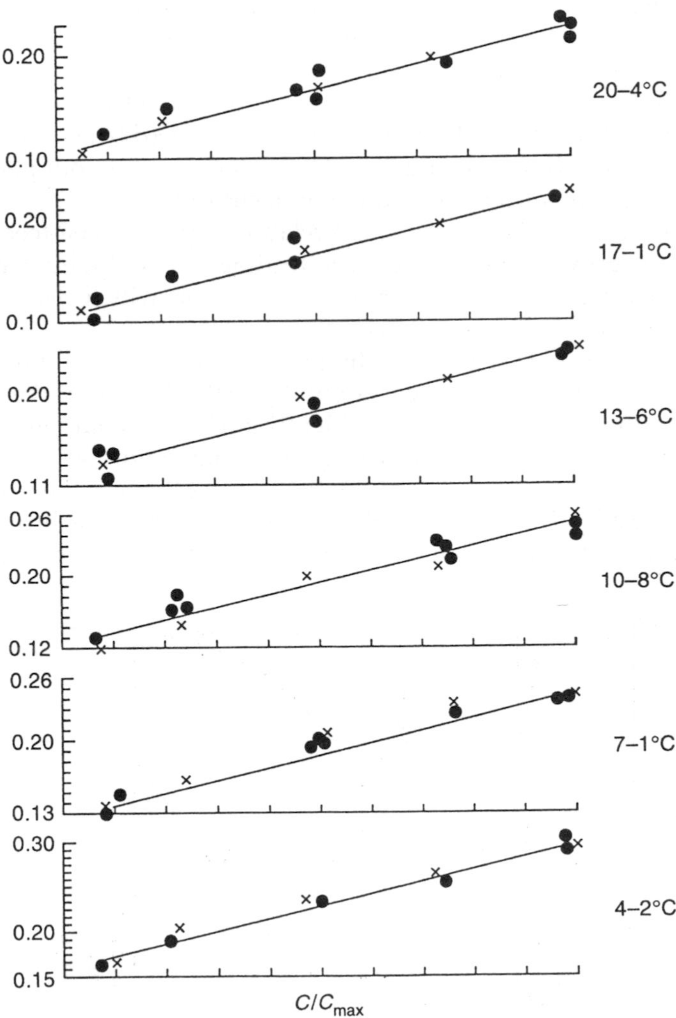

Figure 2.6 Relationship between the proportion of the daily energy intake lost in the faeces (Pf = FE/IE) and the energy intake expressed as a proportion of the maximum ration (C/C_{max}) at different temperatures (°C) for trout of live weight close to 50 g (×) and 11 g, 90 g, 250 g (•). Regression lines are fitted to the data. (From Elliott, 1976.)

Figure 2.7, taken from the same study by Elliott, reveals that the energy lost in the excretory products (IE + UE; both were measured in the study by Elliott) as a proportion of daily energy intake actually decreases with increasing ration size. Again, this relationship occurs at a range of temperatures (Figure 2.7).

As stated earlier, HE is the sum of heat energy produced by basal metabolic rate, activity, waste formation and digestion. Basal metabolic rate will not change with ration size. The energy requirement for basal metabolic rate is always the first to be satisfied and the ration required to increase the standard growth rate to zero includes (but is greater than) the ration required to satisfy basal metabolic rate. Having satisfied the basal metabolic rate, energy consumed by the animal can be devoted to other purposes. The heat of activity (H_jE) is likely to increase with increasing ration since the animal will spend more time searching for or

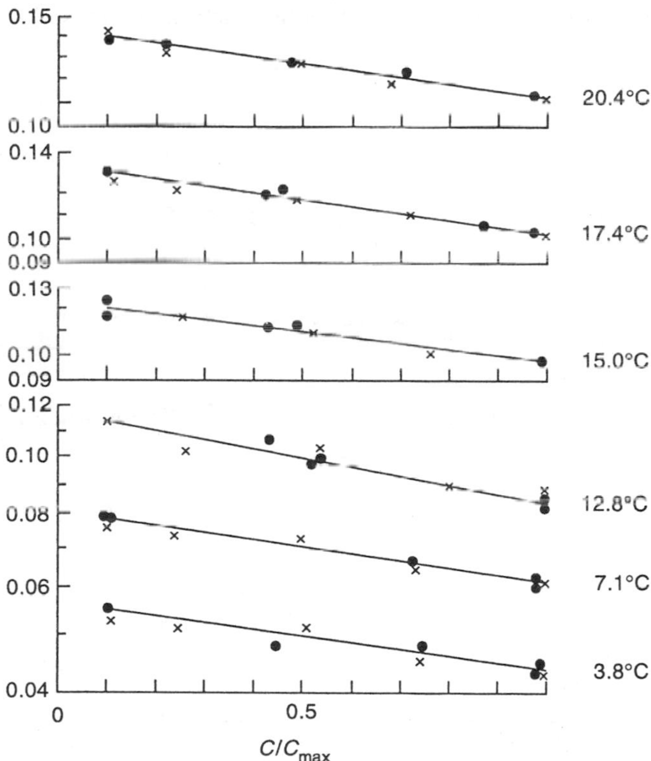

Figure 2.7 Relationship between the proportion of the daily energy intake lost in the excretory products [P_u = (ZE + UE)/IE] and the energy intake expressed as a proportion of the maximum ration (C/C_{max}) at different temperatures (°C) for trout of live weight close to 50 g (×) and 11 g, 90 g, 250 g (•). (From Elliott, 1976.)

gathering food. Heat of waste formation, H_wE, will reflect the production of excretory products and so will decline with increasing ration. Heat increment due to digestion will remain as a constant proportion of the intake energy.

Therefore, the decline in the rate of increase in growth rate as ration size increases is largely due to two factors, increases in H_jE and FE.

The relationship between growth rate and ration size is greatly affected by the temperature at which the animal is cultured. Figure 2.8 shows the relationships between SGR and temperature at different ration levels for the three-spined stickleback (*Gasterosteus aculeatus*) (Allen and Wootten, 1982). By drawing a line across the figure at the point where SGR = 0, it can be seen that, at 3°C, the maintenance ration is equal to 1% body weight per day. At 15°C, maintenance ration is equal to 2% body weight per day. Clearly basal metabolic rate is increasing with increasing temperature. This occurs because all fish are poikilotherms and consequently their body temperatures are affected by the temperature of their environment. As the body temperature increases, all metabolic reactions increase in rate. This

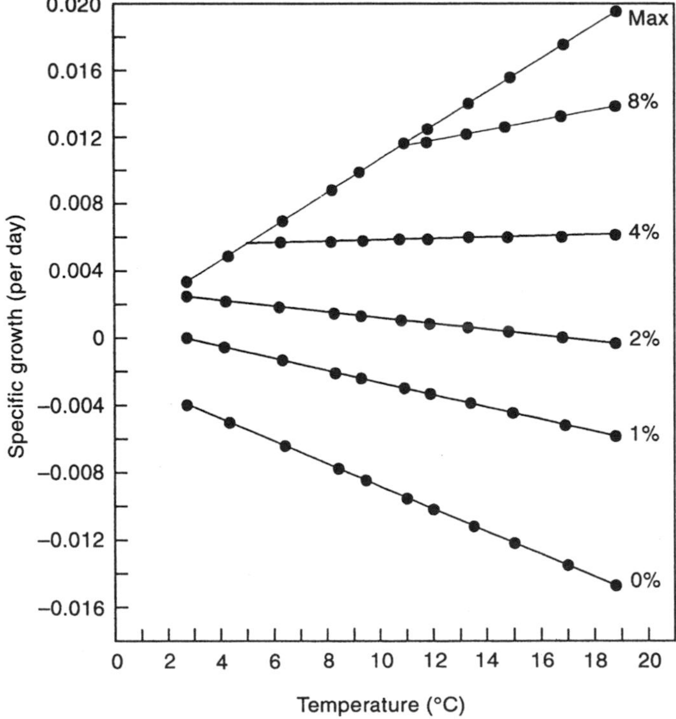

Figure 2.8 Predicted relationship between specific growth rate, ration and temperature for a 250-mg stickleback. Ration size is expressed as percentage of wet body weight. (From Allen and Wootten, 1982.)

has a number of effects. Basal metabolic rate increases, and the rates of reactions that result in growth of the animal also increase. The increased requirement for energy is compensated for by the animal increasing its total food intake. Figure 2.8 shows that the maximum feed intake per day of a 250-mg stickleback is 8% at 11°C and about 2.5% at 3°C. Therefore, while it requires more food to maintain and obtain growth from fish at higher temperatures than at lower temperatures, the increased ration is rewarded with increased growth rates. At a point above optimal temperature, however, growth rate is limited firstly by the need of the animal to reduce its body temperature and secondly by increased metabolism due to a stress response. The relationship between temperature and growth is shown in Figure 2.9, in which it can be seen that the maximum growth rate of sockeye salmon (*Oncorhynchus nerka*) is obtained by feeding to excess at about 14°C (Brett *et al.*, 1969).

The other important parameter affecting the relationship between growth and food intake is the size of the animal. Observed growth rates are usually greater in younger than older fish, but this is because younger fish have a proportionally larger total energy intake. As discussed earlier, an effect called scaling means that, as a fish grows larger, the proportion of energy required for metabolic processes per unit mass decreases and consequently an increased proportion of energy can be devoted to growth.

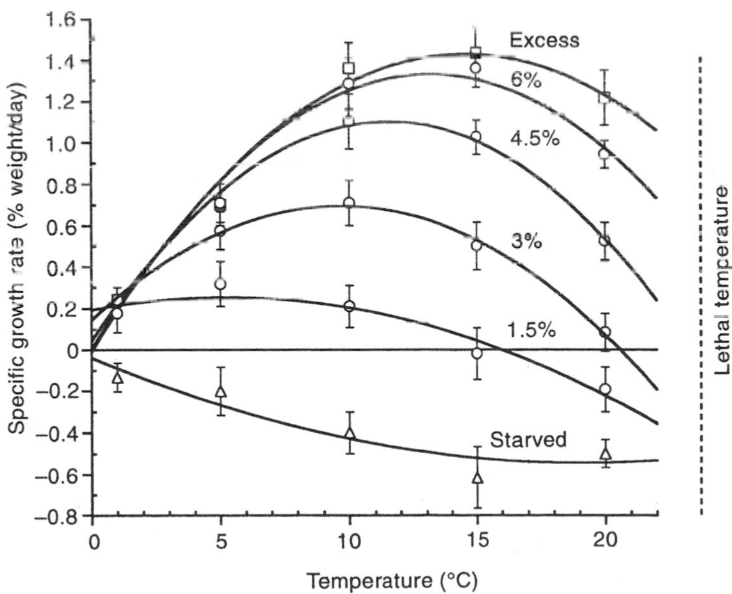

Figure 2.9 Effect of reduced ration on the relation between growth rate (±2 s.e.) and temperature for 7- to 12-month-old sockeye salmon. (From Brett *et al.*, 1969.)

2.5.1 Factors affecting food intake

The amount of food consumed by an animal is restricted by the amount that it can fit into its stomach or foregut. Consequently, food intake is affected by the rate of gastric evacuation. There is considerable evidence to support the hypothesis that gastric evacuation is regulated by the energy content of the food: the greater the energy content, the slower the evacuation rate (Dos Santos and Jobling, 1988). Some evidence has also been presented which suggests that the size of the food particle affects the rate of gastric emptying (Dos Santos and Jobling, 1988). However, because the diets usually fed to cultured species are compounded diets that generally dissociate readily in the stomach or foregut to particles of a similar size, the role of particle size in gastric evacuation of most cultured fish is likely to be minimal. The exception is likely to be those species fed trash fish. However, it can generally be assumed that the most important regulator of gastric emptying is the energy content of the food.

2.5.2 Relationship between feeding frequency and growth

The relationship between feeding frequency and growth rate varies between species. However, a generalization is that increased feeding frequency will not result in reduced growth rates and may result in improved growth rates. Evidence presented by Omar and Günther (1987) indicates that with increased frequency feeding of carp, SGR, PER, %PPV and %ERV are all increased. In that study, feeding frequency was increased from four times daily to six times daily. However, in a study of the effects of feeding frequency on the growth of plaice (*Pleuronectes platessa*), there was no difference between feeding once per day or having food continuously available (Jobling, 1982).

2.6 DETERMINATION OF FEEDING RATE

From the preceding discussion, it can be seen that to determine the appropriate feeding rate for a particular aquaculture venture requires knowledge of the size of the fish, water temperature, biomass in the culture vessel and the relationship between the specific growth rate and the ration for that size of fish at that temperature. For those species that have been well studied, e.g. salmonids, tables available from feed manufacturers give recommendations regarding the optimal ration levels and feeding frequency for that species. An example of such a table is given in Table 2.4 and shows the ration level, the number of days per week and the number of feeds per day that should be given for salmon. For species that have been cultured for a much shorter period, simple tables are available. An example is given in Table 2.5, which shows the feeding frequencies and ration sizes for shrimps.

Table 2.4 Recommended feeding rate (*R*) and frequency (*F*) for salmon

Temperature (°C)	Fish size (g) 0.5 – 1.5 *R*[a]	*F*[a]	1.5 – 2.5 *R*	*F*	2.5 – 3.5 *R*	*F*	3.5 – 5.0 *R*	*F*	5.0 – 7.5 *R*	*F*	7.5 – 11.5 *R*	*F*	11.5 – 18.0 *R*	*F*	>18.0 *R*	*F*
2.2	2.8	7/5	2.4	7/4	1.9	7/2	1.8	6/1	1.4	5/1	1.4	E/1				
3.3	3.0	7/5	2.6	7/4	2.1	7/2	2.0	6/1	1.7	5/1	1.8	E/1				
4.4	3.4	7/5	2.8	7/4	2.3	7/2	1.9	7/1	1.6	6/1	1.3	5/1				
5.5	3.8	7/5	3.0	7/4	2.5	7/2	2.1	7/1	1.9	6/1	1.4	5/1	1.4	E/1	1.0	E/1
6.5	4.2	7/5	3.3	7/2	2.7	7/2	2.1	7/1	2.1	6/1	1.7	5/1	1.8	E/1	1.2	E/1
7.7	4.6	7/5	3.7	7/4	2.9	7/2	2.5	7/1	2.3	6/1	2.0	5/1	2.0	E/1	1.4	E/1
8.8	5.0	7/5	4.1	7/4	3.2	7/2	2.7	7/1	2.6	6/1	2.2	5/1	2.4	E/1	1.6	E/1
9.9	5.6	7/5	4.5	7/4	3.6	7/2	2.9	7/1	2.8	6/1	2.1	6/1	1.8	5/1	1.8	E/1
11.1	6.2	7/5	4.9	7/4	4.0	7/2	3.2	7/1	3.0	6/1	2.3	6/1	2.1	5/1	2.2	E/1
12.1	6.8	7/5	5.4	7/4	4.4	7/2	3.6	7/1	3.3	6/1	2.6	6/1	2.4	5/1	2.6	E/1
13.2	7.5	7/5	6.0	7/4	4.8	7/2	4.0	7/1	3.7	6/1	2.8	6/1	2.7	5/1	3.0	E/1
14.3	8.3	7/5	6.6	7/4	5.3	7/2	4.4	7/1	4.2	6/1	3.0	6/1	2.9	5/1	3.4	E/1
15.4	9.1	7/5	7.2	7/4	5.9	7/2	4.8	7/1	4.7	6/1	3.3	6/1	3.2	5/1	3.8	E/1

[a] *R* means daily feed allowance in g/100 g fish weight. *F* means days fed weekly/number of feedings daily.
Source: Lovell (1989a).

Table 2.5 Recommended feeding rates and frequencies for various sizes of shrimps

Stage/size	*Daily feeding rate (percentage of body weight)*	*Times fed daily*
P_{15}–P_{30}	30–20	6
P_{30}–0.5 g	20–15	4
0.5–2 g	15–12	3–4
2–5 g	12–8	3
5–10 g	8–6	3
10–20 g	6–4	2–3
>20 g	4–3	2–3

Source: Lovell (1989a).

In the absence of such data, aquaculturists should maintain their own set of records regarding growth rates, rations fed, feeding frequencies, water temperature and fish size; over a number of seasons this will allow the refinement of feeding rates to optimize efficiency.

CHAPTER 3

Metabolism

3.1 INTRODUCTION

Nutrients consumed by fish are digested in the gut, are absorbed by the gut lining and appear in the bloodstream as their component molecules (Figure 3.1). These molecules circulate in the body and are taken up by a variety of tissues, where they are subjected to a number of different chemical reactions. The end-point of these reactions is either degradation to liberate the energy contained in the molecules or growth of the organism as evidenced by the production of tissue. Breakdown of molecules is termed catabolism, whilst the synthesis of molecules is termed anabolism. In this chapter the nature and regulation of the catabolic and anabolic reactions that occur in fish are discussed and related to the requirement for various nutrients in the diet.

3.2 METABOLISM

The word 'metabolism' describes the reactions, both catabolic and anabolic, occurring within an organism that result in nutrients being used for energy or growth.

Three major nutrients make up the bulk of food consumed. These are proteins, lipids and carbohydrates. Proteins are digested to release their component amino acids, which are subsequently used either for synthesis of new proteins or for energy. Energy from amino acids is obtained from the carbon skeleton, which is degraded after deamination (Figure 3.2). Lipid is broken down to its constituent fatty acids in the gut. Following absorption, fatty acids are resynthesized into lipid, which then form micelles or droplets. Lipid droplets often circulate in the blood system of fish, but in order to be used by the organism they must again be broken down to their constituent fatty acids. Fatty acids are then used for the synthesis of membranes or are further degraded for energy. Breakdown of lipids or fats occurs much more slowly than breakdown of amino acids. Carbohydrates are generally consumed as complex

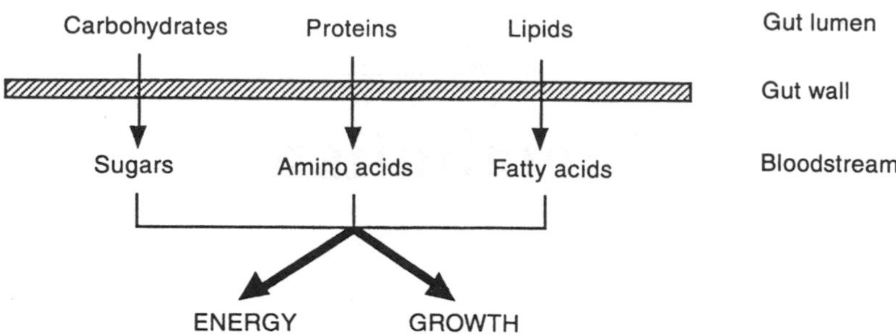

Figure 3.1 The fate of gross nutrients.

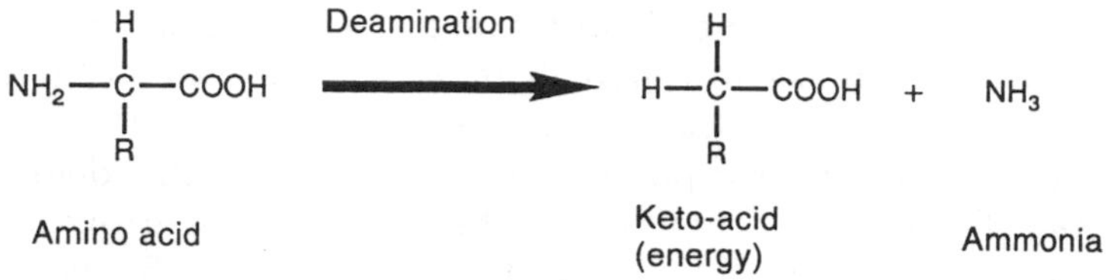

$$\underset{\text{Amino acid}}{NH_2-\overset{\displaystyle H}{\underset{\displaystyle R}{C}}-COOH} \xrightarrow{\text{Deamination}} \underset{\text{Keto-acid (energy)}}{H-\overset{\displaystyle H}{\underset{\displaystyle R}{C}}-COOH} + \underset{\text{Ammonia}}{NH_3}$$

Figure 3.2 Deamination of an amino acid producing ammonia and a carbon skeleton (keto acid) subsequently used to provide energy.

molecules, the most common forms being starch and cellulose. In general, however, cellulose is not digested by fish. Starch is broken down to produce glucose, which again is further degraded to provide energy. The degradation of amino acids, fatty acids and simple sugars is mediated by catabolic pathways, each of which is under a variety of regulation. Most of these pathways were initially described in studies of mammals, but most have been confirmed in studies of teleosts.

A summary of the catabolism of the three major nutrients is shown in Figure 3.3. The degradative pathways of amino acids, fatty acids and simple sugars all reach a common intermediate compound – acetyl coenzyme A (acetyl CoA). Acetyl CoA enters the citric acid cycle, which in turn is linked to the process of oxidative phosphorylation. The result is the production of CO_2, the consumption of O_2 and the liberation of energy, which is then stored as high-energy phosphate molecules, generally adenosine triphosphate (ATP).

Other nutrients are required for optimal growth of aquatic organisms. These include minerals, vitamins, purines and pyrimidines. These compounds are generally used as cofactors in enzyme-mediated reactions, and their importance and roles will be discussed in detail in sections 3.8 and 3.9.

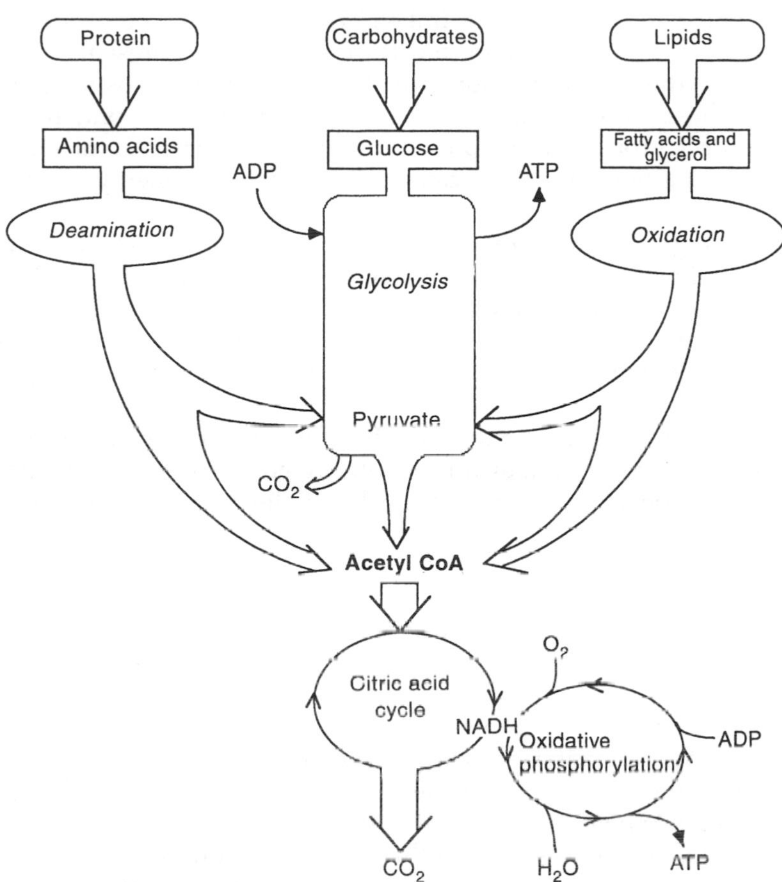

Figure 3.3 The interrelationships between protein, lipid and carbohydrate metabolism. All pathways interconnect at acetyl CoA and citric acid cycle.

3.2.1 High-energy storage molecules

The reactions that maintain a living organism often result in the use or release of energy. Examples of energy-releasing reactions are the metabolism of glucose or amino acids to carbon dioxide. An example of an energy-consuming reaction is the synthesis of a protein from a number of amino acids. In order to couple the energy-releasing reactions to the energy-consuming reactions, high-energy molecules are used as intermediates in the reactions. These intermediates provide a sort of energy 'currency'.

ATP is the high-energy molecule which constitutes the most common cellular energy currency. It is composed of an adenosine moiety to which three phosphoryl groups are linked. The bonds between each of the phosphoryl groups contain most of the energy present in these molecules, thus the conversion of ATP to adenosine diphosphate (ADP) releases the energy contained in the bond between the second and third phosphoryl group. By coupling this energy release to an enzyme-mediated reaction, the energy can be used to drive an energy-consuming reaction.

Other high-energy storage molecules commonly involved in metabolic reactions are guanosine triphosphate (GTP) which acts similarly to ATP, and nicotinamide adenine dinucleotide (NAD) and nicotinamide adenine dinucleotide phosphate (NADP), both of which are readily converted between high-energy reduced forms and low-energy oxidized forms.

This process is fundamental to the functioning of living organisms and you will notice throughout this chapter reference to the production of ATP or the consumption of ATP or other high-energy molecules in each of the reactions.

3.3 CARBOHYDRATES

Carbohydrates, also known as sugars or saccharides, are essential components of all living organisms, having roles as readily metabolized energy stores, as molecules which facilitate transfer of energy throughout the organism and as structural components. The basic units of carbohydrates are known as monosaccharides. Monosaccharides in biological systems are generally either synthesized as a result of a process called gluconeogenesis (section 3.4.2) or are the products of photosynthesis. Monosaccharides are found as principal components of nucleic acids or linked together in chains to form polymer complexes.

The polymer complexes fall into two distinct groups. Oligosaccharides consist of only a few linked monosaccharide units; they are often associated with proteins, forming glycoproteins, and with lipids, forming glycolipids. Polysaccharides consist of many covalently linked monosaccharide units and are extremely large. Their usual function is either as structural molecules (e.g. cellulose in the cell walls of plants or chitin in the exoskeleton of invertebrates) or as dense high-energy storage molecules such as starch in plants and glycogen in animals.

3.3.1 Monosaccharides

Monosaccharides occur in an enormous variety of forms. The basic form is that of a carbon chain backbone, about which oxygen and hydrogen atoms are arranged. The backbone may consist of three, four, five, six or even seven carbon (C) atoms. Occasionally there are sulphur (S) and/or amine (NH_2) groups present as side chains. Monosaccharide molecules with the same chemical composition may be arranged differently, giving the molecule a different shape. These variations are known as isomers.

The most commonly occurring monosaccharide in nature is D-glucose. It and several other monosaccharides, including D-glyceraldehyde, D-ribose, D-mannose and D-galactose are important components of larger biological molecules. The structures of these monosaccharides are shown in Figure 3.4.

Another important feature of monosaccharides is their capacity to interconvert readily between the linear form and a circular form. The interconversion of glucose is shown in Figure 3.5.

Many metabolic reactions involve hydrolysis (breakdown) or synthesis of one form of carbohydrate or another. A discussion of the nature of these molecules is too great to go into in detail here but may be found in most general texts on biochemistry. However, the important molecules will be described when discussing the various metabolic reactions.

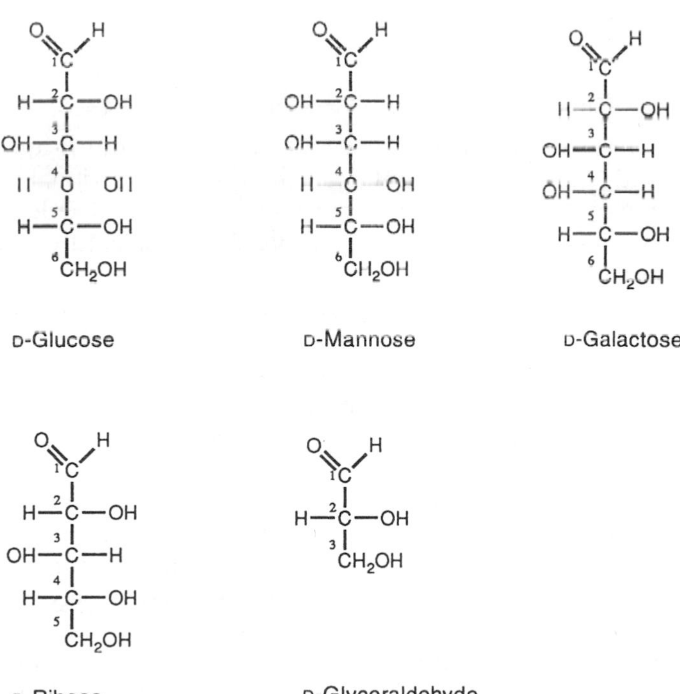

Figure 3.4 The chemical structure of several major monosaccharides.

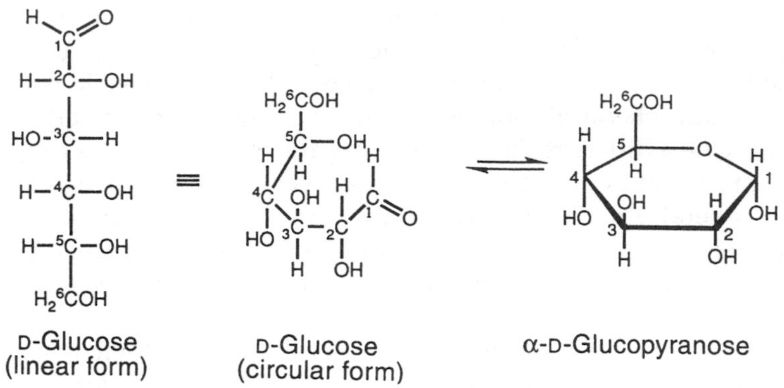

Figure 3.5 The reaction of D-glucose associated with its conversion from a linear form to a cyclic form called D-glucopyranose.

3.3.2 Polysaccharides

Monosaccharides occur in nature primarily in their circular form and in polysaccharide complexes. The physical characteristics of polysaccharides are determined by the exact monosaccharides of which they are composed and the way these are joined together. To understand this a basic knowledge of the nomenclature of polysaccharide bonding and of the effect on the three-dimensional structure of different types of bonds is necessary. Two types of bonds occur – α-linkages and β-linkages. A polysaccharide linked by an α-linkage has a three-dimensional structure that is irregular, with branches and spaces within the molecule, while β-linkages produce a regular three-dimensional structure with the sacchrides tightly packed.

Any organic molecule consisting of more than one carbon atom has those atoms numbered. The convention for numbering the more important saccharides is shown in Figure 3.4. Several common disaccharides and their constituent monosaccharides are shown in Figure 3.6. These will be used as examples to demonstrate the variety of bonds called glycosidic bonds that can occur between monosaccharides to produce polysaccharides with different structures and functions. Sucrose is composed of glucose and fructose, which are joined with an α(1–2) linkage, that is the number 1 carbon of glucose is linked to the number 2 carbon of fructose by an α-linkage. Sucrose occurs in large amounts in some plants and is a common form of carbohydrate supplement in pig and cattle diets. Lactose, which is found only in mammalian milk, is composed of galactose and glucose linked by a β(1–4) linkage. Maltose, isomaltose and cellobiose are all composed of two glucose monosaccharides. Although these molecules differ only in the way that they are linked together (Figure 3.6), these linkages result in an enormous difference in their properties. Maltose is linked by an α(1–4) linkage, isomaltose by an α(1–6) linkage and cellobiose by a β(1–4) linkage. These glycosidic bonds are broken to liberate the two monosaccharides by enzymes called carbohydrases. There are numerous examples in all organisms of α-carbohydrases, but carbohydrases

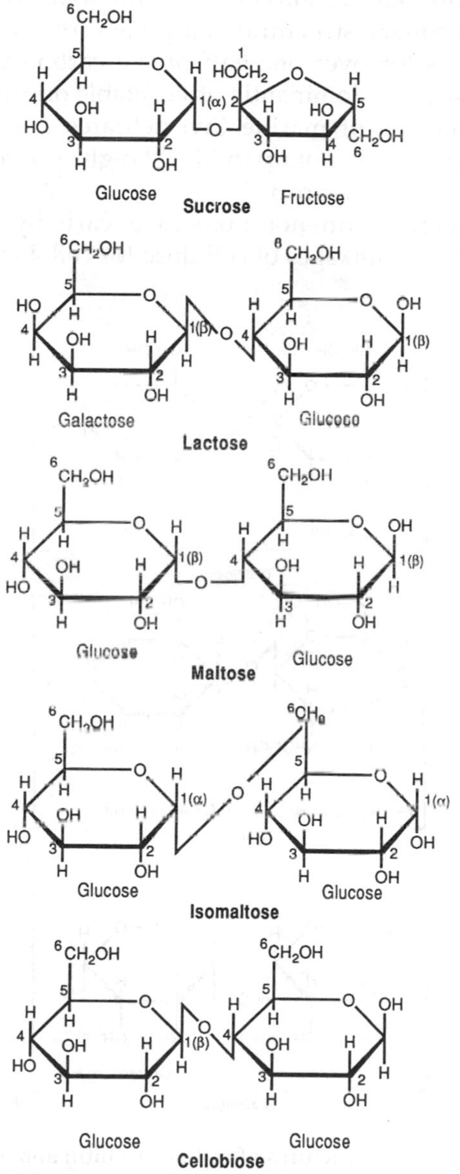

Figure 3.6 The chemical structure of some common disaccharides.

capable of hydrolysing the β(1–4) linkage in cellobiose only occur naturally in bacteria, while a carbohydrase capable of breaking the β(1–4) link in lactose is found only in neonatal mammals. As a result, α-linked sugars are much more useful for inclusion in diets as carbohydrate sources than β-linked sugars.

Four particularly important polysaccharides will be discussed here. These are cellulose, chitin, amylose (or amylopectin) and glycogen. Cellulose is the primary structural component of terrestrial plant cell walls and accounts for over one-half of the carbon in the biosphere. Although cellulose is predominantly of vegetable origin, it also occurs in the stiff outer mantles of marine invertebrates known as tunicates. Cellulose is a linear polymer of up to 15 000 D-glucose residues linked by β(1–4) glycosidic bonds (Figure 3.7).

Although vertebrates do not possess a carbohydrase capable of hydrolysing the β(1–4) linkages of cellulose (or cellobiose), the digestive

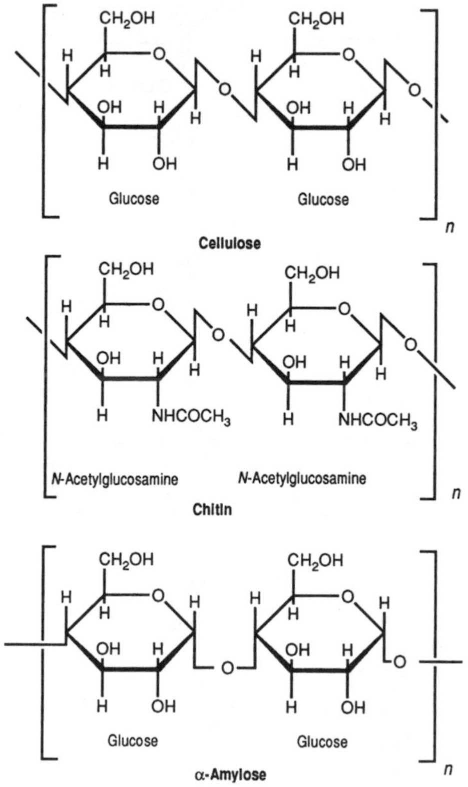

Figure 3.7 The repeating structures of cellulose, chitin and amylose (n may be as great as 10^6).

tracts of mammalian herbivores contain symbiotic microorganisms that secrete a series of carbohydrases collectively known as cellulases. Nevertheless, the degradation of cellulose is a slow process because the molecules are tightly packed and the bonds are not easily accessible. Although cellulase has not been identified in large amounts in any of the finfish (Horn, 1989), under certain conditions materials from plant cell walls are absorbed (Anderson, 1987; Horn, 1989). The nature of these conditions (e.g. food deprivation) and the time taken to observe significant uptake mean that, for practical purposes, cellulose is considered to be unavailable to fish.

Chitin is the principal structural component of the exoskeletons of invertebrates such as crustacea. It is also present in the cell walls of most fungi and many algae. Chitin is a linear polymer of β(1–4) linked *N*-acetyl-D-glucosamine residues (Figure 3.7). It differs chemically from cellulose only in that each C(2)-OH group is replaced by an acetamido group ($NHCOCH_3$).

As a major structural element of crustacea, chitin is an important nutritional component of the diets of carnivorous fish. It is particularly important in the diets of larval and juvenile fish, since it is usual practice to feed *Artemia*, *Daphnia* and other crustacea to larval finfish. Endogenous chitinase, the enzyme that degrades chitin, has been described in the digestive tracts of a number of finfish (Lindsay and Gooday, 1985) and so it can be considered that chitin is available to these animals. However, the efficiency with which it is digested is unknown.

Starch, the carbohydrate storage molecules found in plants, is deposited in the cytoplasm of plant cells in soluble granules composed of α-amylose and amylopectin. α-Amylose is a linear polymer of several thousand glucose residues linked by α(1–4) bonds (Figure 3.7). Although α-amylose is an isomer of cellulose, it has very different structural properties. This is because the β-glycosidic linkages in cellulose cause each successive glucose residue to flip 180° with respect to the preceding residue, so that the polymer assumes an easily packed fully extended conformation. In contrast, the α-glycosidic bonds of α-amylose cause it to adopt an irregular helical conformation. The bonds between glucose residues in α-amylose are therefore much more readily accessible to degradative enzymes and the breakdown of starch is correspondingly faster than the breakdown of cellulose. As a result, starch is a useful component for inclusion in artificial diet preparations.

Amylopectin is a complex of amylose chains having α(1–6) branch points every 24–30 glucose residues. Amylopectin molecules contain up to 10^6 glucose residues and are readily digested by fish.

Glycogen is the storage polysaccharide of animals. It is present in all cells, but is most prevalent in skeletal muscle and liver, where it occurs as cytoplasmic granules. The primary structure of glycogen resembles that of amylopectin, but glycogen is more highly branched with branch points occurring every 8–12 residues. The synthesis and degradation of glycogen are described in section 3.4.3.

A number of other polysaccharides occur in animals. Principal amongst these are mucopolysaccharides and glycoproteins. Glycoproteins vary in carbohydrate content from less than 1% to greater than 90% by weight, and have functions that span the entire spectrum of protein activities. In general, the carbohydrates in mucopolysaccharides and glycoproteins are readily degraded by digestive enzymes and are therefore available when included in diets.

3.4 CARBOHYDRATE METABOLISM

Following digestion (covered in detail in Chapter 4), carbohydrates are absorbed through the wall of the digestive tract. They are absorbed and enter the bloodstream as monosaccharides (i.e. glucose, fructose, galactose, etc.). In vertebrates the absorbed nutrients are transported to the liver where the initial processes of metabolism occur. However, all the processes of metabolism that are to be described here occur in all tissues. Of the monosaccharides, glucose is the most important, acting as a major metabolic energy source circulating in all vertebrates. Some tissues (e.g. brain) use only glucose as an energy source and so maintenance of blood glucose levels is a very important process. Four reaction pathways exist to do this: glycolysis – the breakdown of glucose liberating energy; gluconeogenesis – the synthesis of glucose from other molecules; glycogen synthesis – the storage of excess glucose in glycogen molecules; and glycogenolysis – the breakdown of glycogen to provide free glucose.

3.4.1 Glycolysis

Glycolysis is the conversion of glucose by means of a complex series of enzyme and coenzyme-mediated reactions into pyruvate. Energy released by these reactions is stored in the form of high-energy ATP molecules (Figure 3.8). Glucose enters most cells by a specific carrier that transports it from the exterior of the cell into the cytosol where the glycolytic enzymes are located.

Glycolysis can be described by the equation:

$$\text{Glucose} + 2\text{NAD}^+ + 2\text{ADP} + 2\text{P}_i \rightleftharpoons 2\text{NADH} + 2\,\text{pyruvate} + 2\text{ATP} + 2\text{H}_2\text{O} + 4\text{H}^+$$

where P_i is inorganic phosphate. The process can be considered to occur in two stages. In the initial stage, the six-carbon glucose is phosphorylated and cleaved to yield two molecules of glyceraldehyde 3-phosphate. This process uses the energy from two ATP molecules to

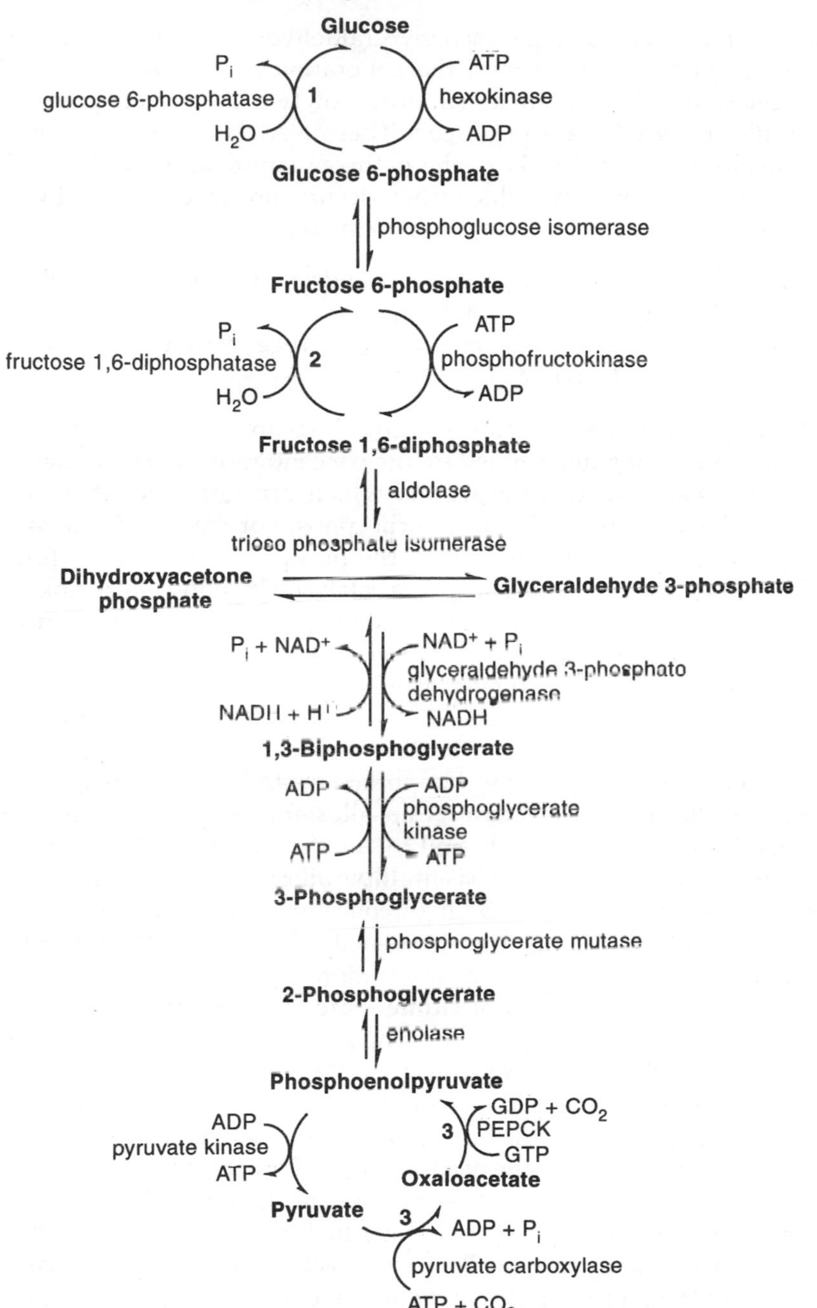

Figure 3.8 The pathways of glycolysis and gluconeogenesis. The three numbered steps are catalysed by different enzymes in gluconeogenesis. Substrates are shown in bold, enzymes in plain text. PEPCK, phosphoenolpyruvate carboxykinase; GDP, guanidine diphosphate; GTP, guanidine triphosphate.

drive it. In the second stage, each glyceraldehyde 3-phosphate molecule is converted to pyruvate with the generation of a total of four ATP molecules. In addition to a net profit of two ATP molecules, two molecules of NADH are produced. These NADH molecules must be continually reoxidized to keep the pathway supplied with NAD^+. The two common ways by which this occurs are determined by the conditions under which glycolysis is occurring:

1. Under aerobic conditions, the mitochondrial oxidation of each NADH to NAD^+ yields three ATP molecules; or
2. Under anaerobic conditions, NAD^+ is regenerated when NADH reduces pyruvate to lactate.

Although most of the reactions of glycolysis are equilibrium reactions with the rates being determined by the concentrations of substrates and products, there are three reactions which are not reversible under intracellular conditions. These are the phosphorylation of glucose to glucose 6-phosphate by hexokinase, the phosphorylation of fructose 6-phosphate to fructose 1,6-diphosphate catalysed by phosphofructokinase and the dephosphorylation of phosphoenolpyruvate (PEP) to pyruvate by pyruvate kinase.

The rate of each of these three reactions is regulated by molecules other than the substrates or products. Hexokinase is inhibited by certain sulphydryl agents. The second of these reactions, that catalysed by phosphofructokinase, is the most important control point in the glycolytic sequence. Phosphofructokinase is an allosteric enzyme. An allosteric enzyme is one which, when bound by an effector molecule, changes its activity. Phosphofructokinase has multiple allosteric modulators or effector molecules; it is inhibited by high concentrations of ATP, citrate and long-chain fatty acids and it is stimulated by ADP or AMP. Therefore, whenever the cell has a high concentration of ATP, or whenever other energy sources – fatty acids or citrate – are available, phosphofructokinase is inhibited and glycolysis is turned off. Conversely, whenever the ATP concentrations are low and thus AMP and ADP predominate, or whenever citrate or fatty acid levels are low, phosphofructokinase activity is stimulated. In this way the breakdown of glucose (or any other hexose, see below) is regulated by the need of the cell for energy.

Pyruvate kinase is also an allosteric enzyme. It is turned off whenever the ATP concentration increases or other fuels such as fatty acids, citrate, acetyl CoA or alanine are available. It is activated whenever there is a build-up of the preceding glycolytic intermediates, particularly fructose 1,6-diphosphate and PEP.

While glucose is the primary end product of the degradation of starch and glycogen, three other hexoses (fructose, galactose and mannose) are

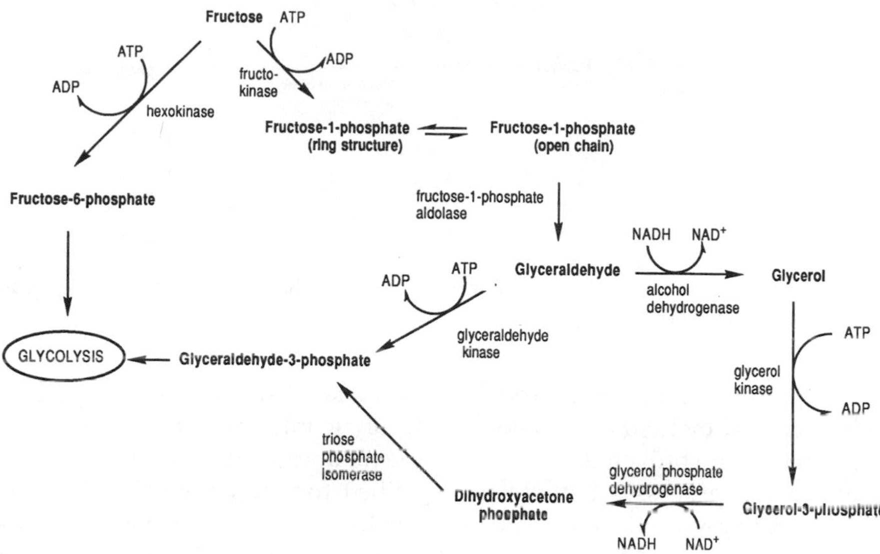

Figure 3.9 The metabolism of fructose showing the intermediate compounds and the enzymes catalysing the reactions.

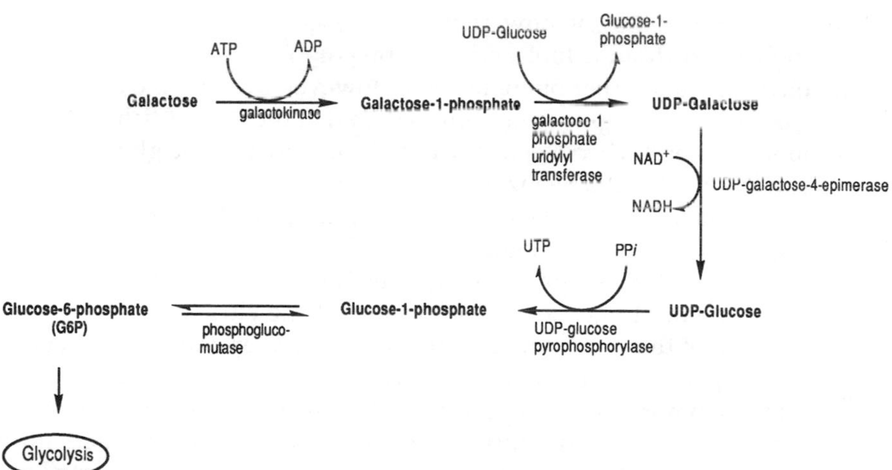

Figure 3.10 The metabolism of galactose showing the intermediate compounds and the enzymes catalysing the reactions. UDP, uridine diphosphate; UTP, uridine triphosphate.

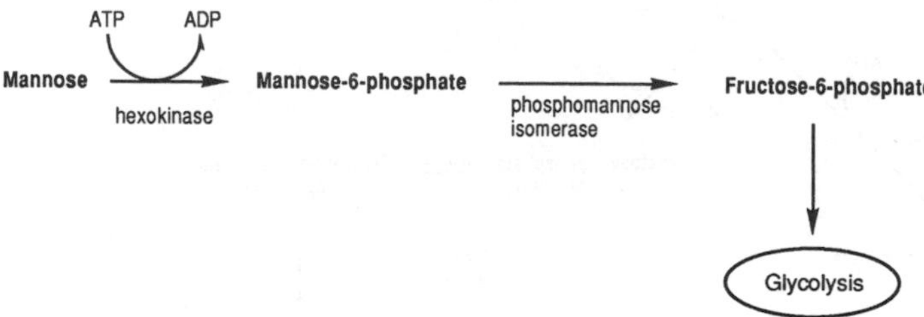

Figure 3.11 The metabolism of mannose showing the intermediate compounds and the enzymes catalysing the reactions.

also prominent digestion products. The metabolism of these compounds is preceded by their conversion to glycolytic intermediates, from which point they are broken down in the same manner as glucose. A number of enzymatic processes have been described for the conversion of these hexoses to glycolytic intermediates. Fructose is either phosphorylated by hexokinase to produce fructose 6-phosphate or it goes through a series of reactions catalysed by six different enzymes to produce glyceraldehyde 3-phosphate (Figure 3.9). Similarly five enzymes catalyse the conversion of galactose to glucose 6-phosphate (Figure 3.10), while two enzymes catalyse the conversion of mannose to fructose 6-phosphate (Figure 3.11).

3.4.2 Gluconeogenesis

You will appreciate by now that glucose occupies a central role in metabolism, both as a fuel and as a precursor of essential structural carbohydrates and other biomolecules. However, because the capacity of an organism to store glycogen is limited and, in the case of fish, glycogen metabolism is perturbed (section 3.4.3), the majority of the glucose needs of a fish are met by gluconeogenesis.

The precursors that can be converted to glucose include the glycolysis products lactate and pyruvate, intermediates of the citric acid cycle (section 3.5) and most amino acids. The amino acids enter gluconeogenesis following their metabolism to one of the citric acid cycle intermediates All of these substances must first be converted to oxaloacetate.

Gluconeogenesis is the reverse process of glycolysis and occurs using the same enzymes as glycolysis with the exception of three reactions (Figure 3.8). These are the three reactions that serve as the regulatory steps in glycolysis. In reverse order, pyruvate is converted to oxaloacetate before conversion to PEP. The two enzymes involved in this process are pyruvate carboxylase, which catalyses the formation of oxaloacetate from pyruvate, and PEP carboxykinase (PEPCK) which converts oxaloacetate to PEP (Figure 3.8). Regulation of pyruvate carboxylase activity occurs as a result of increases in the concentration of two compounds – acetyl CoA (which acts as an allosteric activator of

Table 3.1 Regulators of some glycolytic and gluconeogenic enzyme activities

Enzyme	*Allosteric inhibitors*	*Allosteric activators*
Phosphofructokinase	ATP, citrate	AMP, fructose 2,6-diphosphate, fructose 6-phosphate, P_i
Fructose 1,6-diphosphatase	AMP, fructose 2,6-diphosphate, fructose 6-phosphate	Glycerol 3-phosphate
Pyruvate kinase	Alanine	Fructose 1,6-diphosphate
Pyruvate carboxylase		Acetyl CoA

Source: Adapted from Voet and Voet (1990).

pyruvate carboxylase) and oxaloacetate (which occurs when the citric acid cycle is inhibited by high concentrations of ATP, section 3.5). When these events coincide, the oxaloacetate undergoes gluconeogenesis. The second enzyme in the process, PEPCK, is unregulated, although the activity of this enzyme increases when concentrations of oxaloacetate increase.

The second regulatory reaction of gluconeogenesis is that of the conversion of fructose 1,6-diphosphate to fructose 6-phosphate, catalysed by the enzyme fructose 1,6-phosphatase. This is also an allosteric enzyme, being inhibited by the presence of AMP and stimulated by the presence of 3-phosphoglycerate and citrate.

The third variation between gluconeogenesis and glycolysis is the reaction between glucose 6-phosphate and glucose catalysed by the magnesium (Mg^{2+})-dependent enzyme glucose 6-phosphatase. This reaction does not occur in all tissues; most free glucose produced by gluconeogenesis is synthesized in the liver.

A summary of the regulators of glycolytic and gluconeogenic activity is given in Table 3.1. It should be noted that glycolytic activity is increased when high-energy storage molecules such as ATP are depleted; gluconeogenic activity is increased when such molecules are abundant in the cell.

3.4.3 Glycogen metabolism

Glucose enters the bloodstream and becomes available for metabolism to release energy in a pulsatile fashion as a result of periodic bouts of feeding and digestion. Thus there are periods when excess glucose is available for ATP production and other periods when there is insufficient free glucose in the bloodstream to meet the needs of the animal. In order to overcome the pulsatile nature of the availability of glucose, much excess glucose is polymerized to form glycogen for storage. In most vertebrates, glycogen is readily available in times of metabolic need. The glucose units in glycogen are mobilized by sequential removal from the branched ends of the glycogen molecule. As a result, the branching structure of glycogen permits rapid degradation through the simultaneous release of glucose units from the end of every branch. Excess glucose can also be converted to fatty acids and stored (section 3.7.3).

Glycogen synthesis requires the actions of three enzymes and follows the conversion of glucose to glucose 6-phosphate in the first step of glycolysis, and the subsequent conversion to glucose 1-phosphate by phosphoglucomutase.

The first enzyme, uridine diphosphate glucose pyrophosphorylase (UDP-glucose pyrophosphorylase), catalyses the reaction

$$\text{Glucose 1-phosphate} + \text{uridine triphosphate} \rightarrow \text{UDP-glucose} + 2P_i$$

The second enzyme, glycogen synthase, catalyses the addition of UDP-glucose to the end of the glycogen molecule with the liberation of UDP. The third reaction, catalysed by the enzyme called glycogen branching enzyme, forms branches in the glycogen chains by the formation of β(1–6) glycosidic bonds.

Glycogen synthase is the rate-limiting enzyme in this process. It is an allosteric enzyme, the activity of which is inhibited by AMP and enhanced by ATP, glucose 6-phosphate and glucose. Thus glycogen synthase is activated when the cellular energy levels are high and inhibited when these levels are low.

Glycogen degradation also requires the actions of three enzymes. The first enzyme, glycogen phosphorylase, catalyses the reaction

$$\text{Glycogen} + P_i \rightleftharpoons \text{glycogen} + \text{glucose 1-phosphate}$$

The second enzyme, glycogen debranching enzyme, removes the branching β(1–6) glycosidic bonds, therefore permitting the glycogen phosphorylase action to act on the whole glycogen molecule. The third enzyme, phosphoglucomutase, converts glucose 1-phosphate to glucose 6-phosphate, which then enters glycolysis.

Glycogen phosphorylase is another allosteric enzyme and is the rate-limiting enzyme in glycogen degradation. Its activity is inhibited by the presence of high levels of ATP, glucose 6-phosphate and glucose, and enhanced by high levels of AMP. As a result, in periods of energy shortage in a cell (i.e. when AMP concentrations are high), glycogen phosphorylase is activated. The reverse is true when energy is abundant in a cell. Glycogen debranching enzyme is also rate limiting in that it has a lower maximal rate than glycogen phosphorylase. However, neither it nor phosphoglucomutase is affected to any great extent by other compounds.

As a result of the allosteric regulation of glycogen synthase and glycogen phosphorylase activities, it is possible for most vertebrates to convert excess glucose into glycogen, and to make the glucose available from the glycogen when it is required.

3.4.4 Carbohydrate metabolism in fish

The inability of fish to utilize dietary carbohydrate is well illustrated by experiments known as glucose tolerance tests. A large amount of glucose is administered orally to the animal and the blood concentration of glucose measured over time. In humans, blood glucose returns to normal about 1–2 h after feeding. Fish respond to these tests much more slowly. For example, trout show increasing blood concentrations of glucose for at least 7 h (Palmer and Ryman, 1972) .

There is an apparent relationship between the natural diet of a species in the wild and its capacity to deal with glucose. Carnivorous fish such as the yellowtail (*Seriola quinqueradiata*) respond less well to glucose tolerance tests than the red sea bream or snapper (*Pagrus australis*). The omnivorous carp (*Cyprinus carpio*) has been found to be able to clear blood glucose most quickly (Furuichi and Yone, 1981), although the response is still very poor relative to mammals. This information indicates that fish are unable to metabolize glucose quickly. When fish are presented with diets high in carbohydrate, the excess glucose appears to be used to synthesize glycogen (Palmer and Ryman, 1972).

Although the enzymes required for glycogen degradation are present in fish, their role in maintaining a constant blood glucose level between bouts of feeding seems to be minimal in both eels (*Anguilla anguilla*) (Dave *et al.*, 1975) and sockeye salmon (*Oncorhynchus nerka*) (French *et al.*, 1983) Extensive periods of starvation (up to 164 days) do not result in the depletion of muscle glycogen in these species, although carp show a 75% reduction of hepatopancreatic glycogen after 100 days of starvation (Nagai and Ikeda, 1971).

Murat *et al.* (1978) showed that the gluconeogenic pathway is more important in maintaining blood glucose levels in carp than the breakdown of glycogen. The effects of glucagon, a hormone responsible for raising glucose concentration in the blood when it is below that required for proper functioning of the animal, is largely removed by an inhibitor of gluconeogenesis. Glucagon thus apparently stimulates gluconeogenesis, not glycogenolysis.

Further demonstration of the relative inefficiency of glucose metabolism in fish can be obtained from the data of Bever *et al.* (1977) and Lin *et al.* (1978). Using ^{14}C-labelled glucose, these workers showed that glucose turnover (a measure of glucose degradation and synthesis) in kelp bass (*Paralabrax* sp.) and Coho salmon (*Oncorhynchus kisutch*) is 5–10% that of omnivorous mammals and that glucose utilization is about 10% that of omnivorous birds and mammals.

From this information it is apparent that carbohydrate in fish diets has to be carefully controlled since the excess deposited as glycogen is subsequently less readily available to the fish for use as energy than other molecules. Feeding studies (Furuichi and Yone, 1980) have shown that optimal levels of dietary carbohydrate for carp are 30–40%, for red sea bream approximately 20% and for yellowtail approximately 10%.

When incorporating carbohydrate into a fish diet, it is preferable to utilize a carbohydrate that requires some degree of digestion, such as starch, rather than monosaccharides such as glucose. This will at least allow a time lag between consumption of the carbohydrate and the appearance of glucose and other monosaccharides in the blood of the fish. The resultant slower increase in the plasma of the fish will result in a greater degree of catabolism of these substrates for energy.

3.5 THE CITRIC ACID CYCLE

The citric acid cycle (also known as the tricarboxylic acid or Krebs' cycle) is a series of reactions that converts the acetyl group (CH_3CO-) of acetyl CoA to two molecules of CO_2 such that the energy released is utilized for ATP generation. A summary of the citric acid cycle is shown in Figure 3.12. The initial reaction occurs when oxaloacetate reacts with acetyl CoA to form citrate. Citrate is converted to isocitrate. The conversion of isocitrate to oxalosuccinate results in a conversion of one NAD^+ to NADH. Oxalosuccinate is then converted to α-ketoglutarate with the liberation of a carbon dioxide molecule. α-Ketoglutarate is transformed to succinyl CoA with the liberation of the second carbon dioxide and the production of a second NADH. Succinyl CoA is converted to succinate (a molecule of GTP is produced in this reaction). Succinate is then transformed to fumarate with the conversion of flavin–adenine dinucleotide (FAD) to $FADH_2$. Fumarate is then transformed to oxaloacetate by an intermediate L-malate with the production of a third NADH.

The citric acid cycle is truly cyclic with the production of oxaloacetate from oxaloacetate and acetyl CoA. An infinite number of acetyl CoA molecules can be oxidized by a single oxaloacetate. The reaction can thus be summarized

$$\text{Acetyl CoA} + 3\ NAD^+ + FAD + GDP + P_i \rightarrow 2\ CO_2 + 3\ NADH + FADH_2 + GTP$$

As stated earlier (section 3.2.1), although NADH and $FADH_2$ are used as sources of energy in some cellular reactions, by far the most common form of energy storage in the cell is ATP. As a result, it is advantageous for the cell to convert the NADH and $FADH_2$ into ATP. It does this by an extremely complex process called oxidative phosphorylation. For the purposes of this discussion it is sufficient to know that, by a series of reactions, two molecules of O_2 are converted to H_2O; during the process three NADH molecules are consumed to produce nine ATP molecules, and one $FADH_2$ is consumed to produce a further two ATP molecules. The process of oxidative phosphorylation and the consequent production of high-energy ATP molecules ceases when an animal is deprived of oxygen and is clearly a very important metabolic process.

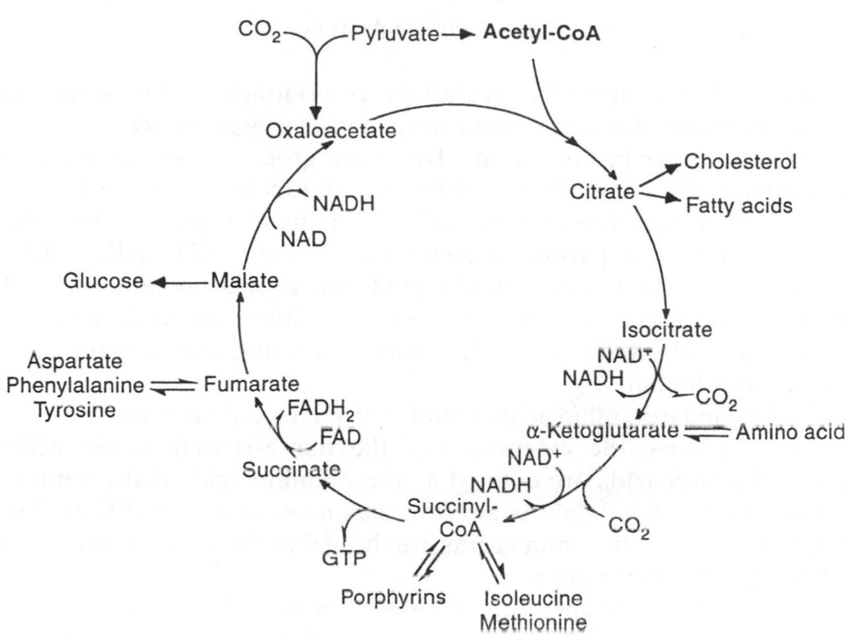

Figure 3.12 The citric acid cycle showing the positions at which intermediates are drawn off for use in anabolic pathways and the points where products of catabolism enter.

The citric acid cycle is regulated by its products in a manner similar to glycolysis. Acetyl CoA, citrate and succinyl CoA all inhibit the reactions of which they are the products. ATP inhibits the production of α-ketoglutarate and NADH inhibits the production of acetyl CoA, citrate, α-ketoglutarate and succinyl CoA. In this way, if there is sufficient NADH and ATP available to the cell, the citric acid cycle is inhibited. When these compounds are limiting, the citric acid cycle is activated.

The citric acid cycle is also important in another way. Most pathways are either catabolic or anabolic, not both. However, the citric acid cycle is capable of performing both functions. It is catabolic because it involves degradation and results in the production of high-energy compounds, notably the intermediates of the cycle are only required in catalytic amounts to maintain the degradative function of the cycle. The anabolic function lies in its production of the starting materials for several biosynthetic pathways, as shown in Figure 3.12. These biosynthetic pathways are discussed in some detail elsewhere in this chapter, but are summarized here. Glucose biosynthesis (gluconeogenesis) utilizes malate, lipid (fatty acid and cholesterol) biosynthesis utilizes citrate, odd-chain fatty acid and porphyrin synthesis utilizes succinyl CoA and amino acid biosynthesis utilizes oxaloacetate, α-ketoglutarate, succinyl CoA or fumarate as substrates.

3.6 AMINO ACIDS

Amino acids are important metabolic compounds in fish since, for reasons discussed later, they serve as the major energy source.

Amino acids can be divided into two major groups: those that are used in protein synthesis and those that have other functions. Table 3.2 shows the molecular structure of those amino acids used in protein synthesis. They conform to a particular general structure – $NH_2CHRCOOH$ – where R is any one of a number of organic side chains having some or all of carbon, hydrogen, oxygen, nitrogen and sulphur atoms (Figure 3.2). Some of the side chains are quite complex, forming ring structures, for example tryptophan.

The 20 standard amino acids found in proteins can be divided further into two groups, the essential and the non-essential amino acids. Essential amino acids are defined as those amino acids that cannot be synthesized by an animal. As a result, they must be obtained from their diet. The non-essential amino acids are those that the animal can synthesize from other compounds.

Amino acids suffer one of three fates in an animal. They can be used for protein synthesis, they can be subjected to structural change to produce another compound or they can be degraded for energy. Most amino acids consumed by fish are degraded for energy.

3.6.1 Amino acid metabolism

The breakdown of amino acids can incorporate three stages which are:

1. deamination (removal of the amino group, which is either converted to ammonia or transferred to become the amino group of a glutamic acid molecule);
2. conversion of amino acid carbon skeletons (the α-keto acids produced by deamination) to citric acid cycle intermediates; and
3. incorporation of ammonia into urea.

In all animals, ammonia build-up in the blood is toxic. In order to prevent this happening in higher vertebrates, ammonia is converted to urea for excretion in the urine. Animals that may suffer a limitation in the availability of water, such as desert-dwelling lizards, may even excrete this ammonia via solid, crystallized uric acid. The bony fish have no need to expend energy converting ammonia to urea since they are bathed in an aquatic medium and have an extremely good transfer mechanism for ammonia across the gills.

The marine bony fish, although living in a solution which is more concentrated than their body fluids, are able to osmoregulate by excreting salt through their gills. The elasmobranchs do not maintain the salinity

Table 3.2 Covalent structures of the 20 amino acids found in proteins

Name	*Structural formula*[a]	*Residue mass (Da)*	*Average occurrence in proteins (%)*
Alanine	$H-C(COO^-)(NH_3^+)-CH_3$	71.0	9.0
Arginine	$H-C(COO^-)(NH_3^+)-CH_2-CH_2-CH_2-NH-C(NH_2)=NH_2^+$	157.2	4.7
Asparagine	$H-C(COO^-)(NH_3^+)-CH_2-C(=O)NH_2$	114.1	4.4
Aspartic acid	$H-C(COO^-)(NH_3^+)-CH_2-C(=O)O^-$	114.0	5.5
Cysteine	$H-C(COO^-)(NH_3^+)-CH_2-SH$	103.1	2.8
Glutamic acid	$H-C(COO^-)(NH_3^+)-CH_2-CH_2-C(=O)O^-$	128.1	6.2
Glutamine	$H-C(COO^-)(NH_3^+)-CH_2-CH_2-C(=O)NH_2$	128.1	3.9

Table 3.2 (Continued)

Name	*Structural formula*[a]	*Residue mass (Da)*	*Average occurrence in proteins (%)*
Glycine	$H-C(COO^-)(NH^+_3)-H$	57.0	7.5
Histidine	$H-C(COO^-)(NH^+_3)-CH_2-$ (imidazole ring, N, NH)	137.1	2.1
Isoleucine	$H-C(^-OOC)(NH^+_3)-C(CH_3)(H)-CH_2-CH_3$	113.1	4.6
Leucine	$H-C(COO^-)(NH^+_3)-CH_2-CH(CH_3)_2$	113.1	7.5
Lysine	$H-C(COO^-)(NH^+_3)-CH_2-CH_2-CH_2-CH_2-NH^+_3$	129.1	7.0
Methionine	$H-C(COO^-)(NH^+_3)-CH_2-CH_2-S-CH_3$	131.1	1.7
Phenylalanine	$H-C(COO^-)(NH^+_3)-CH_2-$ (benzene ring)	147.1	3.5

Table 3.2 (Continued)

Name	*Structural formula*[a]	*Residue mass (Da)*	*Average occurrence in proteins (%)*
Proline	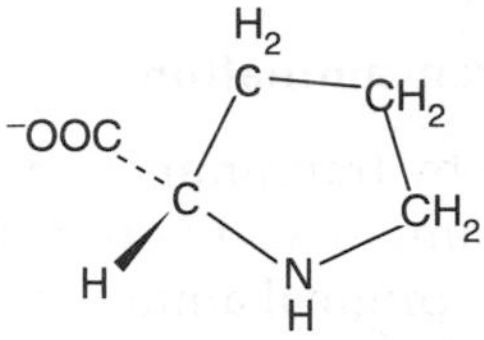	97.1	4.6
Serine	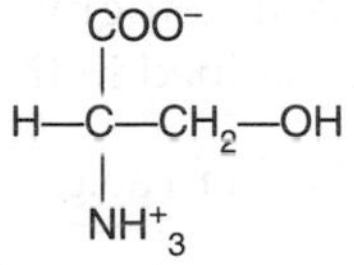	87.0	7.1
Threonine	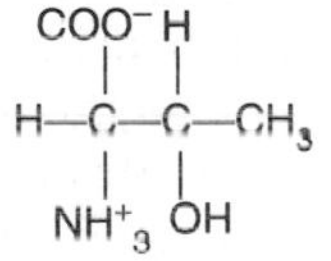	101.1	6.0
Tryptophan	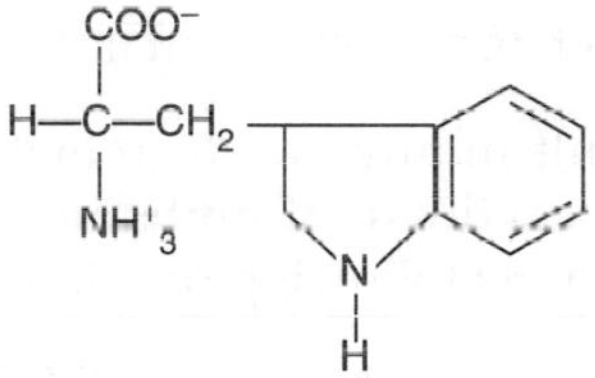	186.2	1.1
Tyrosine	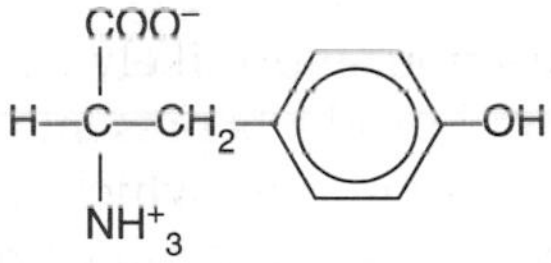	163.1	3.5
Valine	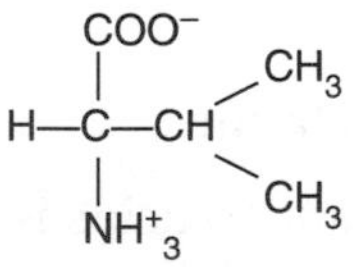	99.1	6.9

[a] The ionic forms shown are those predominating at pH 7.0. For the molecular masses of the parent amino acids, add 18.0 Da, the molecular mass of water, to the residue masses.
Source: Adapted from Voet and Voet (1990).

of their body fluids in this way. Rather, many of these types of fishes produce urea from the ammonia liberated by deamination and use high concentrations of urea in their body fluids to prevent dehydration and salt accumulation.

3.6.2 Transamination

Most amino acids are deaminated by transamination. In this process, the amino group of the deaminated amino acid is transferred to an α-keto acid to yield the α-keto acid of the original amino acid and a new amino acid. These reactions are catalysed by enzymes called transaminases. The predominant amino group acceptor is α-ketoglutarate producing glutamic acid as the new amino acid, as described in the following reaction:

$$\text{Amino acid} + \alpha\text{-ketoglutarate} \rightleftharpoons \alpha\text{-keto acid} + \text{glutamic acid}$$

This reaction requires a substance called pyridoxal 5′-phosphate, a derivative of pyridoxine, also known as vitamin B_6, as a cofactor.

Each transaminase has a specific amino acid substrate, and most accept only α-ketoglutarate or oxaloacetate as the α-keto acid substrate. This ensures that only glutamic acid or aspartic acid is the major amino acid product of transamination. Glutamic acid and aspartic acid are interconverted by the reaction

$$\text{Glutamic acid} + \text{oxaloacetate} \rightleftharpoons \alpha\text{-ketoglutarate} + \text{aspartic acid}$$

Transamination does not result in any net deamination. Deamination occurs largely through the breakdown of glutamic acid by glutamate dehydrogenase, as described in the following reaction:

$$\text{Glutamate} + NAD^+ + H_2O \rightleftharpoons \alpha\text{-ketoglutarate} + NH_3 + NADH$$

The regulation of deamination reactions most likely occurs in response to the concentrations of substrates and products present in the cell.

Two non-specific amino acid oxidases which also catalyse the deamination of amino acids have a relatively minor role in amino acid catabolism.

3.6.3 Degradation

The second stage of the breakdown of amino acids is the degradation of the carbon backbones. The process of degradation converts these to citric acid cycle intermediates or their precursors, allowing them to either be metabolized to CO_2 and H_2O or used in gluconeogenesis. Amino acids have widely differing carbon skeletons, and their conversions to citric acid cycle intermediates follow correspondingly diverse pathways. The degradative pathways, however, can be grouped according to the intermediate at which they enter the citric acid cycle. The important thing is to understand the features which give commonality to groups of amino acids rather than the detail of the mechanisms of breakdown, and so only summary pathways will be described.

The breakdown of a number of amino acids to produce pyruvate is shown in Figure 3.13. Degradation of cysteine, alanine and serine produce pyruvate. Glycine is converted to serine, which is then deaminated to produce pyruvate, whilst threonine is cleaved to produce glycine and acetaldehyde, subsequently converted to acetyl CoA. The pyruvate produced as a result of these reactions is converted to oxaloacetate as

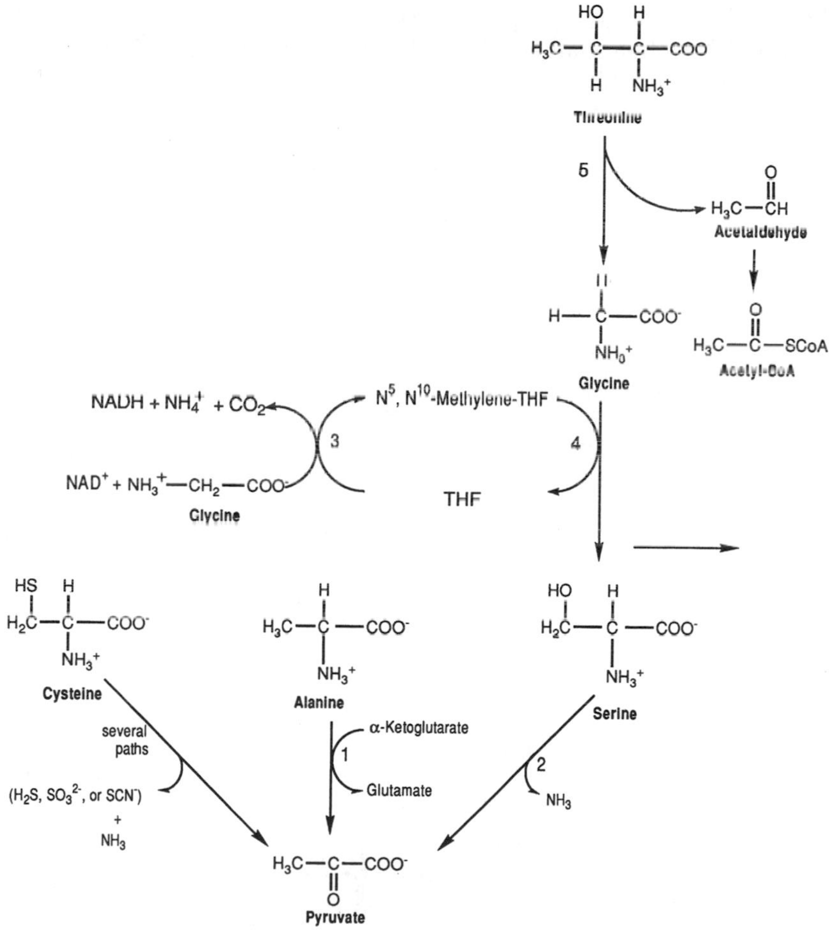

Figure 3.13 The degradative pathway converting alanine, cysteine, glycine, serine and threonine to pyruvate. The enzymes involved are (1) alanine aminotransferase, (2) serine dehydratase, (3) glycine cleavage system, (4) and (5) serine hydroxymethyltransferase.

previously described (section 3.4.2), which, with the acetyl CoA, enters the citric acid cycle.

The degradation of asparagine and aspartic acid to oxaloacetate following a series of deaminations is shown in Figure 3.14.

The degradation of arginine, glutamic acid, glutamine, histidine and proline to α-ketoglutarate is shown in Figure 3.15. The amino acid glutamic acid (or glutamate) serves as a key to this process. Glutamic acid is converted to α-ketoglutarate following deamination. The other four amino acids are converted to glutamic acid by a variety of processes. Glutamine is deaminated to form glutamic acid. Arginine passes through the intermediate ornithine (also an amino acid, though not one of the 20 standard amino acids) during conversion to glutamate 5-semialdehyde. Proline also takes a two-step pathway to glutamate 5-semialdehyde. This compound is then converted to glutamate. Histidine follows a more

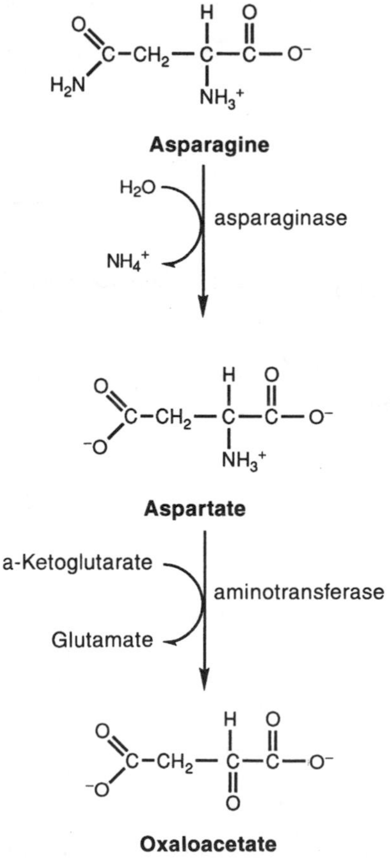

Asparagine

H_2O → NH_4^+ asparaginase

Aspartate

a-Ketoglutarate → Glutamate aminotransferase

Oxaloacetate

Figure 3.14 The degradative pathway of asparagine and aspartate.

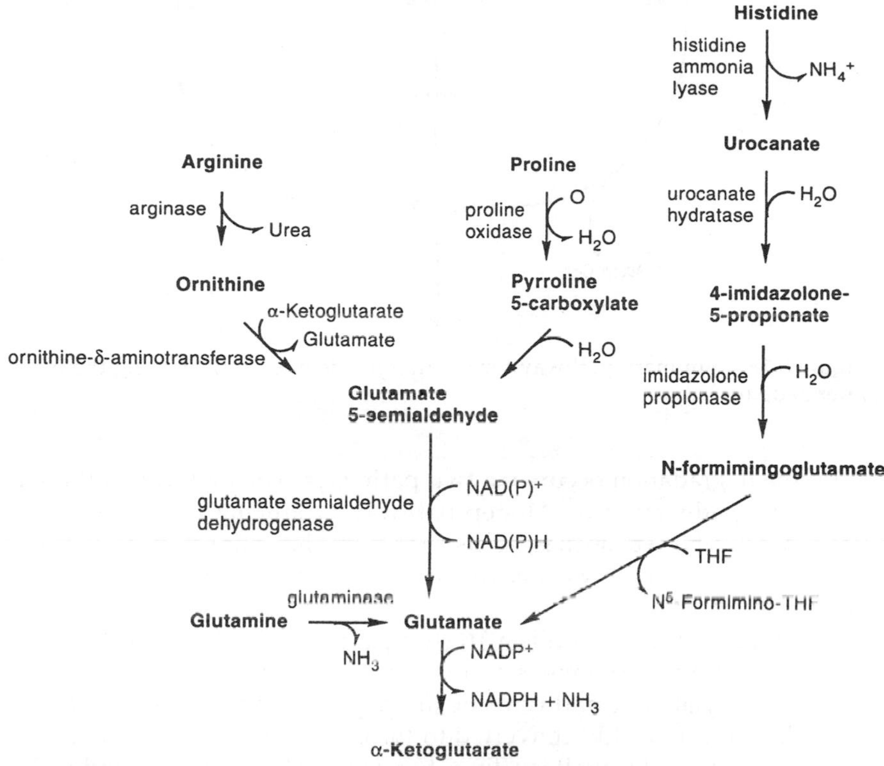

Figure 3.15 The degradation pathways of arginine, glutamate, glutamine, histidine and proline to produce α-ketoglutarate.

complex pathway, having three intermediates before its conversion to glutamic acid.

Methionine, isoleucine and valine are degraded to succinyl CoA via the intermediate propionyl CoA. The reaction converting propionyl CoA to succinyl CoA involves the participation of a coenzyme, vitamin B_{12}. Figure 3.16 shows the important intermediates in the degradation of methionine. Note that one of the by-products of methionine degradation is the production of the amino acid cysteine. Isoleucine and valine are degraded by a six-step or a seven-step pathway respectively to propionyl CoA. The degradation of isoleucine involves the production of acetyl CoA as well as propionyl CoA. Leucine is broken down by a very similar pathway to that followed by isoleucine and valine (isoleucine, valine and leucine are all branched-chain amino acids), the difference being that a six-step pathway produces acetyl CoA and acetoacetate. The acetoacetate is readily converted to a second acetyl CoA molecule.

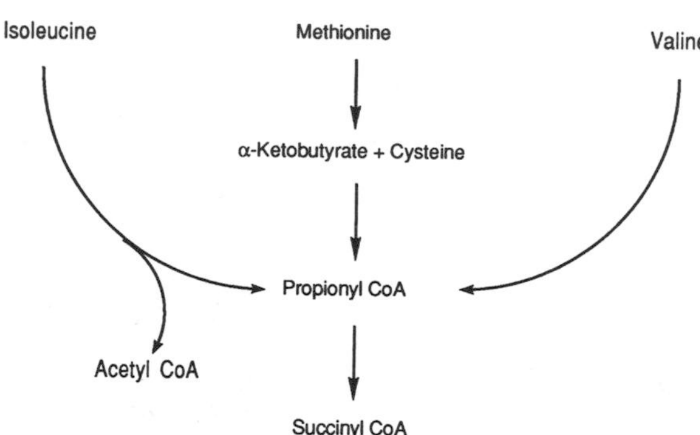

Figure 3.16 Summary pathway for methionine degradation showing the major intermediates.

Lysine degradation occurs via two pathways. The first, called the saccharopine pathway, is an 11-step pathway producing acetoacetate and acetyl CoA. The second has, as its first step, the action of L-amino acid oxidase. The two pathways have a common intermediate and the same end products.

Tryptophan degradation is a 16-step pathway which has acetoacetate as its end product. Alanine is an intermediate by-product of this degradation. Phenylalanine and tyrosine are degraded by the same pathway. Phenylalanine is readily converted to tyrosine by the addition of an OH group to its phenyl ring. It is a five-step degradative pathway and yields fumarate and acetoacetate.

3.6.4 Amino acid synthesis

Many amino acids are synthesized by pathways that are only present in plants and microorganisms. Some, required by animals for protein synthesis, must therefore be obtained in their diets. These substances are known as essential amino acids. The remaining amino acids, those which can be synthesized by the animal, are termed non-essential amino acids. Listed in Table 3.3 are the generally accepted essential and non-essential amino acids for most fish. Meeting the requirement for essential amino acids is not as straightforward as it would seem; some essential amino acids are readily converted to non-essential acids if these are lacking. As a result, the dietary requirement for an essential amino acid often reflects the requirement for it and at least part of the requirement for its non-essential relatives. Two examples of this are methionine and cysteine and phenylalanine and tyrosine. As was shown earlier, part of the degradative pathway of methionine involves the production of cysteine (Figure 3.16). Therefore, when considering the dietary requirement for methionine, it

Table 3.3 Essential and non-essential amino acids in fish

Essential	*Non-essential*
Arginine	Alanine
Histidine	Asparagine
Isoleucine	Aspartate
Leucine	Cysteine
Lysine	Glutamate
Methionine	Glutamine
Phenylalanine	Glycine
Threonine	Proline
Tryptophan	Serine
Valine	Tyrosine

must be remembered that it also includes some of the requirement for cysteine. Inclusion of cysteine in the diet can alleviate some, but not all, of the methionine requirement. Similarly, tyrosine is readily produced from phenylalanine and the dietary requirement for phenylalanine may include an amount which may be used to produce tyrosine. As was the case above, inclusion of tyrosine in the diet can alleviate some, but not all, of the requirement for phenylalanine. The relationship between phenylalanine and tyrosine is tighter than that between methionine and cysteine since cysteine can be synthesized from molecules other than methionine.

All the non-essential amino acids except tyrosine can be synthesized by simple pathways leading from one of four common metabolic intermediates: pyruvate, oxaloacetate, α-ketoglutarate or 3-phosphoglycerate. Alanine, asparagine, aspartate, glutamate and glutamine are synthesized by simple one- or two-step amination reactions from their organic precursors, pyruvate, oxaloacetate or α-ketoglutarate. Proline and arginine are synthesized from glutamate. Cysteine and glycine are derived from serine, which is in turn synthesized from 3-phosphoglycerate.

3.6.5 Protein turnover

Apart from providing a major energy source for fish, amino acids are used as components of protein. Proteins, among the largest molecules present in cells, are linear polymers of amino acids. The amino acids are linked by peptide bonds between the carboxyl group of one amino acid and the amino group of the next (Figure 3.17), thus they are also known as polypeptides. The structure of a protein is determined by the sequence of amino acids of which it is composed, which in turn is determined by the gene coding for the production of that particular protein.

Synthesis and degradation of protein occur continuously in tissues. When the rate of protein synthesis is greater than the rate of protein

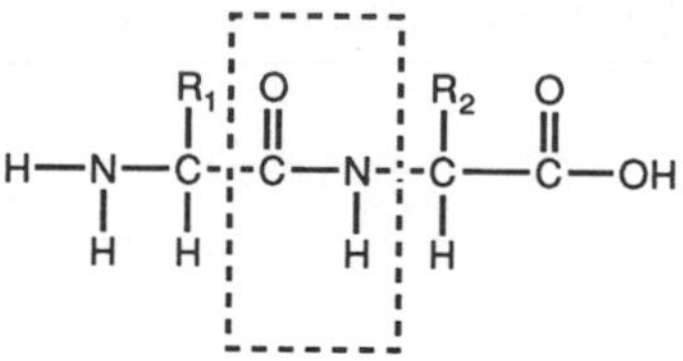

```
        R1   O        R2   O
        |    ||       |    ||
H—N—C—C—N—C—C—OH
  |    |        |    |
  H    H        H    H
```

Figure 3.17 A peptide bond outlined in broken lines which links two amino acids.

degradation, the animal is able to grow. This requires that essential amino acids be available from the diet. If there is insufficient energy or essential amino acids for growth, then protein degradation will be greater than protein synthesis and the animal will lose weight.

The synthesis of protein, described by the pathway

$$\text{DNA} \rightarrow \text{mRNA} \rightarrow \text{protein}$$

is regulated in many ways and during both processes. Protein synthesis is also regulated by the concentration of amino acids available for inclusion in protein.

No reliable direct method for measuring protein breakdown in fish is known. An estimate can be made from the differences between the rate of synthesis and growth (protein deposition). A number of enzymes are capable of catabolizing tissue proteins. The notable enzymes are the cathepsins and acid, neutral and alkaline proteinases.

3.6.6 Amino acid pools

Of great importance in a consideration of protein synthesis is the partitioning of amino acids within an organism between different pools. The body pools of free amino acids are relatively small compared with the pool of amino acids held in protein and are derived from three sources. The principal sources are from the diet and from the catabolism of body proteins. Catabolism of body proteins provides less than 50% of free amino acids in fish (Cowey and Luquet, 1983) and they are therefore highly dependent on dietary sources for free amino acids. The third source is the synthesis of non-essential amino acids.

The location of amino acids in fish is shown in Figure 3.18. There is continual exchange of amino acids that are incorporated into proteins and those that are free in tissues, and of those that are free in tissues and those that are free in blood. Some amino acids in the blood are transported preferentially in erythrocytes (red blood cells). Thus the large muscle mass of a fish acts as an effective reservoir for amino acids.

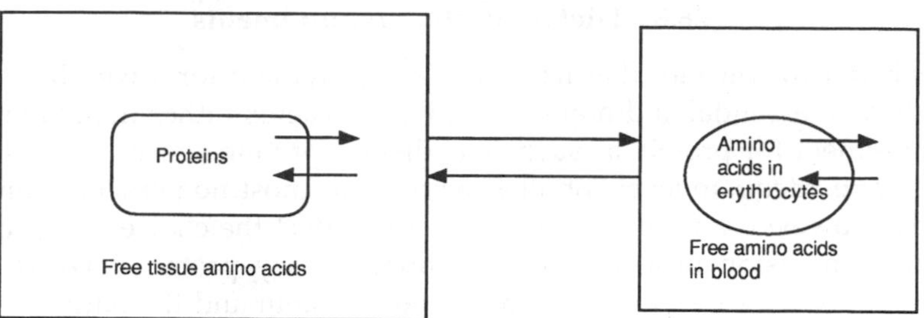

Figure 3.18 The interconnection of the amino acid pools of protein, tissue, blood and erythrocytes.

3.6.7 Amino acid metabolism and diet

The amount of protein consumed by a fish in its diet affects its metabolic state. High dietary protein levels result in increased body free amino acid concentrations, ammonia excretion, protein synthesis, gluconeogenic enzyme activities and decreased glycolytic enzyme activities. There is little effect on the level of the activities of amino acid-catabolizing enzymes.

One of the common methods of determining the requirements of a species for an essential amino acid involves feeding a diet that is nutritionally complete except for the essential amino acid in question. This is then added at various levels, ranging from deficiency to excess, and results in a number of biochemical responses. With increasing levels of the essential amino acid up to sufficiency, no change in the plasma level of that amino acid is observed. However, with greater than sufficient levels of the amino acid, there is a proportional increase in plasma concentrations. Generally the enzymes responsible for catabolizing amino acids are not increased by increasing dietary amino acid concentrations. Those enzymes that do show an increase do so at less than twofold. This may explain the increase in plasma amino acid concentrations observed in fish fed diets containing excess amino acids. In contrast, similar changes in dietary amino acid cause greater than 10-fold changes in the activities of amino acid-catabolizing enzymes in rats (Wilson, 1989).

During prolonged starvation, proteolytic activity in fish tissues increases, with the amino acids produced being used for energy. Although all amino acids would be liberated by the catabolic activity, studies of the tail muscle of dogfish indicate that only alanine was released into the plasma in significantly large quantities (Leech *et al.*, 1979). Similar data were obtained in a study of spawning salmon by Mommsen *et al.* (1980). This suggests that the amino acids released by the catabolism of protein are metabolized to produce alanine in the muscle before release into the bloodstream.

3.6.8 Dietary protein requirements

It is better to consider that fish have a requirement for a well-balanced mixture of essential and non-essential amino acids rather than having a requirement for protein as such. The dietary protein source must therefore have adequate levels of all essential and most non-essential amino acids. However, a number of other factors affect the choice of a protein and its inclusion in a diet. The optimal dietary protein level is also affected by the digestibility of the dietary protein and the nature of the non-protein energy sources in the diet.

Table 3.4 shows the estimated protein requirements of juvenile fish of a variety of species. It can be seen that the protein requirements range from 30% to 56% of the diets. These values can be considered to be estimated levels of protein in diets since a number of the assumptions made in determining these values are in question. They are likely to be overestimates because of inadequate consideration of:

- the energy concentration of the diet – the metabolizable energy values of protein, fat and carbohydrate are not generally determined but are assumed to be similar to other animals;
- the amino acid composition of the dietary protein; and
- digestibility of the dietary protein, which varies considerably between protein sources, and even between different batches of the same protein source. It was not determined in many of the studies listed in the table.

The increasing cost of traditional protein sources and the associated use of non-traditional protein sources (discussed in Chapter 10) together with the environmental cost of using excess protein in fish diets (discussed in Chapter 11) mean that more precise estimates of protein will be required in the future.

3.6.9 Amino acid requirements

All studies on finfish to date have shown that they need the same essential amino acids (Table 3.3) as most other animals. Apart from the method described in section 3.6.7, two techniques have been used to determine the essential amino acid requirements of fish. The most common method measures growth of fish fed diets containing graded levels of one amino acid. The diet with the lowest level of the essential amino acid that produces the maximum or near maximum level of growth is considered to contain the minimum requirement for that particular essential amino acid.

An alternative technique for determining the essentiality of amino acids is to use ^{14}C labelling studies. A radiolabelled substrate (e.g. glucose) is injected into the fish. The radiolabel is incorporated into those amino acids that the animal is able to synthesize. All other amino acids (not radiolabelled) are considered to be essential and are therefore required in the diet. This method identifies the essentiality of an amino acid, but not the dietary level required.

Table 3.4 Estimated protein requirements of juvenile fish

Species	*Protein source(s)*	*Estimated requirements (%)*
Channel catfish (*Ictalurus punctatus*)	Whole egg protein	32–36
Common carp (*Cyprinus carpio*)	Casein	38
Estuary grouper (*Epinephelus salmoides*)	Tuna muscle meal	40–50
Gilthead bream (*Pagrus aurata*)	Casein, FPC[a] and amino acids	40
Grass carp (*Ctenopharyngodon idella*)	Casein	41–43
Japanese eel (*Anguilla japonica*)	Casein and amino acids	44.5
Largemouth bass (*Micropterus salmoides*)	Casein and FPC	40
Milkfish (fry) (*Chanos chanos*)	Casein	40
Plaice (*Pleuronectes platessa*)	Cod muscle	50
Puffer fish (*Fugu rubripes*)	Casein	50
Red sea bream (*Chrysophrys major*)	Casein	55
Salmonids		
Chinook salmon (*Oncorhynchus tshawytscha*)	Casein, gelatin and amino acids	40
Coho salmon (*Oncorhynchus kisutch*)	Casein	40
Rainbow trout (*Oncorhynchus mykiss*)	Fish meal	40
	Casein and gelatin	40
	Casein, gelatin and amino acids	45
Sockeye salmon (*Oncorhynchus nerka*)	Casein, gelatin and amino acids	45
Smallmouth bass (*Micropterus dolomieui*)	Casein and FPC	45
Snakehead (*Channa micropeltes*)	Fish meal	52
Striped bass (*Morone saxatilis*)	Fish meal and SP[b]	47
Tilapias		
Tilapia aurea (fry)	Casein and egg albumin	56
Tilapia aurea (juvenile)	Casein and egg albumin	34
Tilapia mossambica	White fish meal	40
Tilapia nilotica	Casein	30
Tilapia zillii	Casein	35
Yellowtail (*Seriola quinqueradiata*)	Sand eel and fish meal	55

[a] Fish protein concentrate.
[b] Soy proteinate.
Source: Wilson (1989).

Table 3.5 Essential amino acid requirements of fish

Amino acid	*Requirement (g/100 g protein)*
Arginine	3.3–5.9
Histidine	1.3–2.1
Isoleucine	2.0–4.0
Leucine	2.8–5.3
Lysine	4.1–6.1
Methionine[a]	2.2–6.5
Phenylalanine[b]	5.0–6.5
Threonine	2.0–4.0
Tryptophan	0.3–1.4
Valine	2.3–4.0

[a] Requirement varies depending upon the amount of cysteine in the diet.
[b] Requirement varies depending upon the amount of tyrosine in the diet.
Refer to Table 7.1 for requirements of individual species.

Table 3.5 summarizes the range of essential amino acid requirements that have been determined for a variety of species of finfish. Further details of the amino acid requirements of individual species are presented in Table 7.1. The requirements for individual amino acids are fairly consistent between species, although variability is apparent both between species and between studies on the same species. A large part of the variability may be explained by differences in the methods used by various workers. An example of this is the arginine requirements as determined for rainbow trout. This requirement has been determined at between 3.3% of the protein (Kaushik, 1979) and about 5.9% of the protein (Ketola, 1983). In both of these studies, semipurified diets were used. Semipurified diets contain a mixture of purified (e.g. an amino acid) and whole ingredient (e.g. fish meal). Kaushik (1979) used a diet containing zein, fish meal and free amino acids as the amino acid source, whilst Ketola (1983) used a diet containing corn gluten meal and free amino acids as the amino acid source. Failure to determine the comparative digestibility of these protein sources or a variation in the amino acid balance may have resulted in the variable data obtained. Although there is no alternative, the use of free amino acids in test diets also confounds the results because of problems experienced with leaching of these amino acids from the diet between feeding and ingestion, or the possibility of differential uptake between free and protein-bound amino acids.

The histidine requirements of fish fall within a range from 1.3% to 2.1% of dietary protein. Histidine metabolism appears to be closely regulated in fish. Serum free histidine levels in channel catfish respond positively to graded dietary histidine intake when dietary histidine levels are in excess of requirement (Wilson *et al.*, 1980). Carnosine, a metabolic product of histidine, is depleted in the muscle of Chinook salmon fed a histidine-deficient diet (Lukton, 1958), also indicating that histidine metabolism is closely regulated.

Isoleucine, leucine and valine are branched-chain amino acids. Their requirements as a percentage of dietary protein are: isoleucine 2–4%, leucine 2.8–5.3%, and valine 2.3–4%. For each of these three amino acids, the requirements for Japanese eel (*Anguilla japonica*) are higher than for other species.

There is considerable interaction between the branched-chain amino acids. In investigations of channel catfish, Wilson *et al.* (1980) found that serum isoleucine levels increased with increasing isoleucine intake and, at the same time, serum leucine and valine concentrations also increased. In channel catfish fed increasing levels of dietary leucine, there was no change in free serum leucine concentration although serum isoleucine and valine levels increased markedly when leucine was increased in the diet from 0.6% to 0.7% of protein. At dietary leucine levels of 1.2% of protein, isoleucine and valine returned to baseline. A similar effect was seen when dietary valine concentrations were increased (Wilson *et al.*, 1980). The relationship between these branched-chain amino acids varies between species. However, it seems likely that tissue uptake, and probably the intracellular metabolism of the branched-chain amino acids, is a cooperative process, a feature which should be considered when determining dietary requirements for these amino acids.

Lysine requirements range from 4.1% to 6.1% of dietary protein. A dietary relationship between lysine and arginine has been well documented in a number of animals, including some fish, and is known as the lysine/arginine antagonism. Kaushik and Fauconneau (1984) found that increasing dietary lysine increased plasma arginine levels in rainbow trout and considered that this was the result of an antagonism between the catabolism of lysine and arginine. As already mentioned, phenylalanine is readily metabolized to tyrosine and there is a sparing effect of dietary tyrosine on phenylalanine. Accordingly phenylalanine requirements are generally determined in diets which are very low in tyrosine. These values are expressed as total aromatic amino acid requirement and vary from 5% to 6.5% of dietary protein.

Similarly the sulphur amino acid requirements can be expressed jointly, since cysteine can be metabolized from methionine. Methionine or total sulphur amino acid requirements vary from 2.2% of dietary protein to 6.5% of dietary protein. In most fish diets made from readily available ingredients, methionine is limiting relative to other amino acids. Careful attention must therefore be paid to ensure that methionine requirements are met.

Dietary threonine requirement varies between 2% and 4% of dietary protein, whilst tryptophan ranges from 0.3% to 1.4% of dietary protein, much lower than the requirement for any other essential amino acid.

3.6.10 Nutritional value of proteins

Nutritional values of proteins are used as a guide to the effectiveness of a particular protein source in supplying the animal's requirements. There are three main methods for quantifying the nutritional value of proteins. These are protein efficiency ratio (PER), net protein utilization (NPU) and the essential amino acid index.

PER is an expression which relates the gram of weight gained to the grams of crude protein fed according to the formula

$$\text{PER} = \frac{\text{g wet weight gain}}{\text{g crude protein fed}}$$

Although this method makes no allowance for protein used for maintenance, it is still widely used as a method of determining appropriate protein sources for fish diets.

Net protein utilization is probably the most satisfactory method and is described by the formula

$$\text{NPU} = \text{biological value} \times \text{digestibility}$$

However, several technical difficulties occur when determining both biological value and digestibility. Accordingly this is often rewritten as

$$\text{NPU} = \frac{\text{Final body protein} - \text{initial body protein}}{\text{Total protein fed}} \times 100$$

The third method of evaluating nutritional value, the essential amino acid index, can only be used if the amino acid requirement for the given species is known. The nutritional value of a protein can then be made by directly comparing the essential amino acid content with the requirement values in an iterative fashion. However, this is a fairly crude technique and takes no account of variations in digestibility and availability of amino acids from different protein sources.

3.6.11 Factors affecting protein and amino acid requirements

Very little is known of the non-dietary factors which affect protein and amino acid requirements of fish. Fish age is an important factor in determining the level of dietary protein and hence amino acid requirements. Generally, protein requirements diminish as the fish grows older.

Environmental parameters are also thought to affect amino acid requirements. Changes in water temperature have also been reported to affect protein requirement of fish. The evidence is contradictory and has little practical relevance to formulating fish diets without further investigation. The arginine requirement of rainbow trout has been reported to decrease as salinity increases (Kaushik, 1979), however the opposite situation was described by Zeitoun *et al.* (1973). Additional studies are needed to clarify these points.

3.6.12 Rapid determination of amino acid requirements

The detailed and accurate determination of amino acid requirements for any species of fish is a lengthy procedure and requires considerable experimentation. One method that has been applied to assist in the initial formulation of diets has been to identify the proportion of each essential amino acid in the body tissue of the species in question. The amino acid composition of the whole body of six different fish species is shown in Table 3.6. Comparison of these data with the requirements for essential amino acids (Table 3.5) indicates that this is generally a good measure of the relative requirements for essential amino acids, although in most cases the value determined by whole-body tissue analysis is greater than that determined by dietary or labelling studies. Nevertheless, it remains an excellent rapid technique for ensuring that a diet is balanced with regard to the essential amino acids for any given species.

3.7 LIPID METABOLISM

Lipids are organic molecules containing many carbon atoms in a variety of chain or ring conformations. They are substances of biological origin and their physical properties result in them being soluble in organic solvents, but only sparingly soluble, if at all, in water. The main function of lipids in animals is either as high-energy storage molecules or as components of cell membranes. There are five major classes of lipids: fatty acids, triglycerides, phospholipids, sterols and sphingolipids.

Fatty acids are carboxylic acids with long-chain hydrocarbon side groups. The more common biological fatty acids are shown in Table 3.7. The nomenclature of fatty acids follows a particular convention. This is

$$C_{x:y(n-z)}$$

where x denotes the number of carbon atoms, y denotes the number of double bonds in the chain and z denotes the carbon at which the first double bond appears numbering from the non-carboxyl (COOH) end. n is often replaced by ω (omega) in popular jargon; hence ω-3 and ω-6 fatty acids. Thus, for example, α-linolenic acid is a C18:3(n-3) fatty acid, while γ-linolenic acid is a C18:3(n-6) fatty acid (Table 3.7).

In higher plants and animals, the predominant fatty acid residues are those of the C_{16} and C_{18} species, of which half are unsaturated (contain at least one double bond between carbon atoms) and are often polyunsaturated (contain more than one double bond).

Fatty acids rarely occur free in nature, but more generally occur in esterified form, called triacylglycerols or triglycerides, so named since they are triesters of glycerol. As can be seen from Figure 3.19, three fatty acids are bonded to the glycerol molecule. Triacylglycerols are an

Table 3.6 Amino acid composition of whole-body tissue of certain fishes expressed as g/100 g amino acids

Amino acid	*Rainbow trout*	*Atlantic salmon*	*Coho salmon*	*Cherry salmon*	*Channel catfish*
Alanine	6.57	6.52	6.08	6.35	6.31
Arginine	6.41	6.61	5.99	6.23	6.67
Aspartic acid	9.94	9.92	9.96	9.93	9.74
Cysteine	0.80	0.95	1.23	1.34	0.86
Glutamic acid	14.22	14.31	15.25	15.39	14.39
Glycine	7.76	7.41	7.31	7.62	8.14
Histidine	2.96	3.02	2.99	2.39	2.17
Isoleucine	4.34	4.41	3.70	3.96	4.29
Leucine	7.59	7.72	7.49	7.54	7.40
Lysine	8.49	9.28	8.64	8.81	8.51
Methionine	2.88	1.83	3.53	3.14	2.92
Phenylalanine	4.38	4.36	4.14	4.63	4.14
Proline	4.89	4.64	4.76	4.33	6.02
Serine	4.66	4.61	4.67	4.48	4.89
Threonine	4.76	4.95	5.11	4.63	4.41
Tryptophan	0.93	0.93	1.40	0.83	0.78
Tyrosine	3.38	3.50	3.44	3.58	3.28
Valine	5.09	5.09	4.32	4.85	5.15

Source: Halver (1989).

Table 3.7 The common biological fatty acids

Symbol[a]	*Common name*	*Systematic name*	*Structure*
Saturated fatty acids			
12:0	Lauric acid	Dodecanoic acid	$CH_3(CH_2)_{10}COOH$
14:0	Myristic acid	Tetradecanoic acid	$CH_3(CH_2)_{12}COOH$
16:0	Palmitic acid	Hexadecanoic acid	$CH_3(CH_2)_{14}COOH$
18:0	Stearic acid	Octadecanoic acid	$CH_3(CH_2)_{16}COOH$
20:0	Arachidic acid	Eicosanoic acid	$CH_3(CH_2)_{18}COOH$
22:0	Behenic acid	Docosanoic acid	$CH_3(CH_2)_{20}COOH$
24:0	Lignoceric acid	Tetracosanoic acid	$CH_3(CH_2)_{22}COOH$
Unsaturated fatty acids			
16:1	Palmitoleic acid	9-Hexadecenoic acid	$CH_3(CH_2)_5CH{=}CH(CH_2)_7COOH$
18:1	Oleic acid	9-Octadecenoic acid	$CH_3(CH_2)_7CH{=}CH(CH_2)_7COOH$
18:2	Linoleic acid	9,12-Octadecadienoic acid	$CH_3(CH_2)_4CH{=}CHCH_2)_2(CH_2)_6COOH$
18:3	α-Linolenic acid	9,12,15-Octadecatrienoic acid	$CH_3CH_2(CH{=}CHCH_2)_3(CH_2)_6COOH$
18:3	γ-Linolenic acid	6,9,12-Octadecatrienoic acid	$CH_3(CH_2)_4(CH{=}CHCH_2)_3(CH_2)_3COOH$
20:4	Arachidonic acid	5,8,11,14-Eicosatetraenoic acid	$CH_3(CH_2)_4(CH{=}CHCH_2)_4(CH_2)_2COOH$
20:5	EPA	5,8,11,14,17-Eicosapentaenoic acid	$CH_3CH_2(CH{=}CHCH_2)_5(CH_2)_2COOH$
24:1	Nervonic acid	15-Tetracosenoic acid	$CH_3(CH_2)_7CH{=}CH(CH_2)_{13}COOH$

[a] Number of carbon atoms:number of double bonds.
Source: Voet and Voet (1990).

efficient form in which to store metabolic energy mainly because they are less oxidized than carbohydrates or proteins and hence yield significantly more energy on oxidation. Since they are not soluble in water, they are stored in an anhydrous form, whereas glycogen binds about twice its weight of water under physiological conditions. Hence, a given weight of lipid (mixture of triacylglycerols) provides about six times the metabolic energy of an equal weight of hydrated glycogen.

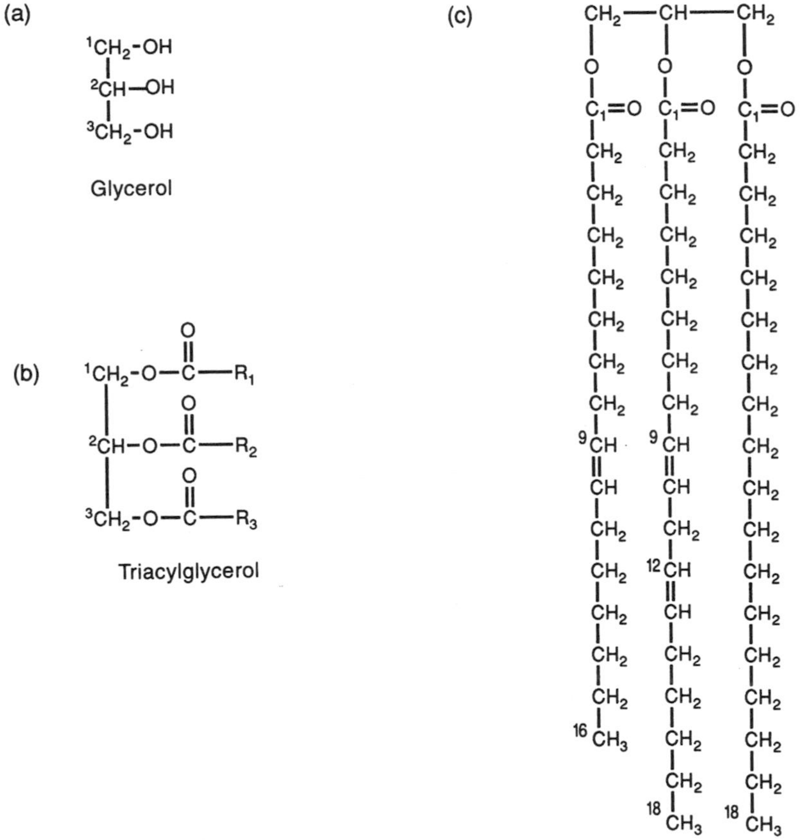

(a) Glycerol

(b) Triacylglycerol

(c) 1-Palmitoleoyl-2-linoleoyl-3-stearoyl-glycerol

Figure 3.19 The structure of (a) the glycerol molecule which forms the backbone of triacylglycerols, (b) the generalized form of triacylglycerols, where R_1, R_2 and R_3 are any fatty acids and (c) an example of a triacylglycerol molecule having R_1 = palmitoleic acid [16:1 (*n*-7)] R_2 = linoleic acid [18:2 (*n*-6)] and R_3 = stearic acid (18:0). (Adapted from Voet and Voet, 1990.)

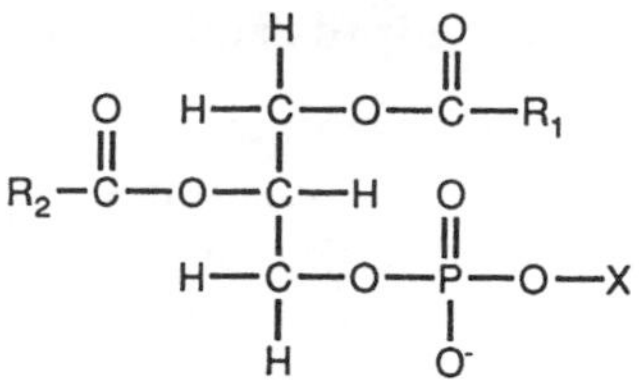

Phospholipid

Figure 3.20 The general formula of the glycerophospholipids. R_1 and R_2 are long-chain hydrocarbon tails of fatty acids and X is derived from a polar alcohol.

Another class of lipids is the glycerophospholipids. Also called phospholipids, glycerophospholipids are the major lipid components of biological membranes. These molecules have a similar structure to triacylglycerols except that one of the fatty acid chains is replaced by a phosphate attached to another organic molecule (Figure 3.20).

The fourth major class of lipids are sterols, identified by their four fused carbon rings. A variety of side chains attached to carbon rings at C_3 and C_{17} give them their individual characteristics. The most notable sterol lipid is cholesterol (Figure 3.21), being a major component of animal cell membranes and a precursor of steroid hormones and bile acids.

The fifth class of lipids is sphingolipids, which are also major membrane components. These molecules, common in the membranes of nerve cells, possess a charged 'head' region, which may be a phosphate group, as in glycerophospholipids, or carbohydrate groups.

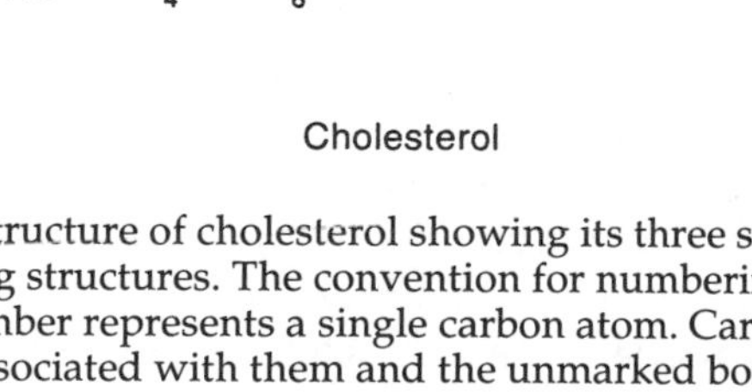

Cholesterol

Figure 3.21 The structure of cholesterol showing its three six-carbon ring and one five-carbon ring structures. The convention for numbering the carbon atoms is shown. Each number represents a single carbon atom. Carbon atoms always have four bonds associated with them and the unmarked bonds are attached to hydrogen atoms.

3.7.1 Lipid catabolism

Lipids are transported in the bloodstream either as lipoprotein complexes called very low-density lipoproteins (VLDLs) or as very small droplets called chylomicrons. Complexing the lipids to proteins allows otherwise insoluble lipid components to be maintained in aqueous solution. The triacylglycerol components of VLDL and chylomicrons are hydrolysed to free fatty acids and glycerol in the target tissues (generally adipose tissue and skeletal muscle) outside of the cell by an enzyme called lipoprotein lipase. After transport across the cell membrane, they are either resynthesized into triacylglycerols or they are oxidized to release energy. Triacyglycerols which are already stored in adipose tissue but are required for metabolism are hydrolysed to glycerol and free fatty acids by a different enzyme, triacylglycerol lipase, before passing out of the cell.

3.7.2 Fatty acid oxidation

Fatty acid oxidation liberates the energy contained in the fatty acid and occurs by two major steps. The first step is priming, or activation, of the fatty acid by linking a CoA to it according to the reaction

$$\text{Fatty acid} + \text{CoA} + \text{ATP} \rightleftharpoons \text{acyl CoA} + \text{AMP} + \text{PP}_i$$

where PP_i is inorganic diphosphate. The second step involves a series of β-oxidation reactions, as shown in Figure 3.22. The process is cyclic, consisting of four steps, each cycle liberating two of the carbon atoms from the fatty acid with the production of acetyl CoA. Note that the acetyl CoA molecule liberated contains a single bond between the two carbon atoms of the acetyl group. The acetyl CoA is then oxidized further in the citric acid cycle to release energy. Since the presence of a double bond in the carbon chain inhibits the action of the enzymes involved in the β-oxidation, a number of enzymes are present to relocate or remove the double bonds to allow the β-oxidation of unsaturated fatty acids to occur.

Most fatty acids have an even number of carbon atoms and they can therefore be completely converted to the two-carbon acetyl group of acetyl CoA. Some terrestrial plants and marine organisms synthesize fatty acids with an odd number of carbon atoms. As a result, in the final round of β-oxidation, these fatty acids form propionyl CoA, which is then converted to succinyl CoA for entry into the citric acid cycle.

Acetyl CoA produced by the oxidation of fatty acids can be further oxidized via the citric acid cycle, as previously discussed (section 3.5). In higher vertebrates, however, a significant fraction of this acetyl CoA has another fate. By a process known as ketogenesis, acetyl CoA is converted into compounds known as ketone bodies, namely acetoacetate, D-β-hydroxybutyrate or acetone. In these animals, ketone bodies serve as

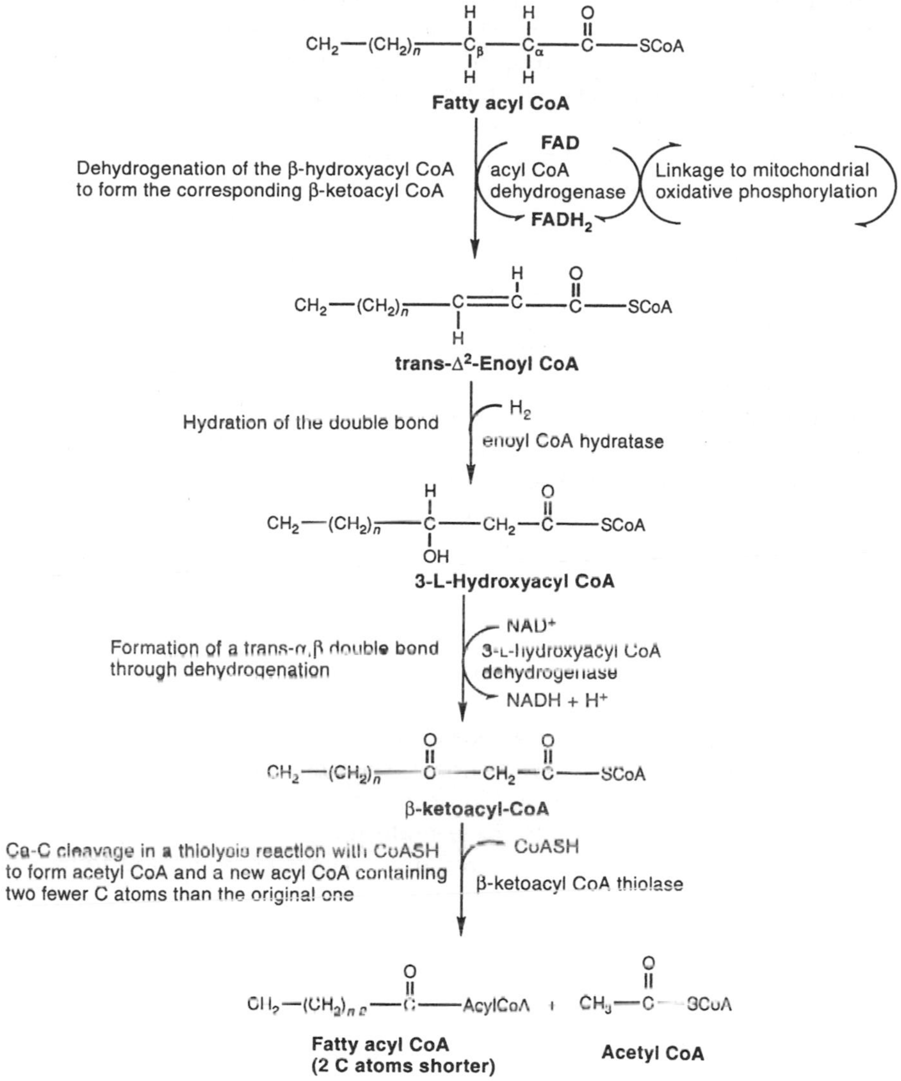

Figure 3.22 The β-oxidation pathway of fatty acyl CoA showing the intermediate compounds and the enzymes catalysing the reactions.

alternatives to glucose as major circulating energy sources. Notably, ketone body formation does not generally occur in fishes.

3.7.3 Fatty acid biosynthesis

Fatty acid biosynthesis occurs through the linkage of C_2 units, that is the reverse of the β-oxidation process just described. There are a number of differences between the two pathways which facilitate their regulation independently of each other. The first major difference is that, rather than the fatty acid conjugating to a CoA molecule, it is attached to acyl-carrier protein (ACP). The second difference is that the C_2 unit donor is malonyl CoA.

The end product of fatty acid synthesis is palmitic acid ($C_{16:0}$). This molecule is the precursor of longer chain saturated and unsaturated fatty acids, produced through the actions of enzymes called elongases and desaturases. Elongases act essentially in the reverse of the β-oxidation pathway; the fatty acid is elongated as its CoA derivative rather than as its ACP derivative.

Unsaturated fatty acids are produced by terminal desaturases. These enzymes catalyse the reaction shown in Figure 3.23. In this figure, the synthesized fatty acid is represented in a simplified form. Here $(CH_2)y$ represents the portion of the fatty acid chain between the CoA group and the site of desaturation. The $(CH_2)x$ represents the portion of the fatty acid chain distal to the site of desaturation. This section may contain double bonds depending upon whether desaturation has already occurred. The value of x in animals is always at least 5. By this method a variety of unsaturated fatty acids may be synthesized by combinations of elongations and desaturation reactions. There are three terminal desaturases in vertebrates designated Δ^9-, Δ^6- and Δ^5-fatty acyl CoA desaturases which desaturate at C_9, C_6 and C_5 respectively. In animals desaturation does not occur at positions beyond C_9. Palmitic acid is the shortest available fatty acid in animals. The rules above preclude the formation of the double bond at C_{12} of linoleic acid or of that at C_{15} of α-linolenic acid. To synthesize these molecules, a Δ^{12}- or a Δ^{15}-fatty acyl CoA desaturase respectively is required. These enzymes are found in

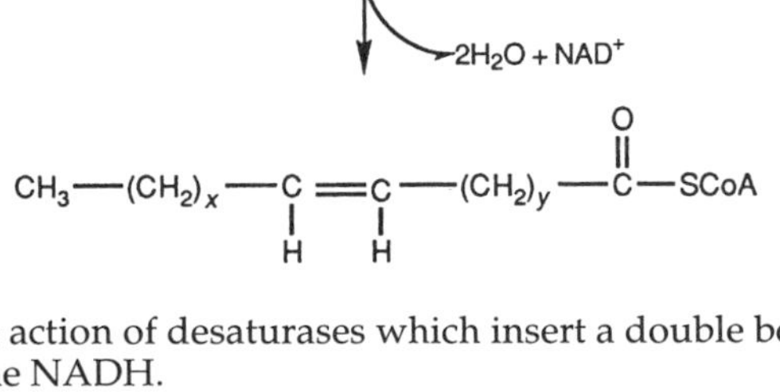

$$CH_3-(CH_2)_x-CH_2-CH_2-(CH_2)_y-C(=O)-SCoA$$

$$\xrightarrow[\text{terminal desaturate}]{NADH + 2H^+ + O_2 \;\rightarrow\; 2H_2O + NAD^+}$$

$$CH_3-(CH_2)_x-CH=CH-(CH_2)_y-C(=O)-SCoA$$

Figure 3.23 The action of desaturases which insert a double bond at $n-(x+2)$ at the expense of one NADH.

plants and some microorganisms and so (n-3) or (n-6) fatty acids must be obtained from the diet. The (n-3) and (n-6) fatty acids play very important roles in the proper functioning of an animal. Fatty acids which are required in the diet are termed essential fatty acids.

Regulation of fatty acid metabolism is under the control of two peptide hormones, insulin, which stimulates the synthesis of fatty acids, and glucagon, which stimulates their degradation. These hormones are secreted by the pancreatic tissue of the fish depending upon the levels of circulating metabolites and hence the overall energy status of the animal.

3.7.4 Synthesis of other lipid classes

Animals are able to synthesize other lipid classes from precursors found in the cell. Cholesterol is synthesized via a complex process which utilizes acetate as a substrate. Because of this there is no dietary requirement for cholesterol. Other lipid classes, triglycerides, phospholipids and sphingolipids, utilize fatty acids and various other molecules as substrates, and thus the dietary requirement is for fatty acids as described above.

3.7.5 Dietary lipid requirements of fish

Apart from satisfying the requirements of a fish for essential fatty acids (section 3.7.6), dietary lipid acts as a source of energy. In general, 10–20% lipid in fish diets gives optimal growth rates without producing an excessively fatty carcass (Cowey and Sargent, 1979). Interspecific variation in the ability of different species to utilize lipid as a source of energy is prevalent. For example, when rainbow trout were fed diets with lipid contents of 5–20% and protein contents of 16–48%, the optimum ratio of protein to lipid was found to be 35% protein, 18% lipid (Takeuchi *et al.*, 1978a, b). However, carp fed diets with a fixed protein level of 32%, with lipid varying from 5% to 15% and with corresponding decreases in carbohydrate, did not show increased growth or food conversion rates (Takeuchi *et al.*, 1979).

Fasting fish generally utilize lipid reserves as an energy source in preference to protein and carbohydrate. A study of Coho salmon indicated that this occurred initially as an increase in the rate of lipid breakdown since the activities of lipid-synthesizing enzymes remained unchanged for a period of 2 days (Lin *et al.*, 1977a, b). However, significant decreases in lipid synthesis were noticeable after 23 days of food deprivation. There is also evidence that lipid mobilization in starved fish is selective, with shorter (C_{18} and C_{16}) and less saturated fatty acids being mobilized first (Sargent *et al.*, 1989). However, again interspecific variation predominates.

3.7.6 Essential fatty acid requirements of fish

When considering the essential fatty acid requirements of fish, it is useful to consider the origin of these compounds in natural systems. The major primary producers in marine ecosystems are unicellular algae. Actively growing algae contain approximately 20% of their dry weight as lipid, and 50% of this is present as (*n*-3) polyunsaturated fatty acids (PUFAs). Notably the red algae can also be rich in C20:4(*n*-6) as well as (*n*-3) PUFAs. The major consumers of phytoplankton are crustacean zooplankton, which are subsequently consumed by planktivorous fish. Throughout this process the (*n*-3) PUFAs are not desaturated although they are generally elongated, being present largely as C_{18} PUFAs in algae and C_{20} and C_{22} PUFAs in zooplankton and fish. Generally, the amount of C_{20} is greater than that of C_{22} in the zooplankton, and vice versa in the fish.

Freshwater microalgae have similar constituent fatty acids, although (*n*-6) PUFAs are more common in freshwater algae. There does appear to be a variation in the fatty acid composition of freshwater fish relative to marine fish, although most of the data for freshwater fish have been derived from studies of salmonids. Freshwater fish contain higher proportions of C18 fatty acids and (*n*-6) PUFAs than do marine fish. However, both are rich in (*n*-3) PUFAs, especially C20:5(*n*-3) and C22:6(*n*-3).

Fish possess a range of elongase and desaturase enzymes capable of modifying dietary fatty acids and the products of endogenous fatty acid biosynthesis. They do not, as stated above in section 3.7.3, possess the desaturases necessary for converting C18:2(*n*-6) and C18:3(*n*-3) from C18:1(*n*-9). Consequently, all (n-3) and (n-6) PUFAs in fish lipids originate from (*n*-3) and (*n*-6) PUFAs consumed in the diet which are therefore essential fatty acids (EFAs).

All fish studied to date appear to require C18:3(*n*-3) PUFAs at about 1–2% of the diet by dry weight. This requirement can be reduced by feeding longer chain (*n*-3) PUFAs such as C20:5(*n*-3), C22:5(*n*-3) or C22:6(*n*-3) to approximately 0.5% of the diet by dry weight. Studies of the nutritional requirements of juvenile turbot indicate that they are unable to efficiently elongate C_{18} fatty acids and thus require C20(*n*-3) or C22(*n*-3) fatty acids in their diet (Sargent *et al.*, 1989). In larger (200 g) turbot, this limitation can apparently be overcome by including C18:3(*n*-3) at 4% of the diet (Leger *et al.*, 1979), although the mechanism by which this occurs is unclear. Sargent *et al.* (1989) consider that a requirement for C_{20} or C_{22} PUFAs is a general feature of marine species. Drawing such conclusions is acknowledged by those authors to be difficult because very small amounts of essential fatty acids are required by fish and the likelihood of other ingredients used in test diets containing PUFAs of unknown composition is high. It is therefore extremely difficult to be certain that a known amount of PUFAs is being fed. Care should be taken not to provide excess short-chain PUFAs. In a study of the effects of dietary lipids on the growth rate of channel catfish, Stickney and Andrews (1972) found that fish diets containing beef tallow (highly saturated) or menhaden oil [high proportion of C20:5(*n*-3) and C22:6(*n*-3)] grew faster and had a lower food conversion ratio and lower percentage body fat than fish fed a diet containing safflower oil [large proportion of C18:2(*n*-6)] or linseed oil [large proportion of C18:3(*n*-3)].

The significance of these requirements to the preparation of diets for use in aquaculture and to fish nutrition generally is:

- that terrestrial herbivorous ruminants break down the (*n*-3) PUFAs present in plants during their digestion of cellulose, therefore (*n*-3) PUFAs are present in very low amounts in terrestrial animal meat meals;
- the abundant seed oils of terrestrial plants rich in (*n*-6) PUFAs do not satisfy the major requirement of fish for (*n*-3) PUFAs;
- the lipids found in aquatic organisms have a high proportion of (*n*-3) PUFAs that is maintained through the food web; as a result, diets for farming fish rely heavily on fish oils and oils in fish meal for provision of the (*n*-3) PUFA requirements of fish.

3.8 VITAMINS

Vitamins are organic molecules that act as cofactors or substrates in some metabolic reactions. They are generally required in relatively small amounts in the diet and, if present in inadequate amounts, may result in nutrition related disease, poor growth or increased susceptibility to infections.

The vitamin requirements of the majority of species of fish in culture have not been determined. As a result, the data obtained from studies of salmonids, carp or catfish are usually applied to other species. Most practical diets, diets formulated and manufactured from ingredients that are readily available, include vitamins at the levels published by the United States National Research Council (NRC, 1983) for warmwater fish unless there is evidence of a different requirement that has been obtained from specific studies. As can be observed from Table 3.8, measured vitamin requirements of different species vary considerably. If data from a different species are used, one of three outcomes may occur. The first is that too great a level of vitamin may be included in the diet, which at the least increases the cost of the feed unnecessarily, or at the most can result in vitaminosis, a disease induced by excess amounts of vitamin in the diet. Vitaminosis is usually caused by excess amounts of fat-soluble vitamins which are difficult for an animal to excrete. If insufficient vitamin is included in the diet, a nutritional disease or poor growth performance is observed. The third alternative is that the right amount of vitamin is included. Most nutritionists hope for the last outcome.

Table 3.8 Summary of the vitamin requirements of various species of fish

	Sea bass	*Atlantic salmon*	*Ayu*	*Channel catfish*	*Common carp*	*Eel*	*Turbot*	*Pacific salmon*	*Rainbow trout*	*Red drum*	*Sturgeon*	*Tilapia*	*Yellowtail*
Thiamine	R		12	1	R	R	0.6–2.6	10–15	40				11.2
Riboflavin	R	R	40	9	7–14	R		20–25	9				11.0
Pyridoxine	5	5	12	3	5–6	R	1.0–2.5	15–20	9				11.7
Pantothenic acid	R	R	50	15	30–50	R		40–50	40				35.9
Nicotinic acid			100	14	28	R		150–200	300				12.0
Biotin			0.3	R	1	R		1–1.5	0.4				0.67
Inositol	R		400	NR	440	R		300–400	510				423
Choline			350	400	4000	R		600–800	11 100		1700–3200		2920
Folic acid			3	R		R		6–10	21				1.2
Ascorbic acid	700	50	300	60	R	R	R	100–150	400	60–75	NR	R	122
Vitamin A			10 000 IU	1000–2000 IU	10 000 IU			2000–2500 IU	7000 IU				5.68
Vitamin D			2000 IU	250–500 IU					3000 IU				NR
Vitamin E	R	35	100 IU	50 IU	100 IU	200		30 IU	200			50–100	119.0
Vitamin K		R	10 IU	R					50				NR
Vitamin B_{12}				R		R		0.015–0.02	0.21			NR	

Data are collated from Halver (1989), Lovell (1989a) and Wilson (1991).
Values are mg/kg diet unless stated otherwise.
R, required, NR, not required, IU, international units.

Determining vitamin requirements in fish, as in other animals, is difficult, since many are produced by microorganisms in the gut. Depending upon the activity of an animal's gut microflora, the requirement for dietary vitamins can vary. It is likely that all vitamins are required by all fish, but whether it is necessary to include them in a formulated diet is a different matter. A species such as carp can obtain many nutrients from decaying organic matter which it will consume in addition to any artificial diet in all but the most sterile of conditions, and so the requirements for vitamins in supplementary feeds for this species are less than for a species which has a carnivorous feeding habit and relies entirely upon the components of an artificial diet for its vitamin intake. Similarly, animals held in tanks have less opportunity to consume natural sources of vitamins than those held in ponds and so require greater levels of vitamins in any artificial feed they may be given.

Vitamins can be divided into two groups, the fat-soluble vitamins, A, D, E and K and the water-soluble vitamins, which are the B group, C and some specific cofactors. The major vitamins, their active forms and physiological functions are listed below. The role of vitamins in the pathways listed here are covered in more detail in the relevant sections earlier in this chapter. The reader is encouraged to refer to those sections for further details.

3.8.1 Individual vitamins

(a) Thiamine (vitamin B_1)

The active form of thiamine is thiamine pyrophosphate. Thiamine pyrophosphate is a coenzyme for reactions which are involved in carbohydrate metabolism. These reactions are the decarboxylation of α-keto acids and the formation and degradation of α-ketols. Magnesium is also required as a cofactor. Of particular importance is the role of thiamine in the conversion of pyruvate to acetyl CoA.

(b) Riboflavin (vitamin B_2)

Riboflavin is a yellow pigmented molecule which in the body forms a component of the molecule flavin adenine dinucleotide (FAD). FAD is a cofactor in a number of oxidation reduction reactions and acts as an energy currency similar to ATP. It is particularly important in the degradation of pyruvate, fatty acids and amino acids, and in the process of electron transport.

(c) Nicotinic acid (niacin)

Nicotinic acid, or niacin as it has been commonly named, is a component of two high-energy molecules – nicotinamide adenine dinucleotide

(NAD) and nicotinamide adenine dinucleotide phosphate (NADP) (section 3.2). Nicotinic acid can be synthesized by most animals from the amino acid tryptophan, and niacin deficiencies can be avoided by feeding a diet high in tryptophan. Tryptophan, however, is only present in very small amounts in most animal meals, and supplementation of the diet with nicotinic acid is therefore advisable. NAD and NADP are, like FAD, important in a number of oxidation and reduction reactions that occur within cells. Of obvious importance are reactions involved in the citric acid cycle, but involvement of NAD or NADP in metabolic reactions is fairly common (see sections discussing biosynthesis and degradation of fatty acids, carbohydrates and amino acids).

(d) Pantothenic acid (coenzyme A)

Pantothenic acid or coenzyme A is the coenzyme of acetyl, acyl or propionic CoA. This molecule serves as a carrier of various carbohydrate groups and is involved in reactions in fatty acid oxidation, fatty acid synthesis, pyruvate oxidation and acetylations. The presence of coenzyme A within a cell is fundamental to the transfer of energy throughout the various reactions.

(e) Pyridoxine (vitamin B_6)

Pyridoxine acts as an important coenzyme in transamination reactions. The full importance of this coenzyme in fish nutrition will be realized when it is recalled that one of the major sources of energy in fish metabolism occurs from the degradation of amino acids.

(f) Biotin

Biotin acts to facilitate carbon dioxide transfer in reactions which require the addition of CO_2 to another molecule. Biotin is made by intestinal bacteria, and a deficiency in this vitamin is not readily produced by withholding the vitamin from the diet. A glycoprotein component of egg white called avidin binds biotin in the gut such that it cannot be absorbed by the intestinal mucosa and thus can be used to induce biotin deficiency. Heat denatures avidin, allowing it to be digested and so any biotin complexed can be absorbed. Clearly care must be taken if including raw eggs into fish feeds.

(g) Folic acid

Folic acid is a cofactor in the transfer of single carbon entities [e.g. methyl (CH_3) groups] to other molecules in the same way that biotin is a carrier for CO_2. The folic acid molecule is composed of three separate parts called pterodine, *p*-aminobenzoic acid and glutamic acid. In some organisms, folic acid can be synthesized if *p*-aminobenzoic acid is provided in the diet.

(h) Cyanocobalamin (vitamin B_{12})

Vitamin B_{12} is essential for normal maturation and development. It is required for the synthesis of choline and the metabolism of single carbon fragments. The requirement for vitamin B_{12} in the diet is only as a trace, and it is difficult to induce a dietary deficiency because vitamin B_{12} is synthesized by microorganisms in the gut. The usual problem with vitamin B_{12} deficiency occurs because a carrier mucoglycoprotein called intrinsic factor is lacking in the gut. The absence of intrinsic factor means that the vitamin B_{12} present in the intestinal contents will not be absorbed by the animal and deficiency results.

(i) Ascorbic acid (vitamin C)

Ascorbic acid is required in the diet of some species of fish. The first recognized function of ascorbic acid is its role in hydroxylating the proline to hydroxyproline for use in cartilage synthesis. However, it acts as a strong reducing agent in a number of other reactions. It is involved in carnitine synthesis and in the detoxification of pesticides and other toxicants in processes involving cytochrome P450. It is a highly labile vitamin easily destroyed by cooking or lengthy or improper storage of food. It is usually added to five- to 10-fold excess to allow for degradation during storage and to provide some shelf-life to the feed.

(j) Inositol

Inositol has no known cofactor activity but, as an important component of phospholipids, is important in the synthesis of membranes in cells.

(k) Choline

Choline also has no known coenzyme function but acts as an important methyl donor in a number of metabolic reactions, such as the production of acetylcholine from acetyl CoA. Choline can be synthesized by animals if there is adequate methionine in their diet. Inclusion of choline into fish diets alleviates some of the requirement for methionine which is present in low amounts in most protein sources.

(l) Retinoic acid (vitamin A)

Vitamin A is a fat-soluble vitamin which is important as a component of the protein rhodopsin, a light-absorbing pigment found in the retina of the eye. Two forms of vitamin A are known. Vitamin A_1 is common in marine fishes, whilst vitamin A_2 is common in freshwater fishes. The two molecules are very similar, the only difference being that there is an additional double bond present in the carbon ring structure of vitamin A_2. Vitamin A is also important in cells other than the retina since deficiency causes damage in epithelial, bony and connective tissues. However, its exact function is unknown. β-Carotene is often used as a dietary source of vitamin A since it is composed of two vitamin A molecules which are readily separated by hydrolysis.

(m) Vitamin D

The vitamin D group of compounds consists of several molecules, although the most important are vitamin D_2 (ergocalciferol) and vitamin D_3 (cholecalciferol). Of these it appears that fish are only able to use vitamin D_3. Vitamin D is a precursor to 1,25-dihydroxycholecalciferol, a hormone important in regulating calcium and phosphate levels in the serum. Vitamin D synthesis occurs in most animals by ultraviolet radiation of 7-dehydrocholesterol. Whilst this reaction is likely to occur in many fishes, the shallow penetration of UV irradiation through water is likely to lead to a requirement for vitamin D in the diet of fish.

(n) Vitamin E

Vitamin E, also known as tocopherol, acts as an antioxidant, particularly protecting polyunsaturated fatty acids. High dietary levels of polyunsaturated fatty acids increase the requirement for dietary vitamin E. An interaction exists between vitamin E and selenium, a metallic antioxidant; vitamin E requirements are greater in selenium-depleted fish.

(o) Vitamin K

Vitamin K is a group of one of several compounds called phylloquinone or menaquinone, which are isolated from plant or animal tissue respectively. A third compound, menadione, is a synthetic compound that has greater vitamin K activity than either of the naturally occurring substances. Vitamin K is important in the synthesis of prothrombin, a protein which is important in blood clotting. Vitamin K is required by all animals, including fish, for normal blood clotting.

3.8.2 Vitamin requirements of fish

As stated earlier, the requirement for vitamins by different species of fish varies greatly according to their usual feeding habit and capacity to synthesize them. The known requirements of those fish that have been investigated are shown in Table 3.8. It can be seen from Table 3.8 that not only do the vitamins required vary, but the dietary requirement also varies. This may be the result of variation between experimental regimes, but is more likely to be the result of variations between species. It is therefore very difficult to determine a generally recommended level of vitamin supplementation that will be satisfactory for all fish species.

Inclusion of vitamins in diets is complicated further since most of them are highly labile molecules and are readily destroyed during processing. It is easy to overcome this problem for the water-soluble vitamins by adding excess amounts since any excess consumed is readily excreted. However, in the case of fat-soluble vitamins, excess amounts accumulate in the body and cause vitaminosis or vitamin poisoning. Adding large amounts of vitamins, even those that are water soluble, has the additional disadvantage of the extra expense, which is undesirable. The common practice of basing inclusion of vitamins on known requirements of other species generally means using data from rainbow trout for coldwater fish, and from catfish or carp for warmwater fish. This strategy probably provides a compromise between determining the exact requirement experimentally and adding vitamins to excess.

3.8.3 Dietary sources of vitamins

Vitamins are contained in the fresh products from which diets are made. However, processing method and time can destroy these compounds, a problem not easily resolved. For example, cooking destroys vitamin C, whilst another vitamin, thiamine, is protected by heat. Heat treating fish meal destroys the thiaminase which breaks down thiamine. Processes such as solvent extraction of oil seeds removes the fat-soluble vitamins and choline. Inositol, on the other hand, is common in seeds but is largely present in a form which is indigestible to fish.

Considerable work is being undertaken to develop methods of protecting vitamins included in animal feeds, particularly vitamin C. The use of techniques, which physically protect (encapsulation) or chemically protect active sites (e.g. esterification of vitamin C; O'Keefe and Grant, 1991), is proving to be successful in delivering accurate doses of vitamins to fish.

3.9 MINERALS

Minerals, or inorganic elements, are needed by animals to maintain many of their metabolic processes and to provide material for major structural elements (e.g. skeleton). Not all elements that are used in metabolism are required in the animal's diet. Fish, particularly marine fish, have an advantage in that their ambient medium contains many of the elements that they need for growth and survival.

The minerals required for normal metabolism can be divided into two groups, major and trace minerals. Major minerals are required in large quantities and include calcium, phosphorus, magnesium, sodium, potassium, chlorine and sulphur. Trace minerals are those required in trace amounts and include iron, iodine, manganese, copper, cobalt, zinc, selenium, molybdenum, fluorine, aluminium, nickel, vanadium, silicon, tin and chromium. A description of the minerals essential for fish is given below. Again the reader is referred back to the sections describing the reactions in which minerals act as cofactors.

3.9.1 Individual minerals

(a) Calcium and phosphorus

The majority of calcium is found in the skeleton and scales of bony fish, and during fasting these structures provide a pool of calcium for resorption (Lovell, 1989a). Calcium is also important for a variety of other physiological processes including metabolism, nerve and muscle function and osmoregulation. Phosphorus is the other major mineral required by fish. Again the majority of phosphorus is found in the bones and scales (Lovell, 1989a). Phosphorus also has roles in a number of metabolic reactions and is a constituent of many important molecules such as ATP. However, relative to the available stores in bone and scales, the amount of phosphorus involved in these molecules is relatively small.

(b) Iron

Iron is required by fish in trace amounts and acts as a constituent of haemoglobin and the cytochromes, proteins that are important in oxidative phosphorylation.

(c) Magnesium and manganese

Like calcium and phosphorus, a large proportion of the magnesium in fish is contained in skeletal tissue. Magnesium is important as a cofactor in a number of metabolic reactions and seems to be particularly important in maintaining muscle tone. Likewise, manganese is important as a cofactor in a number of metabolic reactions and is important in maintaining proper function in nerve cells. In a number of instances, manganese and magnesium can replace each other.

(d) Copper and zinc

Copper and zinc are important components of a number of metalloenzymes. Approximately 20 different enzymes have been found to contain zinc. The functions of these metalloenzymes encompass a wide variety of metabolic processes.

(e) Iodine

All animals have a requirement for iodine, which is an important component of thyroid hormones. Thyroid hormones, as discussed later, are important in regulation of the growth and development of fish.

(f) Selenium

Selenium is an important inorganic element and constitutes an integral part of the enzyme glutathione peroxidase. Glutathione peroxidase destroys hydrogen peroxide and other hydroperoxides, which oxidize various molecules in the cell such as unsaturated fatty acids. Selenium also imparts a protective effect against the toxicity of heavy metals such as cadmium and mercury.

(g) Sodium, potassium and chlorine

Sodium, potassium and chlorine are amongst the most common inorganic elements found in fish. Appropriate levels of these ions are required for proper functioning of all cells, particularly in maintaining ion gradients between the inside and outside of cells, and for maintaining nerve function. These elements are extremely common in the environment, especially marine environments, and tissue levels of these ions are maintained as an inevitable result of osmoregulation. They are also extremely plentiful in ingredients used for preparation of diets, and as such supplementation of these elements is not required.

(h) Other trace elements

A number of trace elements apart from those listed above are required for adequate fish nutrition. Cobalt acts as a component of vitamin B_{12}. Chromium is important in normal carbohydrate and lipid metabolism, whilst sulphur is required for the synthesis of cysteine.

Most of the minerals required by fish are found commonly either in the environment or in a component of their diet. Animal meat meals are rich sources of iron, magnesium, zinc, iodine, selenium, calcium and phosphorus, while plant meals contain adequate amounts of manganese and copper. Selenium, sodium, potassium, chlorine and magnesium are also commonly found dissolved in water, and can be absorbed by fish through their gills.

In general, the dietary requirements for minerals are poorly understood because of the difficulty in devising mineral-deficient diets and the need to deplete tissue mineral stores. Often generalizations from studies of higher vertebrates must be made. A summary of the known mineral requirements of fish is contained in Table 3.9.

Table 3.9 Summary of the mineral requirements of various species of fish

	Sea bass	*Atlantic salmon*	*Channel catfish*	*Common carp*	*Eel*	*Sole*	*Pacific salmon*	*Rainbow trout*	*Red drum*	*Tilapia*
General	2% UPS XI[a]									
Phosphorus		13 000	4500	6000–7000	2500–3200	7000	R	6000	8600	9000
Manganese		20	2.40	13			R	20		12
Copper		6	5	3				5		3–4
Iron		73	30	150	170			60		
Selenium		R	0.25				R	0.3		
Iodine and fluorine		R					0.6–1.1 μg/kg			
Potassium		NR					8000			
Keto-carotenoids		R								
Calcium			4500		2700					6500
Magnesium			400	400–500	400			500		590–770
Zinc			20	15–30			R	30	20–25	10
Cobalt				0.1						

[a] Proprietary mineral mix.
Data are collated from Halver (1989), Lovell (1989a) and Wilson (1991).
Values are mg/kg diet unless stated otherwise.
R, required, NR, not required.

3.10 MANIFESTATION OF NUTRITIONAL DISEASES

Although the balance of protein, carbohydrate and lipid in fish diets is important, the effects of an imbalance are difficult to observe without measuring such parameters as growth or feed conversion.

Features of diets that are likely to predispose an animal to disease include an imbalance in the protein–energy ratio of a diet, and inadequate provision of essential amino or fatty acids. The symptoms of such deficiencies include development of fatty livers, excess visceral fat, reduced feeding activity and failure to grow. These deficiencies do not generally result in large-scale mortalities over short periods, but do have economic consequences associated with poor food conversion rates, low growth rates and less desirable body composition. If uncorrected over long periods, total loss of stock may result.

Vitamin and mineral deficiencies are the most readily observed forms of nutritional disease. The requirements for most vitamins and minerals are common to many tissues, however individual tissues are able to preferentially retain vitamins or minerals within them. Disease that appears as a result of a deficiency of a particular vitamin or mineral usually appears in that type of tissue least able to maintain or hold the vitamin or mineral concerned. The manifestations of vitamin and mineral deficiencies observed in fish are shown in Table 3.10. The most common form of symptom is loss of appetite and failure to grow, which can be observed in most species with practically any kind of deficiency. The two other common symptoms of nutritional deficiency diseases are malformations of skeletal structures (indicating the importance of vitamins and minerals in the development of the skeleton) and haemorrhage of the skin and fins.

Poor nutrition can affect the well-being of the animal and thus make it more susceptible to opportunistic infections by various disease-causing agents. In these cases it is impossible to perform controlled experiments to determine whether a poorly composed diet predisposed the fish to infection or whether the infectious agent was virulent enough such that the fish would have been infected no matter what its diet. The only reliable method to determine if a diet predisposes fish to disease is to compare growth and survivorship of those fish to others fed a different diet. On occasion, however, it may be possible to prevent low-level mortalities (1–5% of the population per week) simply by changing diet.

3.11 REGULATION OF METABOLISM AND GROWTH BY HORMONES

As well as the obvious importance of nutrients in regulating metabolism via the concentration of substrates and products, a number of hormones circulate within an animal to regulate growth.

Table 3.10 Summary of the manifestations of vitamin and mineral deficiencies observed in fish

Vitamins	*Signs of deficiency*
Thiamine	Poor growth, susceptibility to post-handling shock, high mortality, loss of appetite, loss of sense of distance, cataracts, fin haemorrhage, convulsions, neurological disorder, trunk winding, low transketolase activity
Riboflavin	Erratic swimming, cataracts, loss of appetite, haemorrhagic eyes, erosion, haemorrhage and ulceration of the skin, dwarfism, reduced erythrocyte glutathione reductase activity, kidney necrosis
Pyridoxine	Avoidance of schooling, erratic swimming, surfacing, lesions of lip, high mortality, convulsions, reduced alanine aminotransferase activity, degenerative changes in liver, kidney and gill, nervous disorders, slow growth, disintegration of erythrocytes, loss of appetite, ascites, exophthalmia
Pantothenic acid	Ventral fin haemorrhage and erosion, clubbed gill, high mortality, loss of appetite, haemorrhagic eyes, exophthalmia, anaemia, skin lesions, ataxia, abnormal swimming, atrophy of pancreatic islets
Nicotinic acid	Loss of appetite, poor growth, erratic swimming, skin and fin haemorrhage, deformed operculum, exophthalmia
Biotin	Anaemia, reduced skin pigmentation, reduced liver pyruvate decarboxylase activity, fatty liver, abnormal swimming, haemorrhage of gastrointestinal tract, ataxia, reduced acetyl CoA levels and pyruvate carboxylase activity
Inositol	Poor growth, abnormal bone formation, dermatitis, white–grey intestine, ataxia
Choline	Fatty liver, haemorrhagic kidneys, white–grey intestine, poor food conversion efficiency
Folic acid	Epithelial and fin haemorrhage and lesions, anaemia, poor growth, large segmented erythrocytes
Ascorbic acid	Bleeding gill, reduced and deformed hard parts and skeleton, exophthalmia, loss of equilibrium, increased mortality, slow growth, lethargy, loss of appetite, haemorrhage
Vitamin A	Exophthalmia, oedema, ascites, anorexia, haemorrhage in liver, skin or gill, opercular deformities, abnormalities of the eye
Vitamin D	Low bone ash, impaired calcium homeostasis, convulsions
Vitamin E	Skin depigmentation, muscular atrophy, susceptibility to disease, anaemia, reduced growth and increased mortality, impaired serum complement function, elevated plasma proteins, increased carcass fat and water composition, erythrocyte haemolysis, increased iron in the spleen and pancreas
Vitamin K	Skin haemorrhage, poor blood clotting, anaemia
Iron	Reduced haematocrit, red cell count and liver, spleen, serum and kidney iron
Copper	Reduced serum copper, cytochrome C oxidase activity and hepatic superoxide dismutase activity, poor growth

Table 3.10 (Continued)

Vitamins	*Signs of deficiency*
Manganese	Reduced haematocrit and haemoglobin, dwarfism, skeletal abnormalities, high mortality, low manganese in bone
Selenium	Lethargy, anorexia, increased mortality, reduced muscle tone and glutathione peroxidase activity
Calcium	Reduced bone ash
Phosphorus	Reduced bone ash, poor growth, skeletal abnormalities, low feed efficiency
Magnesium	Muscle flaccidity, lethargy, reduced bone, serum and whole-body magnesium
Zinc	Reduced serum alkaline phosphatase activity, reduced bone and serum zinc, poor growth, high mortality, skin and fin erosions

Data are summarized from Halver (1989), Lovell (1989a) and Wilson (1991).

3.11.1 Insulin and glucagon

Insulin and glucagon play extremely important roles in the metabolism of carbohydrates in mammals. Insulin is secreted by cells in the pancreas in response to high glucose concentrations in the blood, and acts to reduce this by increasing the rates of glycolysis, glycogen synthesis and lipogenesis, particularly in the liver and the adipose tissue. Glucagon is also secreted by the pancreas, but only when blood glucose levels fall. Glucagon stimulates the rates of gluconeogenesis and glycogen and lipid breakdown in mammals.

In fish, it seems that insulin particularly, but also glucagon, is more important in amino acid metabolism than in carbohydrate metabolism. Amino acids serve as better stimulants of insulin secretion than does glucose. Insulin also serves to increase the incorporation of amino acids into protein (Plisetskaya, 1989). There is also a consistent positive correlation between the blood levels of amino acids and insulin. A high-protein diet increases blood insulin, while a starvation diet decreases blood insulin. Although one study has indicated that there is an increase in muscle RNA levels associated with insulin administration, it is generally accepted that the stimulation of protein deposition rates by insulin is mediated by the alleviation of gluconeogenesis, reduced levels of amino acid breakdown and the associated increase in the availability of precursors for protein synthesis.

3.11.2 Thyroid hormones

There are two other major groups of hormones or factors that are known to be involved in the natural processes involved in growth regulation in fish. Both involve a series of releasing and inhibiting factors regulating the levels of the molecule that has the growth-promoting effect. These two systems are also known to interact with each other (MacLatchy and Eales, 1990).

The first system comprises the thyroid hormones and their regulatory hormones. The first hormone involved is thyroid releasing hormone (TRH), which is released from the hypothalamus, a neural control centre at the base of the brain. The release of TRH is stimulated or inhibited by many factors such as photoperiod or temperature. TRH acts directly on the pituitary gland (a neuroendocrine organ located adjacent to the hypothalamus). The pituitary gland is then stimulated to release thyroid-stimulating hormone (TSH), also known as thyrotropin. Increased blood levels of TSH stimulate the thyroid gland to release thyroxine (T_4). T_4 is modified in peripheral tissues such as the liver by an enzyme called monodeiodinase to produce triiodothyronine (T_3). T_3 has the somatotrophic or growth-promoting effects but the mechanism involved is not understood. Both T_4 and T_3 circulating in the blood have a negative feedback effect on the release of TSH by the pituitary gland.

The effects of administered T_4 and T_3 vary depending on the dose. T_4, when administered to Coho salmon, *Oncorhynchus nerka*, has effects that include increased body weight, body length, appetite and utilization efficiency. In a study of *Ophicephalus punctatus*, Medda and Ray (1979) found that low doses of T_4 induced anabolic effects as evidenced by both reduced ammonia excretion, indicating reduced deamination, and higher muscle and liver RNA and protein levels. This evidence is all indicative of an increased level of protein synthesis. A similar study by Matty *et al.* (1982) demonstrated increased incorporation of radioactive amino acids in fish treated with T_3, thereby supporting the hypothesis that T_3 increases protein synthesis.

However, when Medda and Ray (1979) used higher doses of T_4, they found an increased nitrogen excretion rate and decreased tissue protein and RNA content, indicating that catabolism was increased. The significance of the biphasic action of the thyroid hormones is not known. It is clear though that they have a stimulatory effect on growth which is mediated by an increase in protein synthesis and associated cell size.

An influence of diet on thyroid hormones has also been described. Leatherland *et al.* (1984) reported that plasma T_3 was low in rainbow trout fed a diet containing no carbohydrate and that dietary lipid levels could influence plasma T_4 and T_3 concentrations. Himick *et al.* (1991), however, reported that dietary carbohydrate levels were the determining factor in nutritional stimulation of plasma T_4 rather than protein, lipid or total energy. Reduced dietary levels of pantothenic acid and vitamin D also reportedly brought about reduced T_4 or T_3 levels respectively (Leatherland *et al.*, 1984). In starved fish, a reduced level of monodeiodinase was correlated with reduced T_3 and T_3–T_4 ratios, indicating that secretion of T_4 from the thyroid gland is maintained during food deprivation (Leatherland *et al.*, 1984). However, considerable work is still required to understand the details of the interaction between diet and thyroid hormones.

3.11.3 Growth hormone

The second major system of growth regulatory hormones comprises growth hormone (GH) and its regulatory and mediating hormones. A number of factors, including gonadotropin-releasing hormone (GnRH) and neuropeptide Y, which are secreted by the hypothalamus reportedly stimulate growth hormone release, whilst somatostatin, also secreted by the hypothalamus, inhibits GH release. These factors act on the pituitary gland, which releases GH. Growth hormone acts upon a variety of tissues, inducing the secretion of insulin-like growth factors (IGFs), also called somatomedins, which have the somatotropic effects.

Reduced growth, induced by removing the pituitary gland, is reversed by administration of GH, and growth rates can be enhanced by GH administered to intact fish. Administration of GH has been reported to decrease levels of plasma protein and excretion of nitrogen (Wilson *et al.*, 1988). Subsequent studies have shown that GH promotes growth by increasing protein synthesis and by mobilizing lipid with a subsequent sparing of amino acids for protein deposition. However, aspects of the mechanism by which GH promotes growth are yet to be clearly defined. A number of workers have described increased levels of plasma GH in fish in which food intake and growth rate were decreased (Bjornsson *et al.*, 1988) while Sumpter *et al.* (1991) found that rainbow trout exhibit increased GH concentrations after 7 days of fasting and suggested that this may be because of reduced IGF levels in starved fish.

3.11.4 Steroids

The hormones that received the earliest interest with regard to their growth-affecting properties are the anabolic steroids. These hormones, derived from cholesterol, take a number of forms and have a variety of functions in an animal. Early studies of the roles of steroids in growth regulation concentrated on determining the effect of steroids on gross parameters such as body weight. It was found, in a number of studies, that inclusion of steroids in the diet increased the growth rate of salmonids. The most effective of these was 17α-methyltestosterone, although interspecific variability occurs (Gannam and Lovell, 1991). Subsequent studies have revealed that the action is due to an increased rate of protein synthesis. Increases in appetite and rates of digestion and absorption have also been reported as effects of 17α-methyltestosterone.

The effects of oestrogens on growth are conclusively negative. Feeding diethylstilboestrol results in retarded growth or increased protein degradation in rainbow trout, brown trout, goldfish, catfish and plaice. This may be of significance, since there is considerable secretion of oestrogens during gametogenesis in the female.

The third group of steroids of interest are the corticoids. These hormones are secreted by the adrenal cortex in response to stress, and again the effects on body growth are negative. Corticoids are thought to increase the rate of protein degradation and the rate of lipolysis.

CHAPTER 4

Digestion and absorption

The earliest step in obtaining nutrients for an animal involves ingestion, processing and absorption of nutrients via the digestive system, and in any study of nutrition it is important to understand these processes. The method by which this occurs in any individual species varies greatly in some aspects, although other aspects are general to vertebrates. Accordingly these principles and variations are discussed in this chapter.

4.1 GENERAL ANATOMY AND ORGAN PHYSIOLOGY

Among vertebrates, both chondrichthyan and actinopterygean fish have the least developed and simplest digestive system. This is to be expected, as fish are placed lowermost among vertebrates in the evolutionary scale. Equally, among 20 000 or so known species of actinopterygean or bony fish, one finds some of the most bizarre feeding habits, which raise questions regarding the digestive physiology of various species. An example is a species of the family Cichlidae found in the African Great Lakes, which feeds on the scales of other species, although this sort of food habit is very much the exception to the rule. The generally accepted subdivision of fish on the basis of feeding habits into predatory and non-predatory groups is equally a misnomer since the majority of non-predatory species also ingest animal organisms.

Based on the nature of the food ingested, fish are usually categorized into:

- herbivores, feeding largely on plant material;
- detritivores, feeding largely on detritus;
- omnivores, consuming a mixed diet; and
- carnivores, consuming only animal matter, usually fish and bigger invertebrates.

Again, such divisions are not strictly valid in most cases. The majority of species depend on a mixed diet, and such a dependence is more common among freshwater species. In addition, changes to feeding habits throughout a fish's life-cycle may occur. A species may feed on zooplankton as a juvenile, but may change to feeding largely on plants as an adult.

Fish may also be broadly categorized into ecological groups according to the feeding conditions, i.e. pelagic plankton feeders, benthos feeders, etc. Pelagic plankton feeders could be further subdivided into surface and columnar feeders. These divisions are important in certain types of aquaculture, in particular in warmwater polyculture, in which a combination of species is used, each occupying a different ecological niche.

Food habits of numerous fish species, and the associated specializations and/or adaptations, have been studied by many workers. Such investigations constituted a major component of the traditional investigations in fish biology studies, and it would be futile to attempt to summarize these. Phylogeny is an important factor in the final construction of the digestive apparatus, and it is known that different species with the same food habits/diet may differ in the structure of the alimentary system (Nikolskii, 1965). However, functional adaptations related to the nature of the food and feeding habits remain similar. As shown in Figure 4.1 the alimentary canal of teleosts can be a straight tube from the mouth to the anus, or it may form loops and be structurally divided into different parts which are functionally different.

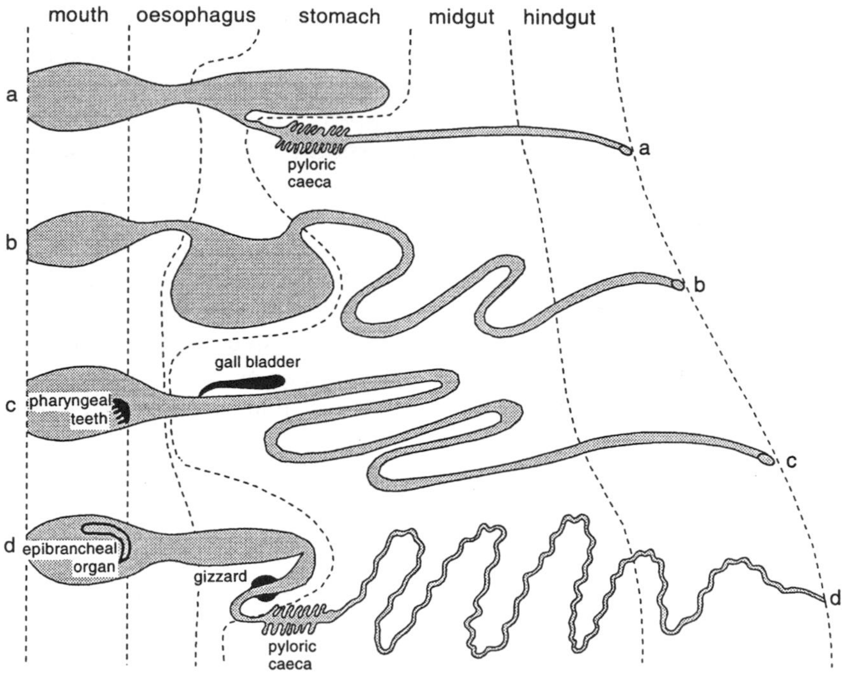

Figure 4.1 The digestive systems of four fish described in the text, arranged in order of increasing gut length. (a) Rainbow trout (carnivore). (b) Catfish (omnivore emphasizing animal sources of food). (c) Carp (omnivore, emphasizing plant sources of food). (d) Milkfish (microphagous planktivore). (From Smith, 1980.)

Some generalities of the anatomy of the alimentary canal of fish need to be considered. Two main groups can be distinguished among the economically and aquaculturally important species:

- fish without a stomach – into this group fall the cyprinid or carp-like fishes comprising a large number of species which are cultured in the tropics, as well as some species which are important to the ornamental fish industry; and
- fish with a stomach – species which are aquaculturally important in this group are the coldwater salmonids and warmwater species (or species groups) such as catfishes, eels, tilapias, groupers and barramundi.

It should be noted that all predatory fish have a stomach, and also generally exhibit teeth to a varying degree.

Table 4.1 Relative gut length of selected cyprinids

Species	*Feeding habits*	*Relative length of the gut*
Labeo horie (Cuv.)	Algae, detritus	15.5
L. lineatus Blgr	Algae, detritus	16.1
L. calbasu (Ham.)	Plants, weeds, algae, diatoms	3.75–10.33
Cirrhina mrigala[a] (Ham.)	Algae, detritus	8
Hypophthalmichthys molitrix[a] (Val.)	Phytoplankton	13
Varicorhinus heratensis Keys	Algae, detritus	6–7.5
Catla catla[a] (Ham.)	Periphyton, plants, insect larvae	4.68
Garra dembensis (Rupp.)	Algae, invertebrates	4.5
Ladislavia taczanowskii Dyb.	Algae, invertebrates	2.0–2.5
Ctenopharyngodon idella[a] (Val.)	Plants	2.5
Amblypharyngodon mola (Ham.)	Plants	2.8
Barbus sharpeyi Gunth.	Plants	2.79–3.18
B. tor (Ham.)	Invertebrates, plants	1.24
B. ticto (Gunth.)	Invertebrates, plants	1.58
Rostrogobio amurensis Tar.	Invertebrates	0.8–1.4
Leptocypris modestus Blgr.	Invertebrates	0.8–0.1
Engraulicypris minutus (Blgr.)	Zooplankton	0.7
Chelethiops elongatus (Blgr.)	Zooplankton	0.75
Erythroculter erythropterus (Bas.)	Carnivorous, insects	0.77–1.50
Elopichthys bambusa (Rich.)	Carnivorous	0.63
Barilius moorei Blgr.	Carnivorous	0.65–0.8
Chela bacaila (Ham.)	Carnivorous	0.88
Ptychocheilis oregonense (Rich.)	Carnivorous	0.78

[a] Cultured species.
Source: Modified from Kapoor *et al.* (1975).

The other generality is the length of the digestive tract relative to body length and the surface area of the mucosa. The ratio of the length of the digestive tract to body length is known as the relative gut length (RGL), and the mean for different species (mRGL) is indicative of their feeding habits. The mRGL does not remain constant for any given species, and within a species it could change with body length, stage of development and changes in feeding habits.

Generally, the mRGL is highest in species that feed on detritus and algae (microphages) which contain a proportion of indigestible material such as sand and fibrous material. The mRGLs of selected cyprinids with different feeding habits are given in Table 4.1 to exemplify this further.

The mRGL is also known to respond to changes in feeding habits of different populations, as has been shown in the case of the green chromid (*Etroplus suratensis*), one of the two Asian cichlids (Figure 4.2). Domestication of the carp is thought to have brought about an increase in its intestinal length. Similarly, a carbohydrate-rich diet induces changes in mean relative intestinal length as a measure of mRGL.

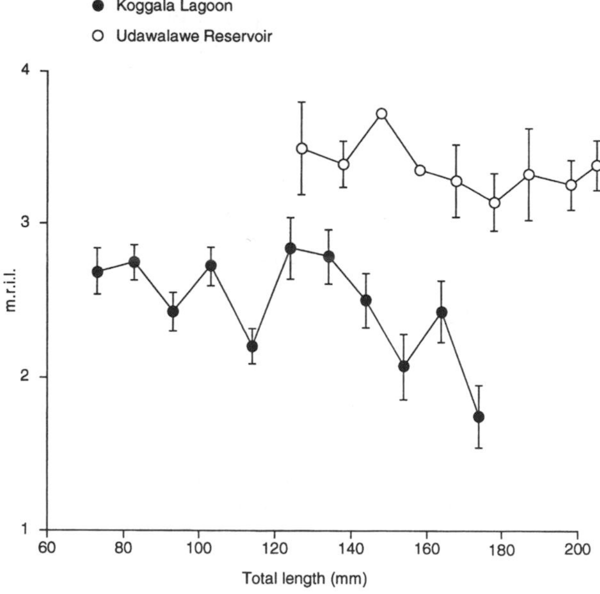

Figure 4.2 The mean relative intestinal length (m.r.i.l.) as a measure of mRGL of two populations of *Etroplus suratensis*: (1) a lagoonal population feeding predominantly on molluscs and (2) a reservoir population feeding predominantly on macrophytes. Numerals on the figure are the number of fish at each length. (From De Silva *et al.*, 1984a.)

4.1.1 Morphology

Detailed morphology of the teleostean alimentary canal has been aptly dealt with by Barrington (1957) and Kapoor *et al.* (1975). In our treatment we will attempt to focus on those features of the gut morphology relevant to aquaculture, and therefore concentrate on modifications and/or features of cultured species wherever possible.

The alimentary canal in teleost fishes is broadly divisible into a *Kopfdarm*, consisting of the mouth, buccal cavity and the pharynx, and a *Rumpfdarm*, which constitutes the remainder of the canal. The latter is generally provided with sphincters and valves. Also, organization of various tissues in the different regions of the alimentary canal is generally uniform in teleosts. The exceptions to this rule are the strata compactum and granulosum, which are not uniformly present in all fishes.

4.1.2 Mouth, buccal cavity and pharynx

The mouth, buccal cavity and pharynx are essentially associated with the predigestive processing of the food – selection, seizure and orientation. Generally, stomachless fish such as the carps do not have teeth on the jaw, as opposed to predatory fish with a stomach. Buccal teeth assist in food capture and holding but do not assist in crushing and/or tearing of the prey, except in unusual cases such as the piranha. The lips are the primary food-procuring organs, and as such the position of the mouth, dentition on the jaws and the buccopharynx (buccal cavity) and the gill rakers show a close relationship to the mode of feeding and the kind of food consumed.

Some fish which consume macrophytes have evolved anatomical adaptations to aid them in ingestion and mastication of plant material. The initial processing of the plant material makes it more suitable for digestive enzymes to act upon it elsewhere in the gut. Notable adaptations are bi- and tricuspid teeth on the jaws of *Tilapia rendalli*, which aid in the cutting and macerating of macrophytes, or the strong and specialized pharyngeal teeth with a flattened, serrated and rasping surface to cut, shred and grind macrophytes found in grass carp (Hickling, 1966).

As an aid to initial processing, most stomachless cyprinids have pharyngeal teeth which are modified to varying degrees depending on the nature of the diet. In Figure 4.3, examples of different types of teeth in three species of tilapias are shown. Such variations in the patterns of dentition are quite common in fish and occur either on the maxillae or on the palate.

Some species commonly cultured in the tropics are filter-feeders, filtering the surrounding water for phyto- and zooplankton, trapping the organisms and ingesting them. The filtering apparatus is composed of the gill rakers of selected posterior gill arches, which are modified to

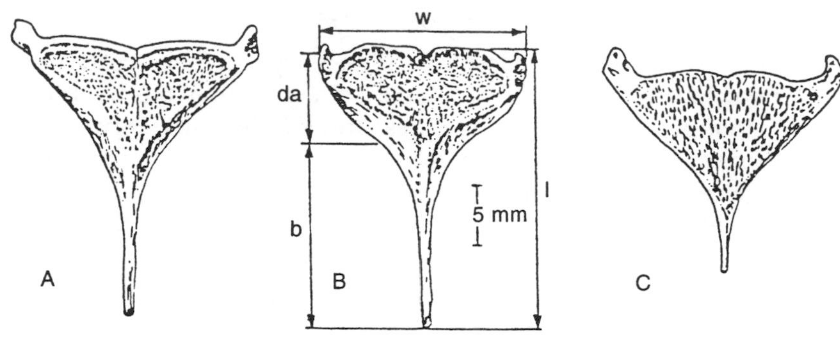

Lower pharyngeal bones of: A *Oreochromis niloticus*
B *Sarotherodon galilaeus*
C *Tilapia rendalli* – da is dentiferous area, b the blade, l the length and w the width

Figure 4.3 Lower pharyngeal bones of three cichlids – *Oreochromis niloticus* (A), *Sarotherodon galileaus* (B) and *Tilapia rendalli* (C) and their dentition. (From Pullin, 1988.)

form a fine sieve. The commonly cultured silver and bighead carps feed in this fashion.

In some species, such as in some mugilids (the mullets), specialized pharyngeal organs (pharyngeal cushions, pads) develop teeth over their outer surface (usually claw shaped). These, together with specialized gill rakers of a modified fifth gill arch, act in selecting food from the ingested material and conveying it into the pharynx, thereby concentrating acceptable food items and rejecting the remainder. An example of the sophistication of this type of adaptation is seen in *Mugil cephalus*, a detrital and microbenthic feeder. This species is known to strain the food contained in the bottom muds by the first and second gill rakers, and drain and concentrate it by the action of the third and fourth rakers, pharyngeal discs and pharyngeal rakers.

The tongue is often not a prominent structure in the buccal cavity of fish, although it is often better developed in carnivorous fish. It is not always sharply demarcated, is not freely movable and it generally has skeletal support, striated musculature and a larger amount of connective tissue. The tongue could supplement the function of teeth in the retention of prey, and may also act as a sensory organ.

The cellular structure of the buccal apparatus (which includes the buccopharynx and associated structures) is a layer of stratified epithelium covering a basement membrane. The epithelium contains mucous cells and taste receptors. Striated muscle is present in the subepithelial tissue providing support and mobility to the structures. However, numerous variations from this basic plan occur in different species, e.g. some teleosts are known to have large club cells in buccopharyngeal epithelium. The primary functions of the epithelium of the buccopharynx are mucus secretion and gustation. No enzymatic activity has been demonstrated in the buccal cavity of any teleostean fish.

4.1.3 Oesophagus

The oesophagus, commonly referred to as the gullet, is short, expandable and may not be clearly demarcated from the stomach, or intestine in stomachless fish. Generally, the oesophagus is lined with stratified epithelium beneath which is a basement membrane. Few club cells, and even taste buds, may be present. It is a muscular organ, the circular muscle layer often being well developed. The mucosal folds are elaborately developed in some species, being composed entirely of mucus cells, as in *Mugil cephalus*. The oesophagus essentially acts as a well-lubricated transit tube for the food, adding fluid and mucus during passage into the stomach or the intestine. The occurrence of oesophageal glands and oesophageal sacs with thick striated muscle coats (inner circular, outer longitudinal) in some species has resulted in additional functions being attributed to the oesophagus. The glands and sacs perform the functions of food storage, trituration and mucus production. The presence of proteolytic activity has been described in the mucosa and contents of the oesophagus of *Girella tricuspidata* (Anderson, 1991).

4.1.4 Stomach

As indicated earlier, not all fishes have a stomach. Species with and without stomachs also occur within the same family (e.g. Blennidae) and even in the same genus (e.g. *Gobius*). The size of the stomach, when present, is related to the duration between meals and the nature of the food. Typically, the stomach wall consists of a number of layers which are characteristic of vertebrates (Figure 4.4). The organization of these layers in order from peritoneal cavity to lumen is

- serous membrane
- muscularis
  - longitudinal muscle layer
  - circular muscle layer
- mucosa
  - submucosa
  - muscularis mucosae
  - stratum compactum
  - lamina propria
  - mucosal epithelium

The stomach can be divided into two regions: the anterior and pyloric regions. The most notable difference between the two regions is that

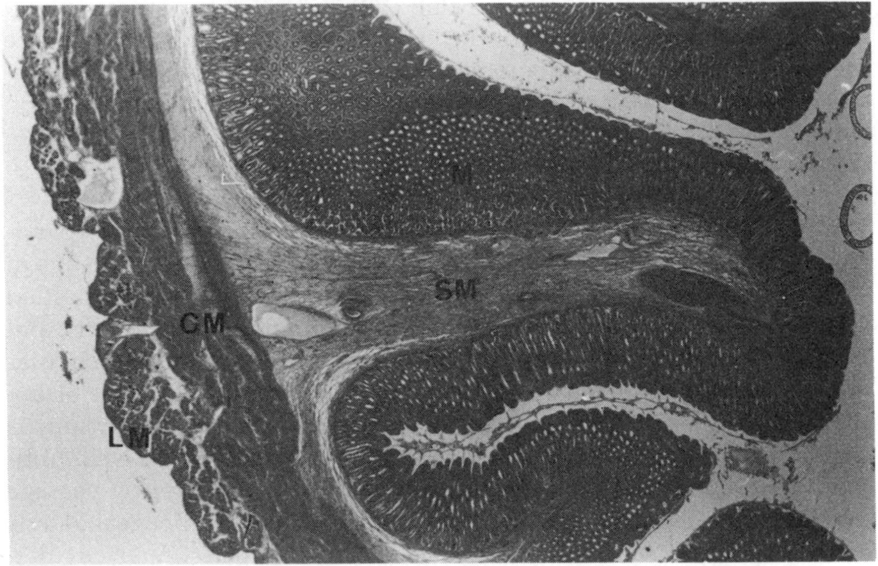

Figure 4.4 A typical histological section of the stomach of a fish, showing mucosa with gastric pits (M), submucosa (SM) and circular (CM) and longitudinal muscle (LM) layers.

gastric glands are more common in the anterior stomach and the musculature is also usually more prominent. Only one type of secretory cells, having a common physiological function, has been identified in the gastric glands of teleosts. Modifications of each of the constituent layers in the wall of the stomach in different species are to be expected. The highest degree of modification is found in the pyloric stomach in several members of the families Clupeoidei, Chanoidei, Characinoidei and Mugiloidei, when the thick muscularis is reduced to a submucosa, and a protective lining enables it to act as a gizzard for trituration and mixing. The development of a 'gizzard' has been attributed to microphagy, and is thought to partly compensate for poor dentition, as demonstrated in the mugilids (Pillay, 1953).

The stomach is a highly extensible organ. For example, in brown trout, enlargement of the stomach by around 30–35% in the longitudinal direction and up to 75% in diameter may occur.

The lack of a stomach in certain species raises the question as to why this should be so. Several hypotheses have been put forward to explain the loss of the stomach, and they are often contradictory and speculative.

It might be noted here that even in predominantly carnivorous groups such as the Siluroidei and Percoidei, stomachless subgroups have evolved (in Loricaridae and Theutidae respectively) and in all instances these are microphagous.

4.1.5 Intestine

The intestine in fishes is a simple, columnar absorbing epithelium lined with a brush border of microvilli, which is typical of absorptive tissues. The detailed structure of the brush border may vary between species. The microvilli play a very important role by substantially increasing the surface area of the cells. This allows greater contact between cells and the nutrients in the lumen, and provides for enhanced cell surface digestion and nutrient uptake than would otherwise occur with a smooth-surfaced cell. Other common constituents are goblet cells and cellular migrants (lymphocytes and various types of granulocytes) which, in fish with stomachs, may be associated with the production of zymogen granules. In stomachless fish, this function is performed by goblet cells of the intestine, which are chiefly mucus producers.

The intestinal epithelium also contains 'pear-shaped' or rodlet cells, the function of which remains controversial. They are considered to be stages in the life-cycle of a goblet cell, unicellular glands or even protozoan parasites by different authors. An enzyme-producing function has been attributed to them by some authors.

In young fish the mucous membrane is mostly smooth, but as the fish ages longitudinal and zig-zag folds appear, finally leading to a sponge-like structure. Unlike the intestinal length (or mRGL), the intestinal surface area relative to body size tends to decrease with growth (Figure 4.5).

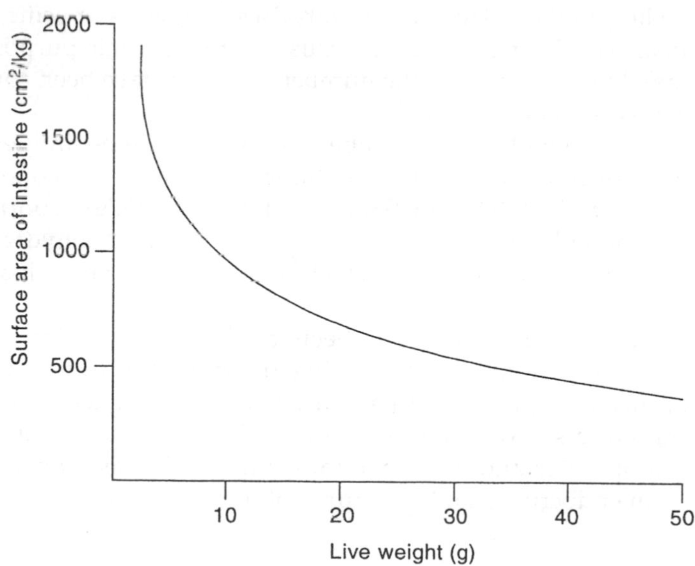

Figure 4.5 Changes in intestinal surface area (cm^2) as a function of body weight (live) for carp. (From Steffens, 1989.)

In most fish, two sections of the intestine may be distinguished with respect to their physiological functions: the anterior section (the area into which the bile and pancreatic juices are received) is often characterized by the presence of large numbers of chylomicrons in the epithelial cells, whilst the distal portion of the intestine is characterized by pinocytotic activity and cells containing granules consisting of absorbed nutrients.

4.1.6 Rectum

The presence of a rectum in the gut of fish also varies with species. The most distal part of the intestine usually possesses a thicker muscular coat and with a marked increase in the number of goblet cells and sometimes granulocytes relative to the more proximal sections. A true rectum delineated from the intestines by an ileorectal valve is only observed in a few species (Anderson, 1986). The number of goblet cells in the rectum is known to change with the feeding habit and starvation.

4.1.7 Pyloric caecae

Also referred to as intestinal caecae, these form auxiliary appendages in teleosts, having a cellular structure similar to that of the intestine. They have a well-developed muscularis, consisting primarily of circular muscles. In some cases the epithelial cells of the pyloric caecae may have cilia. Presence or absence and the number of pyloric caecae varies with species. The number of caecae is not always species specific, but in certain instances this feature has been used for taxonomic purposes (e.g. Mugilidae). In certain species the number of caecae have been correlated with the bulk of food consumed.

It is unclear whether the number of pyloric caecae influence the digestibility of individual nutrients. Buddington and Diamond (1986), De La Noüe *et al.* (1989) and Bauermeister *et al.* (1979) reported that pyloric caecae enhance absorption of amino acids, carbohydrates and lipids, whereas De La Noüe *et al.* (1989) failed to observe such an effect on protein and energy.

A number of functions have been attributed to pyloric caecae. Proposed uses include: supplementing the digestive functions of the stomach, increasing the intestinal surface area; acting as an accessory food reservoir, a site of absorption of carbohydrate and fats or a site of resorption of water and inorganic ions; functioning as a space-saving device or an endocrine gland (precursor of the pancreas).

4.2 DIGESTION

Digestion is the process by which the ingested food material is broken down to simple, small, absorbable molecules. This task is performed primarily by the digestive enzymes. The digestive enzymes, which are secreted into the lumen of the alimentary canal, originate from the oesophageal, gastric, pyloric caecal and intestinal mucosa and from the pancreas.

4.2.1 Digestive fluids and enzymes

Digestive fluids and enzymes are part of gastric secretions, pancreatic secretions, bile, and intestinal secretions.

Acid gastric fluid production probably occurs in most fish, except in those without a stomach when neither HCl nor pepsin is formed in the gut. The fluids and enzymes secreted in teleost fish are summarized in Table 4.2.

Table 4.2 Digestive fluids and enzymes secreted in teleosts

Site/type	*Fluid/enzyme*	*Function/notes*
Stomach		
Gastric secretions	HCl	Reduces gut pH and allows pepsinogens to act
Gastric glands	Zymogen, pepsinogen, HCl, pepsin	Proteolytic enzymes; cleave peptide linkages at the NH_2 groups of aromatic and acidic amino acids. Attack most proteins
	Amylase	Acts on carbohydrates
	Lipase	Acts on lipids
	Esterases	Acts on esters (class of lipids)
	Chitinase	Acts on chitin
Pancreas	Enzymes	Enzymes are stored as zymogens. Proteases produced by the intestine convert trypsinogen into trypsin, which in turn activates others.
	HCO^-_3	Neutralizes HCl entering intestine and prepare intestine for alkali digestion
	Proteases (trypsin, chymotrypsin, carboxypeptidases and elastase)	Optimal action of enzymes at pH 7.0
	Trypsin	Cleave peptide linkages at carboxy groups of lysine or arginine
	Chymotrypsin	Attacks peptide with carbonyls from aromatic side chains

Table 4.2 (Continued)

Site/type	*Fluid/enzyme*	*Function/notes*
Pancreas – *continued*		
	Elastase	Attacks peptide bonds on elastin
	Carboxy peptidases	Hydrolyses the terminal peptide bond of their substrates
	Amylase	Carbohydrate digestion at non-acidic pH
	Chitinase (+NAGase)	Splits chitin into dimers and trimers of *N*-acetyl-D-glucosamine (NAG), which is further broken down by NAGase
	Lipases	Hydrolyse triglycerides, fats, phospholipids and wax esters
Bile (secreted by the liver)	Bile salts, organic anions, cholesterol, phospholipids, inorganic ions	Make the intestinal medium alkaline, emulsify lipids; most bile salts are reabsorbed from the intestine and returned to the liver in the enterohepatic circulation
Intestinal enzymes (secreted by the brush border of the epithelium, but could be partly pancreatic in origin)	Aminopeptidases (alkaline and acid)	Split nucleosides
	Polynucleotidase	Splits nucleic acids
	Lecithinase	Splits phospholipids into glycerol, fatty acids
	Various carbohydrate-digesting enzymes	

The digestive enzymes are hydrolases, compounds capable of catalysing hydrolytic reactions. They are water-soluble proteins. Based on the physiological function they are divided into proteases, lipases and esterases, and carbohydrases.

Apart from the principal sites (Table 4.2), enzymes may be produced in other tissues, e.g. amylase is also produced in the liver. It is also known that enzymes present in the animals which form part of the diet of a fish may enhance endogenous enzyme activity. This is particularly the case in very young fish (Chapter 5).

Most digestion of the food is extracellular, taking place in the lumen of the alimentary canal. It is also likely that enzymes immobilized on the cell surface membrane are also responsible for a certain amount of digestion. This process is largely involved with the intermediate and final stages of the digestive process, and is often regarded as a link in the absorption mechanism.

4.2.2 Protein digestion

Proteolytic enzymes arise from inactive precursors known as zymogens. Zymogens are processed in the lumen of the gut either by acid hydrolysis or proteolytic action. This eliminates the risk of self-digestion of the secreting tissue. Proteases break down peptide links of proteins. Different enzymes are capable of acting on peptide bonds either at the end of the protein (exopeptidases) or at a point within the protein (endopeptidases).

The different points at which endopeptidases and exopeptidases attack during the hydrolysis of a simple protein chain can be illustrated by the following diagram.

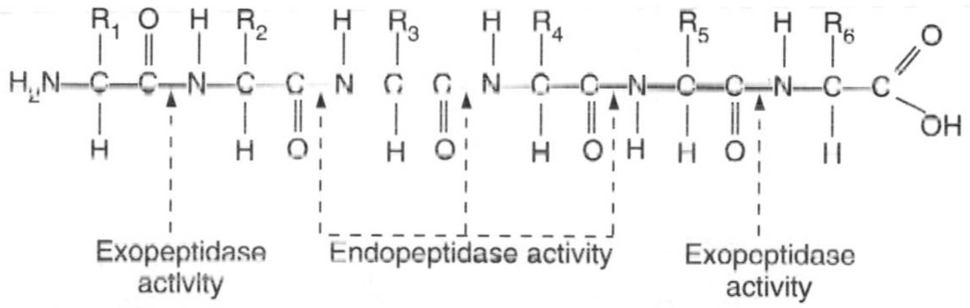

Exopeptidase activity — Endopeptidase activity — Exopeptidase activity

Endopeptidases are very specific in their action and exert their effect only at a particular point within the protein molecule. The susceptibility to attack by a particular endopeptidase is basically determined by the nature of the chemical groups on either side of the bond concerned; a bond that is readily hydrolysed by one endopeptidase may be totally resistant to hydrolysis by another. Hence, the way in which a protein is split up and the chemical nature of the products is determined by the type of endopeptidases present. This is further illustrated by the following example, which shows the hydrolysis of a synthetic protein, benzyloxycarbonyl-L-glutamyl-L-tyrosylglycine amine.

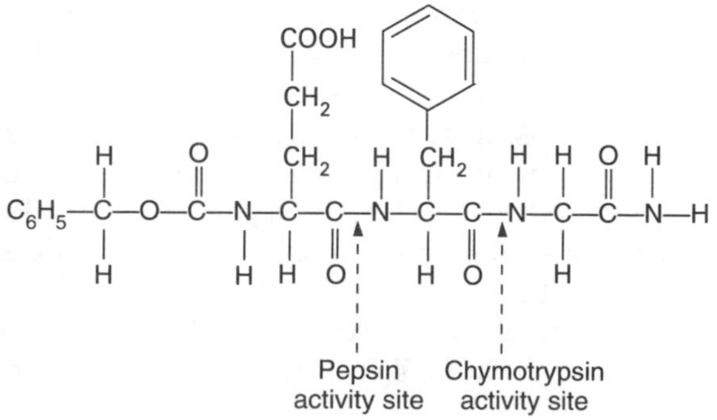

Pepsin activity site — Chymotrypsin activity site

It can be seen that pepsin attacks (hydrolyses) the bond on the amino side of the aromatic radical, while chymotrypsin hydrolyses the bond on the carboxyl side.

Trypsin, on the other hand, acts on peptide bonds between arginine and lysine. These three endopeptidases are probably the most important in the digestion of protein. Between them these can break down to polypeptides the majority of proteins normally occurring in a diet.

There are three groups of exopeptidases: carboxypeptidases, amino peptidases and dipeptidases. Each of these exhibits specificity for a substrate or group of substrates. As with endopeptidases, this specificity is determined by the nature of the groups on each side of the peptide bond to be hydrolysed. Thus, carboxypeptidases remove terminal amino acids in which the carboxyl radical is free (shown as A in the diagram below), while amino peptidases act at the other end of the polypeptide chain and remove the terminal amino acid possessing free amino groups (shown as B).

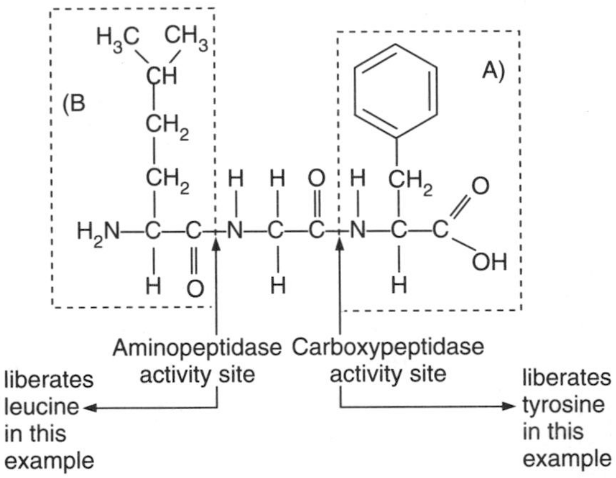

(B
A)
Aminopeptidase activity site
Carboxypeptidase activity site
liberates leucine in this example
liberates tyrosine in this example

Some fish have the capability to absorb dipeptides or even larger polypeptides which are broken down into the individual amino acids intracellularly.

Most of the digestion of protein occurs in the stomach as a result of action of pepsin. Pepsinogen is broken down to pepsin by acid hydrolysis. This occurs as a result of secretion of HCl into the stomach in response to the presence (or sometimes absence) of food in the stomach. Pepsin is also most active in acidic solutions. It is believed that the acidity in the stomach also causes lysis of plant cell walls in macrophyte feeding fish (e.g. tilapias). Hydrolysis of the cell walls by HCl, facilitated by partial crushing of the ingested material by pharyngeal teeth, allows the plant cell contents to be subjected to the actions of the proteolytic enzymes.

In stomachless fish, the role of pepsin is taken by alkaline proteases which are most active in alkaline environments. The effect of pH on protease activity is shown in Figure 4.6. In all fish species, the stomach takes

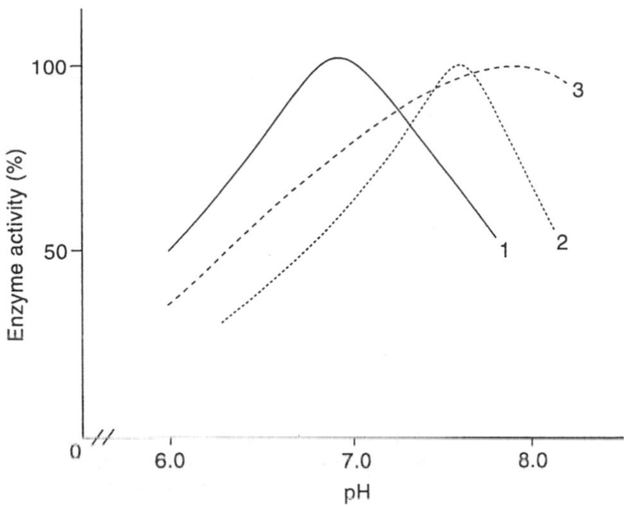

Figure 4.6 Intestinal enzyme activity (%) in rainbow trout as a function of pH. 1, amylases; 2, esterases; 3, proteases. (From Steffens, 1989.)

time to develop. Until it does, there will be inadequate digestion of proteins. Similarly, in stomachless fish, there is initially low activity of alkaline proteases (Table 4.3). This phenomenon provides some explanation for the problems associated with the use of dry feeds during the earliest stages of development. In general, protein digestion in the early juvenile stages is heavily dependent on the alkaline tryptic enzymes rather than on the acidic peptic enzymes.

Activity of some enzymes in carp has been shown to be related to the dietary composition. A fall in protease activity was correlated with a decrease in the proportion of fish meal in the diet. A corresponding increase in amylase activity was observed with an increasing amount of starch in the diet.

Protease activity in the pyloric caecae of rainbow trout has been shown to be dependent on the temperature and feed quality. However, with low-protein diets the temperature effect is not seen. Changes in protease

Table 4.3 Intestinal proteolytic enzyme activity in common carp fry

Age (days)	*t*	*Live weight (mg)*	*Protease activity (μg/g)*
8	5	10.5	42
17	14	57	219
22	19	114	266

t, time from the start of protein intake in days.
Source: Steffens (1989).

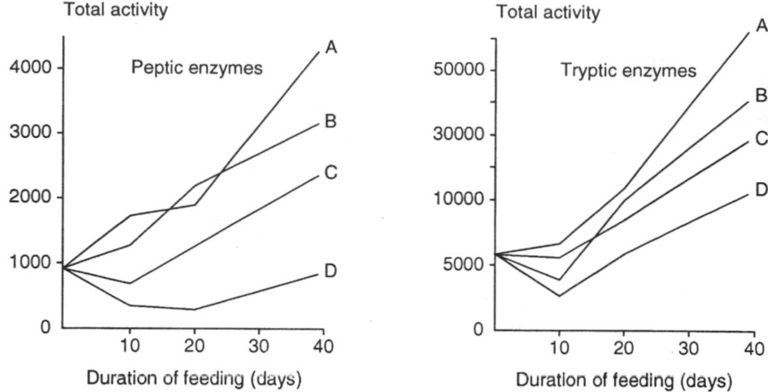

Figure 4.7 Effect of feed composition on the total level of activity of proteolytic enzymes in young rainbow trout. (A) Feed containing 80% fish meal, 0% α-cellulose. (B) Feed containing 60% fish meal, 20% α-cellulose. (C) Feed containing 40% fish meal, 40% α-cellulose. (D) Feed containing 20% fish meal, 60% α-cellulose. (From Kawai and Ikeda, 1973.)

activity in relation to dietary composition in rainbow trout are shown in Figure 4.7.

4.2.3 Fat digestion

The liver plays a very important role in fat digestion; the bile that is produced in the liver and stored in the gall bladder is released when food arrives in the intestine. It contains gallic acids of high surface activity, and these emulsify the fats, breaking large fat droplets into very small droplets, thereby increasing the surface area and making them more accessible to fat-splitting enzymes. All fat-digesting enzymes are classed as lipolytic enzymes or lipases.

Lipase activity has been found in extracts of the pancreas, pyloric caecae and upper intestine, but not necessarily in all three sites in all species. Lipase activity is almost non-existent in the stomach, and in the intestine the principal site for lipase is the mucosal layer.

A major difference between fat hydrolysis and that of proteins or carbohydrates is that lipases show relatively little substrate specificity, and many will catalyse hydrolysis of almost any organic ester. Thus, the progressive breakdown of a fat through various intermediate stages is often catalysed by a single lipase and there is not a precise succession of different enzymes as in proteolysis.

The other major difference between lipases and proteases and carbohydrases is that lipases fairly readily catalyse the synthesis (as well as the hydrolysis) of fats and other esters. Therefore, the products need to be removed (absorbed) if the reaction is to proceed after reaching equilibrium.

The fats normally used by animals are esters of organic acids and higher alcohols (usually glycerol, a trihydric alcohol). The end result of lipolysis of a triglyceride molecule is three molecules of fatty acids and one of glycerol, and the progressive steps involved can be summarized as follows:

CH_2OH
$CHOOCR_2$
CH_2OOCR_3
2, 3-Diglyceride

R_3COOH

CH_2OH
$CHOOCR_2$
CH_2OH
2-Monoglyceride

R_1COOH

CH_2OOCR_1
$CHOOCR_2$
CH_2OOCR_3
Triglyceride

H_2O

H_2O

R_1COOH

H_2O

CH_2COOCR_1
$CHOOCR_2$
CH_2OH
1, 2-Diglyceride

H_2O

R_2COOH

H_2O

R_3COOH

CH_2OH
CH_2OH
CH_2OH
Glycerol

All fat-digesting enzymes act in alkaline media, the pH optimum being slightly variable from group to group. Lipase from the intestinal mucosa has an optimum between pH 7.0 and 7.5, while intestinal esterases are most active between pH 8 and 9.

4.2.4 Carbohydrate digestion

A large number of different carbohydrate-digesting enzymes (=carbohydrases), each with very specific actions, are present in the gut of fish. Carbohydrases, like lipases, have been demonstrated in pancreatic juice, in the stomach, in the intestine and in the bile, but not necessarily at all sites in all species investigated. In most species, however, the pancreas is the main producer of carbohydrases.

Carbohydrases have a broad temperature tolerance (20–40°C) and optimal activity occurs at pH 6–8.

Examples of a few specific carbohydrase actions are the hydrolysis of starch or glycogen to oligosaccharides or maltose by α-amylase or the hydrolysis of di- and oligosaccharides to mono- and polysaccharides. Depending of the structure of the substrata, these molecules are broken down by glucosidases, galactosidases or fructosidases.

Carbohydrase activity responds to the level of dietary carbohydrate. Amylase activity in rainbow trout increases in response to feeding, a rise in the temperature or salinity of the water, or when the fish are given a

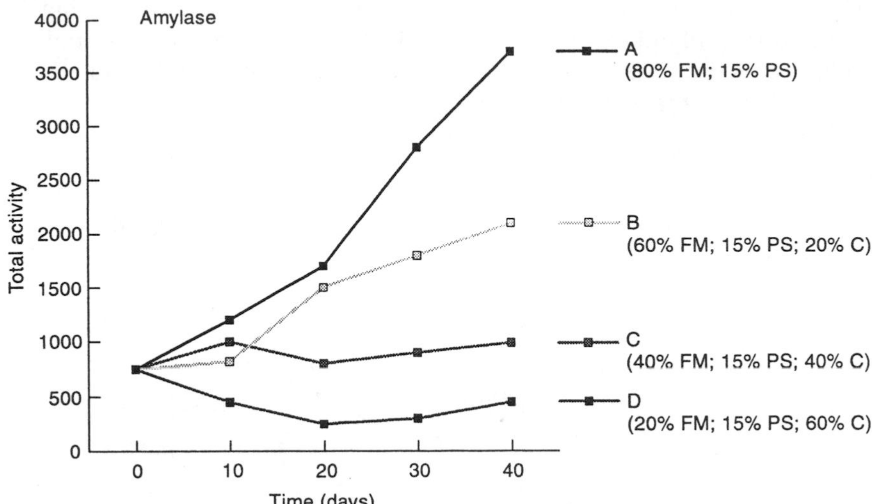

Figure 4.8 Amylase activity in rainbow trout in response to different diets: fish meal (FM); potato starch (PS); α-cellulose (C). (From Kawai and Ikeda, 1973.) The increasing levels of FM in the diets reflect increasing levels of protein.

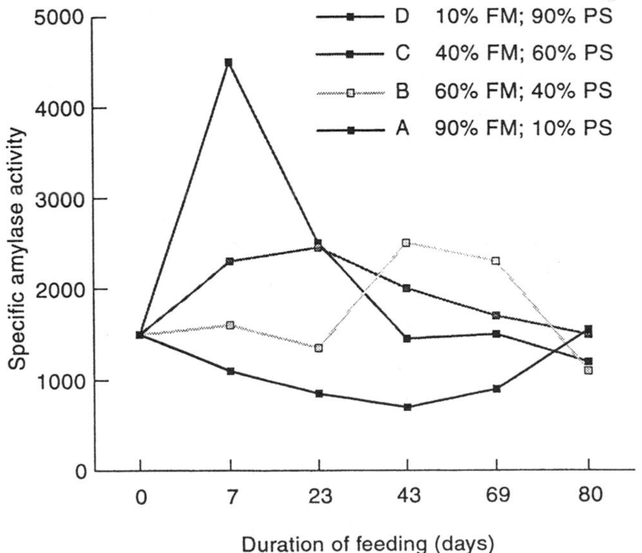

Figure 4.9 Effect of diet composition on amylase activity in common carp. Abbreviations used similar to those for Figure 4.8. (From Kawai and Ikeda, 1972.)

Table 4.4 Relative activities (max = 100) of amylase, α-glucosidase and β-galactosidase in the digestive tract of different fish species

	Amylase			*α-Glucosidase*		*β-Galactosidase*	
Species	*I*	*S*	*P*	*I*	*S*	*I*	*S*
Carassius carrassius	100					34	
Ctenopharyngodon idella	84			100		61	
Oreochromis niloticus	44	31				59	
Cyprinus carpio	35					8	
Hypophthalmichthys molitrix	31					100	
Salmo gairdneri	8	<1	16			2	<1
Anguilla japonica	1	<1		15	<1	20	11
Seriola quinqueradiata	1	<1				22	6

I, intestine; S, stomach; P, pyloric caecae.
Source: Nagayama and Saito (1968).

protein-rich diet (Figure 4.8). On the other hand, tilapia (*O. mossambicus*) shows higher amylase activity when changed to a starch-rich diet. The related species, *O. niloticus* (Nile tilapia), shows a decrease in carbohydrase level in response to high dietary levels of starch but an increase in α-glucosidase and β-galactosidase in response to elevated dietary lactose levels. Common carp respond in an opposite fashion to rainbow trout in relation to dietary starch levels (Figure 4.9). The activity of carbohydrases in general, and of amylase in particular, differs from species to species, and appears to be related to their feeding habits. Similarly, other carbohydrases such as chitinase activity are also known to differ between species (Table 4.4).

In general, both carbohydrase and protease activity are related to the feeding habit of fish, the naturally carnivorous fish species having greater levels of enzymes that digest protein and non-carnivorous species having lower levels. However, the magnitude of difference in proteolytic enzyme activity between species is much lower than that of carbohydrases such as amylase. It is also appropriate when making comparisons between species to consider only closely related species rather than to make broad comparisons.

4.3 MICROBIAL DIGESTION

Microbial digestion plays an important role in cellulose digestion and protein synthesis in some animals, cellulase being formed by the bacteria harboured in the gut. Fish, on the other hand, have been inadequately studied in this respect. Stickney and Shumway (1974) studied 62 species (from 35 families), and 17 species were shown to exhibit some cellulase activity. The occurrence of the enzyme, however, was not related to the gut morphology or the feeding pattern. All evidence indicates that any cellulase activity present in fish is due to the microbial flora.

Generally, bacteria are abundant in the gut of detrital and macrophyte feeders, and they also exhibit considerable proteolytic and amylolytic activity when isolated. In addition, the microbial flora is also known to exhibit chitinase and lecithinase activity. In this respect, Prejs and Blaszczyk (1977) demonstrated that in cyprinids there is a positive correlation between cellulase activity and the amount of plant detritus in the stomach, which again can be attributed to bacteria and fungi colonizing the feed.

4.4 ABSORPTION

The main mechanism of intestinal absorption in fish appears to be similar to that of mammals. Products of digestion are absorbed by diffusion and active transport. An example of active transport is the uptake of glucose involving a carrier. It is an energy-requiring process which moves glucose across a membrane into the epithelial cells even if there is already a high concentration of glucose within that cell.

Diffusion may occur either as facilitated or as simple diffusion. Facilitated diffusion occurs where there is a carrier system that allows the compound to move across an otherwise impermeable membrane. Fructose is an example of a compound absorbed into the intestinal epithelium in this manner. Facilitated diffusion does not require energy and will not move the compound up a concentration gradient.

Simple diffusion does not require a carrier or energy. Fatty acids (lipids) are examples of compounds that are absorbed into intestinal epithelia by simple diffusion. As the micelles or droplets of fatty acids and bile salts approach the epithelial cell surface, lipids are released to pass through the cell membrane. Once inside, the lipids reform droplets, this time called chylomicrons, which have a very thin protein coat to make them water soluble. The chylomicrons are moved through the cell to be released into the circulation and transported to the liver for further processing.

As stated earlier, the processes of absorption are common to all animals with only minor variations, many of which have little relevance to aquaculture. Those that are important are discussed in later chapters. It should be mentioned that antibiotics can influence food absorption either through physiological alteration in the gut or through physical and/or chemical interactions between drugs and ingested material. Antibiotics are frequently used in aquaculture. Administration of certain antibiotics, for example, is known to increase the absorption of unsaturated fatty acids in rainbow trout (Cravedi *et al.*, 1987).

4.5 REGULATORY ACTIVITIES

4.5.1 Secretion of digestive juices

Most animals do not feed continuously, and fish are no exception to this rule. Generally, secretion of digestive juices is elicited so that its presence will correspond with the presence of food in the alimentary canal (note that fish do not possess an equivalent of salivary glands and there is no buccal digestion). Unfortunately, very little is known about the mechanisms and regulatory processes involved in the secretion of digestive juices in fish.

The presence of food in the alimentary canal stimulates both the gland cells in the immediate vicinity and other nervous and/or hormonal mechanisms. These stimuli are relayed to other gland cells or organs so that glands and the processes of enzyme secretion and acid or alkaline juice secretion in regions not immediately adjacent to the food are coordinated. The initiation and transmission of the stimuli causing secretion of digestive fluids and the coordination of the muscles responsible for the movement of food through the alimentary canal are not under direct and voluntary control of the animal, irrespective of whether nervous and/or hormonal mechanisms are concerned. When secretion is controlled by nervous mechanisms, it is mainly the sympathetic and parasympathetic systems which are involved excluding the vagus (10th cranial). Some pathways involved in digestive juices secretion are diagrammatically presented in Figure 4.10. There is evidence from some teleosts that gastrin

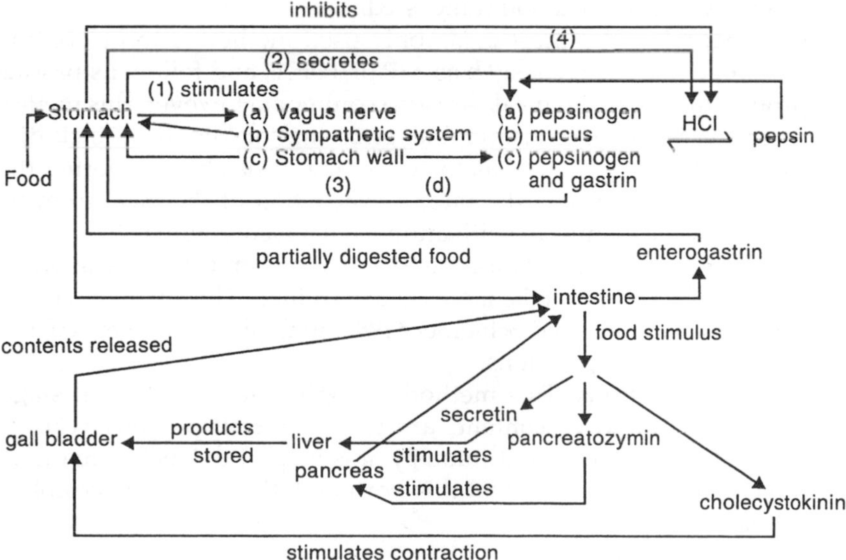

Figure 4.10 Some of the regulatory pathways of the digestive process(es).

is not secreted and that its place is taken by histamine, which has the physiological function of regulation of acid secretion.

4.5.2 Rate of digestion

The specific enzyme activity and the volume of the digestive juices are important for the process of digestion, as are the rate at which the gut fills and the transit times. The terminology used to quantify the rate of digestion is varied.

Gut transit time	Time interval between ingestion and first appearance of faeces
Gut evacuation time	Time taken for the entire quantity of food ingested to be voided
Gastric evacuation time	Time taken between ingestion and emptying of the stomach

The difference between evacuation time and transit time is referred to as 'the retention time'.

All these parameters are equally valid and are sufficiently good and reliable indicators of the rate of digestion. The basic methods utilized are as follows.

- Sacrifice method. Fish are fed a predetermined quantity of food (usually and preferably voluntarily), sacrificed at predetermined intervals and the amount of food remaining in the stomach estimated. The amount remaining in the stomach can be estimated as a percentage of the volume, weight (dry or wet), ash-free dry weight or in calorific value of the amount ingested.
- X-radiography. The most commonly used method is to incorporate an isotope into the feed, such as 32P or 144Ce, and follow its passage through time. Barium meals are also common. However, this method generally involves considerable stressing of the fish, including constant handling as well as force feeding. It has been reported that force feeding decreases the rate of evacuation of the meal from the stomach when compared with animals that feed voluntarily.
- Use of dyes. Dyes are incorporated into diets, and the time at which the dye appears first in the faeces is determined. This method has also been used by Hofer and Schiemer (1983) in their study on evacuation rates of natural populations.
- Direct observations. This method is applicable in the larval stages when the gut and its contents are visible in transparent larvae. As most fish tend to feed continuously, it is helpful to label a 'quantum' of the diet in such a way that this portion of the meal is detectable in faeces.

There have been a large number of investigations into the rate of digestion in a variety of species. It is known that the rate is dependent on a number of exogenous and endogenous factors. The quantitative relationship between gastric emptying time and fish size (weight) differs slightly between species, e.g. in *Limanda* the gastric emptying rate varies as (fish weight)$^{0.386}$ and in *Megalops* as (fish weight)$^{0.41}$.

These quantitative relationships offer support for the relationship(s) proposed in the equations of Fange and Grove (1979) given later, and also offer an explanation for the pattern of feeding by fish in their natural environment. It has been suggested by many workers that the appetite of a fish returns as its stomach empties. Therefore, it would be expected that (for a given species) fish of different sizes which show a regular periodicity of feeding must voluntarily regulate their meal size so that digestion is completed in time for the next feeding. This is known to occur in nature; the mass of the stomach contents of freshly captured fish shows a negative relationship with body weight. Equally, experimental results have shown that voluntary food intake as a proportion of body weight decreases with increase in the fish size.

(a) Meal size

Observations on the effect of meal size on gastric emptying rate and/or digestion are often contradictory. According to some workers, a larger meal size (fed to same size fish) does not increase the time of digestion. However, some others, find that when the meal size increases the time required to empty the stomach increases but the increase in emptying time is not directly proportional to the increase in meal size.

Since digestion requires the secretion of digestive enzymes and acids, digestion and/or gastric emptying does not necessarily commence immediately after ingestion. This time lag is known to be temperature dependent, and may also be influenced by the type of meal.

On initiation of digestion, parts of the ingested material begin to move along the gut, and the weight of the residual material in the stomach begins to decrease. Most researchers have attempted to construct emptying curves, using the methods described earlier. Studies suggest that gastric emptying could be described, for different species, by one of two forms: a straight line, e.g. *Gadus* spp. (Jones, 1974), *Lepisosteus* (Hunt, 1960), *Katsuwonus* (Magnuson, 1969), *Megalops* (Pandian, 1967); or an exponential curve, e.g. *Pleuronectes* (Jobling, 1981b), *Salmo trutta* (Elliot, 1972), *Oncorhynchus* (Brett and Higgs, 1970).

A relationship of the latter type suggests that the rate of emptying is proportional to the instantaneous bulk of food in the stomach and is expressed as

$$W_t = W_0 e^{-b(t-a)}$$

where W_t is the content of the stomach at time t, W_0 is the size of the meal, a is the delay before digestion begins and b is the instantaneous rate of digestion.

Fange and Grove (1979) proposed a general model based on the available data. They suggested that the rate of digestion is primarily related to the surface area of the food bolus in the stomach and therefore

$$\frac{dv}{dt} = -KV^{0.67}$$

where V is the volume of food in the stomach at time t and K is the instantaneous digestion rate (varies with temperature, fish size, etc.).

It follows from this equation that the gastric emptying time (GET) will vary with the ingested meal size as

$$\text{GET (h)} = K'V_o^{0.33}$$

where K' is a constant.

Fange and Grove (1979) argued that, since enzymes typically attack the food bolus at the surface, the rate of digestion will obviously be proportional to the surface area and hence

$$\frac{dv}{dt} = -aV^{0.67}$$

Accordingly, a larger meal has a faster digestion rate (g/h).

Jobling (1981b) reviewed the literature and observed that stimulation depends on stretching the stomach circumference as food enters so that emptying rate depends on the square root of the food volume:

$$\frac{dv}{dt} = KV^{0.5}$$

when GET = $K'V_o^{0.5}$. However, if initial delays in the onset of emptying are allowed for, exponential curves (as described earlier) describe stomach emptying very well.

A number of factors are known to influence the rate at which food moves through the alimentary canal. Amongst these are temperature, species, age size and sex, stocking density, meal size, the type of food and its presentation and the feeding rate.

(b)Temperature

Fish are poikilothermic, and therefore a temperature influence on the speed at which food is processed is to be expected. Temperature influence has also been one of the most studied aspects of rate of digestion. The available information was summarized by Fange and Grove (1979), who defined the relationship between gastric emptying/evacuation time and temperature (Figure 4.11).

Figure 4.11 also gives a good indication of the relationship of gastric emptying time to the feeding niche. Fange and Grove (1979), based on a

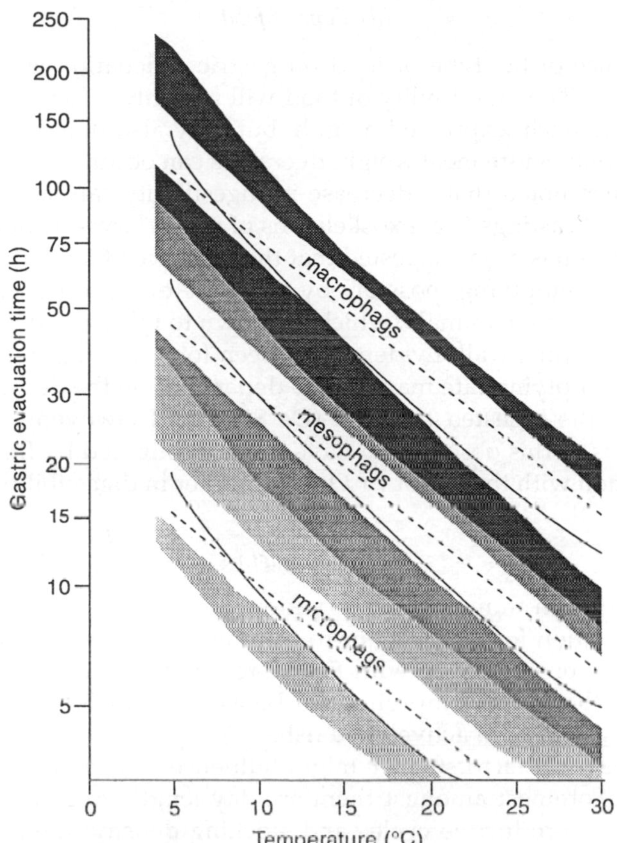

Figure 4.11 Relationship between gastric or foregut emptying times and temperature in fishes. The effect of temperature on digestion rate in each group is represented by Q_{10} of 2.6 (continuous line) or in proportion to $10^{0.035t}$ (broken line). (From Fange and Grove, 1979.)

comparison of gastric evacuation rate of species which are fed diets not normally consumed in nature and their natural food habits, suggested that fish have evolved an inherent rate of gastrointestinal mobility and digestion to suit their natural diets.

(c) Fish size

All experimental observations available have indicated that digestion rate and/or gastric emptying time is influenced by size, ranging as the exponent of the wet weight of the meal between 0.46 and 0.75, and in a few species as high as 1.0. Only one study has indicated that digestion rate decreases as the meal size increases.

(d) Type of food

The influence of the type of food on gastric evacuation is well known (Figure 4.11). The digestibility of food will not only affect emptying rate from the stomach, expressed as g/h, but may also determine the time after ingestion before meal weight decreases can occur.

It has been noted that a decrease in digestibility can be attributed to thick or inert castings (e.g. exoskeletons of insect larvae, mollusc shells). In addition, it has been suggested that the presence of fat in the food may delay gastric emptying, possibly by a release of a hormone, similar to enterogastrone in mammals (which is known to inhibit gastric mobility), from the intestinal wall. Evidence has been forthcoming to indicate that the gastric emptying rate may also be dependent on the density (specific gravity) of the ingested material (De Silva and Owoyemi, 1983). The implications of this observation have been highlighted by Bowen (1978) in association with the use of Cr_2O_3 as a marker in digestibility studies.

(e) Other factors

There have been numerous observations, on a variety of species, that food deprivation for a time prior to feeding tends to slow down gastric emptying when compared with fish tested under continual feeding. In such cases the rate of emptying can be as much as 50–68% less when compared with that of actively fed fish.

There are other factors which might influence digestion and/or gastric emptying. Foremost amongst them are day length, degree of parasitic infestation, reproductive cycles and stocking density, the last possibly exerting its influence via hierarchical effects. However, these aspects have been little studied and are of marginal importance in aquaculture.

4.6 DIGESTIBILITY

Digestibility is the quantification of the digestive processes. It gives a relative measure of the extent to which ingested food and its nutrient components have been digested and absorbed by the animal.

The total and/or dry matter digestibility refers to the degree of digestibility of the complete diet and/or the ingredient in question. Nutrient digestibility refers to a specified nutrient such as protein, lipid, amino acid or carbohydrate of the diet and/or the ingredient.

From an aquacultural point of view, a thorough knowledge of digestibility, i.e. of the methods of evaluation of digestibility and of the influence of age, size, sex, stocking density, time and frequency of feeding and feed quality and quantity, of culturable species is more important than a knowledge of digestive physiology, even though digestibility is dependent on and determined by physiology.

Only a proportion of ingested food is digested and its nutrients absorbed. The rest is voided as faeces, which may be contaminated with small quantities of endogenous enzymes and mucous membrane (gut mucosa) and even with some nitrogenous excretory products.

In providing a cultured organism with a cost-effective diet, a diet that is well assimilated by the cultured organism, it is not sufficient to consider only the nutrient requirements of the organism. A nutritionally balanced diet will only be a good diet if it is also easily and effectively digested and utilized. The digestibility of a diet depends on the nature and type of ingredients which go into the diet, and the final physical form (such as hardness, palatability, stability in water) of the product.

Therefore, when formulating a diet it is essential to have a knowledge of the digestibility (total and individual nutrient) of each of the ingredients of the diet, as well as that of the final product. All digestibility values are expressed as percentages of the original ingested material. Generally, fish meal is digested efficiently, often exceeding 90%, while the apparent protein digestibility of various plant products tested is extremely variable. Equally, the digestibility of macrophytes by different species is variable (Table 4.5). These variations have a number of practical implications. If, for example, it is necessary to formulate a diet which contains 30% protein by dry weight, the required protein can be supplied by a variety of ingredients ranging from animal products such as fish meal to feather meal, and plant products such as soybean meal to oil seed cakes, pulses, etc. Diets compounded by incorporating these different ingredients, although they could be formulated to have 30% protein, are unlikely to be equally digestible with respect to protein as well as to the total dry matter of the diet. Accordingly, when fed on diets containing 30% protein but made up of these different ingredients, the performance of the cultured organism will be different.

Therefore digestibility studies clearly constitute an important and an integral part of diet formulation and preparation. Ideally, digestibility studies should precede feed formulation. However, the digestibility of most common ingredients is presently known for the most commonly cultured species or related species. These values are used in the initial formulation and the formulated diets tested for digestibility later.

4.6.1 Determination of digestibility

Digestibility of a diet/feed can be determined directly or indirectly. Unlike comparable studies on terrestrial animals, those on fish have an inherent difficulty by virtue of the medium in which they live. Faecal traps, for example, are impossible to use, and the voided faeces lose nutrients immediately on discharge. Therefore, all digestibility estimations

Table 4.5 Digestibility coefficients of some non-conventional plant ingredients which have been tested for their suitability for incorporation in fish diets

Ingredient	*Species of fish*	*Level of ingredient used*	*Control diet*	*Digestibility coefficient*	*Growth response*	*Culture system*
Water hyacinth *E. crassipes* (dried)	*Labeo rohita*	20% and 40% of total dietary protein	Fish meal supplying 100% of the dietary protein	Apparent protein digestibility (APD) coefficients were 71% and 63% for the 20% and 40% inclusion levels respectively, and 79% for control	SGRs obtained were 79% and 68% of the SGR on the control diet for the 20% and 40% inclusion levels respectively. The SGR for the control diet was 3.13% per day	Static water indoor glass aquarium
Water hyacinth *E. crassipes* (5 weeks composting)	*O. niloticus*	50% of dietary protein (37.5% of the diet)	Dietary protein supplied by fish meal, groundnut and rice bran	APD coefficients were 46% and 65% respectively	SGRs obtained were 76% and 81% of the SGR on the control diet in a static water system and a recirculating water system respectively. The SGR for the control diet was 1.64% per day	Outdoor concrete tanks in static water or recirculating water systems

Table 4.5 (Continued)

Ingredient	*Species of fish*	*Level of ingredient used*	*Control diet*	*Digestibility coefficient*	*Growth response*	*Culture system*
	O. niloticus (1–$1\frac{1}{2}$ years)	50% of dietary protein (37.5% of the diet)	Dietary protein supplied by fish meal, groundnut and rice bran	APD coefficients were 33% and 36%	SGRs obtained were 77% and 94% of the SGR on the control diet in a static water system and a recirculating water system respectively. The SGR for the control diet was 1.64% per day	Outdoor concrete tanks in static water
Ipil-ipil *Leucaena leucocephala* (leaf meal)	*L. rohita*	20% and 40% of total protein	Fish meal supplying 100% of the dietary protein	APD coefficients were 68% and 63% respectively	SGRs obtained were 79% and 70% of the SGR on the control diet for the 20% and 40% inclusion level respectively. The SGR for the control diet was 2.34% per day	Indoor static water glass aquaria
	O. niloticus	25%, 50% and 100% of dietary protein	Fish meal supplying 100% of the dietary protein	APD coefficients were 72%, 66% and 40% respectively with increasing amount of leaf meal	SGRs obtained were 66%, 36% and 18% of the SGR on the control diet for the 25%, 50% and 100% inclusion levels	Indoor recirculating system and concrete tanks

Table 4.5 (Continued)

Ingredient	*Species of fish*	*Level of ingredient used*	*Control diet*	*Digestibility coefficient*	*Growth response*	*Culture system*
					respectively. The SGR for the control diet was 3.03% per day	
Ipil-ipil *Leucaena leucocephala* (soaked in water for 48 hs)	*O. niloticus*	25%, 50% and 100% of dietary protein	Fish meal supplying 100% of the dietary protein	APD coefficients were 75%, 65% and 41% respectively	SGRs obtained were 89%, 73% and 2.3% of the SGR on control diet at the 25%, 50% and 100% inclusion levels respectively. The SGR for the control diet was 3.03% per day	
Ipil-ipil *Leucaena leucocephala* (soaked in water for 24 hs)	*L. rohita*	20% and 40% of total protein	Fish meal supplying 100% of the dietary protein	APD coefficients were 71% and 63% respectively	SGRs obtained were 86% and 75% of the SGR on the control diet at the 20% and 40% inclusion levels respectively. The SGR for the control	Indoor static water system, glass aquaria

Table 4.5 (Continued)

Ingredient	*Species of fish*	*Level of ingredient used*	*Control diet*	*Digestibility coefficient*	*Growth response*	*Culture system*
					diet was 2.34% per day	
Cassava leaf meal *Manihot esculenta* (soaked in water for 48 hs)	*O. niloticus*	20%, 40%, 60% and 100% of dietary protein	Fish meal supplying 100% of the dietary protein	APD coefficients were 64% 50%, 35% and 18% with increasing amount of the leaf meal	SGRs obtained were 79%, 71%, 44% and 6% of the SGR on the control diet with increasing substitution level. The SGR for the control diet was 2.62% per day	Indoor, recirculating water system. Concrete tanks
Cassava leaf meal *Manihot esculenta* (sun-dried)	*O. niloticus*	20%, 40%, 60% and 100% of dietary protein	Fish meal supplying 100% of the dietary protein	APD coefficients were 64%, 50%, 35% and 18% respectively	SGRs obtained were 82%, 65%, 7% and 8% of control diet with increasing substitution level. The SGR for the control diet was 2.62% per day	Indoor, recirculating water system. Concrete tanks
Mustard oil cake *Brassica juncea*	*C. carpio*	25% and 50% of dietary protein	Fish meal supplying 100% of protein	APD coefficients were 34% and 81% respectively	SGRs obtained were 85% and 67% of the SGR on the control diet for the 25% and 50% inclusion levels respectively.	Indoor, recirculating water system

Table 4.5 (Continued)

Ingredient	*Species of fish*	*Level of ingredient used*	*Control diet*	*Digestibility coefficient*	*Growth response*	*Culture system*
					The SGR for the control diet was 3.58% per day	
Linseed meal *Linum usitatissimum*	*C. carpio*	25% and 50% of dietary protein	Fish meal supplying 100% of protein	APD coefficients were 85% and 78% respectively	SGRs obtained were 86% and 66% of the SGR on the control diet at the 25% and 50% inclusion levels respectively. The SGR for the control diet was 3.58% per day	Indoor, recirculating water system
Sesame meal *Sesamum indicum*	*C. carpio*	25%, 50% and 70% of dietary protein	Fish meal supplying 100% of protein	APD coefficients were 81%, 78% and 78% respectively	SGRs obtained were 74%, 54% and 36% of the SGR on the control diet at the 25%, 50% and 70% inclusion levels respectively. The SGR for the control diet was 3.58% per day	Indoor, recirculating water system

Source: Modified from Wee (1991).

on fish, whichever method one chooses, are subject to some degree of error, except perhaps when the fish is killed and rectal contents are dissected out. The aim is to minimize the possible errors in the experimentation.

In the direct method, the quantity ingested (total or nutrient) and faecal matter voided are determined, and the ratio gives the percentage digestibility of the feed and/or the nutrients under consideration.

However, it is difficult to determine the quantity ingested and/or the total amount of faeces voided accurately as such determinations are subject to many errors. Foremost amongst these is the leaching of nutrients from the faeces and the difficulty in collecting all the faecal material, which often tends to break up into small particles with time, a process facilitated by aeration and the movements of the experimental fish. Also, there are fine faecal particles in suspension which are almost impossible to collect by normal siphoning. The latter can be resolved to some degree by using automatic, continuous faecal collection devices incorporated into an experimental system (Figure 4.12). Attempts to use the direct method for digestibility studies in aquaria without such devices are not recommended. The errors that could arise are many: leaching from faecal material, reingestion of faecal material and incomplete collection of all faecal matter voided.

Such systems, however, are expensive to build and maintain and are generally used by researchers who are working exclusively on aspects of digestibility. Generally, mechanisms which allow representative faecal samples to be collected are more than adequate. These are simple and are easily incorporated into an experimental system (Figure 4.13).

The indirect method of estimation of digestibility was first used by the Swedish scientist Edin in 1918, when he used markers in experiments on cattle. A marker is a material, thought to be indigestible, which is introduced in small quantities and distributed evenly in the test diet, or an indigestible component of the diet itself. These are known as external and internal markers respectively. The marker will concentrate relative to the digestible material in the faeces, and the relative quantities provide a measure of the extent of digestibility of the diet and/or its nutrient components.

A marker, however, will have to be indigestible, should not influence the physiology of digestion of the experimental animal, should move along the gut at the same rate as the rest of the food material and should not be toxic.

Commonly used external markers, i.e. those introduced into the diet, are Cr_2O_3, FeO, SiO_2, polypropylene, etc. By far the most commonly used external marker is Cr_2O_3 in fish as well as in terrestrial animals.

Endogenous or internal markers commonly utilized for digestibility estimations are crude fibre (CF), hydrolysis-resistant organic matter (HROM) and hydrolysis-resistant ash (HRA). Occasionally ash has been

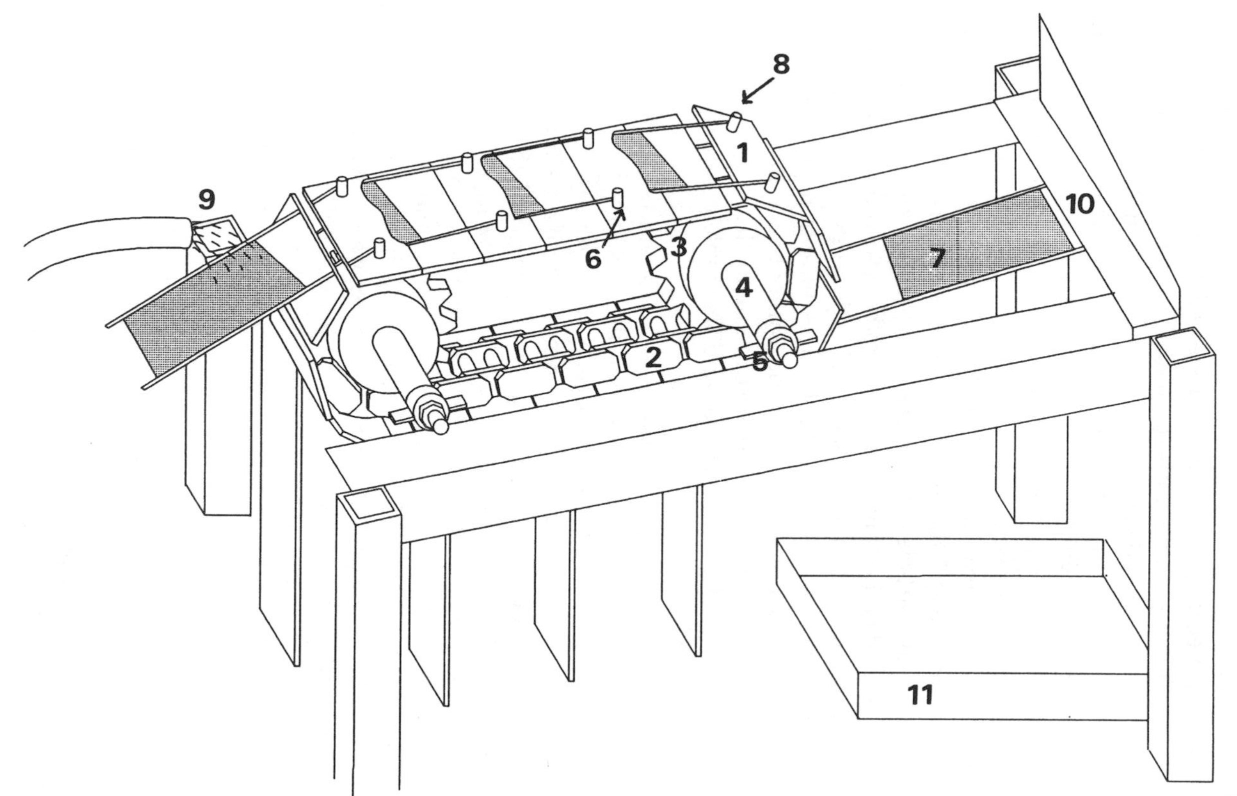

Figure 4.12 Automated faecal collector. 1, nylon pads; 2, nylon straps; 3, nylon drums; 4, axle; 5, waterproof roller bearing; 6, props; 7, brass screens; 8, motor; 9, drainage port; 10, check bar; 11, faecal collection pan. (From Choubert *et al.*, 1982.)

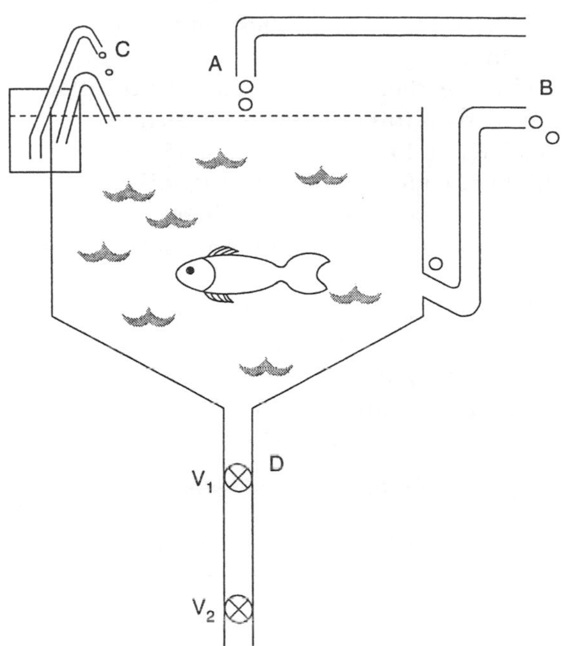

Figure 4.13 Experimental tank. (A) Inflow of seasoned water. (B) Outflow of water from the tank. (C) Filter and aeration. (D) Faeces collection tube. (V_1 and V_2) Control valves. Operation of the faeces collection tube: open V_1 for 5 min and close it again; open V_2 and drain the faeces into a beaker; repeat the procedure 2–3 times.

used, particularly with respect to digestibility estimations of naturally ingested food material. CF and HROM refer to the same group of materials; cellulose and chitin (when present) are the chief constituents of HROM, and cellulose and lignin are the major components of the CF fraction. HRA is the fraction of mineral ash resistant to acid digestion.

Digestibility is estimated from the determined values as follows:

$$\text{Dry matter digestibility}(\%) = 100 - 100\left(\frac{\%\text{ marker in diet}}{\%\text{ marker in feed}}\right)$$

(or total digestibility)

$$\text{Nutrient digestibility }(\%)$$

$$= 100 - 100\left(\frac{\%\text{ marker in diet}}{\%\text{ marker in faeces}} \times \frac{\%\text{ nutrient in faeces}}{\%\text{ nutrient in feed}}\right)$$

where nutrient refers to any nutrient such as protein or lipid.

However, these digestibility estimates are only apparent digestibility estimations, and therefore they are correctly expressed as the percentage apparent dry matter and/or total digestibility and percentage apparent nutrient digestibility.

It was pointed out earlier that faeces are contaminated with endogenous material and therefore, unless a correction factor is introduced and/or allowance made for such material, we would not be estimating the true digestibility of a feed and/or an ingredient.

In order to determine the true protein digestibility of a diet, we would have to make a control diet without protein while the other components remain similar to that of the test diet. Digestibility estimations are made for the two diets and the difference gives us the true percentage digestibility of protein of the test diets. In normal practice, for evaluation of diet and ingredient suitability, estimations of apparent total and/or nutrient digestibilities are sufficient.

(a) Ingredient digestibility

In feed formulation and manufacture, it is essential to have a knowledge of the digestibility of the main ingredients of a diet, as well as of the whole diet. It may appear that if the ingredient in question could be treated as a 'diet' the method described above could be used to determine its digestibility. This is not desirable for many reasons. For example, the ingredient by itself might not behave in the same way as it would as a component of a compounded diet. Alternatively, the fish may not accept (ingest) the ingredient by itself.

The procedure that is presently accepted for estimating digestibility of an ingredient is to:

- use a diet for which apparent total and nutrient digestibilities are known for the species/organism that is being investigated – this diet is referred to as the reference diet;
- prepare a test diet mixing 20–30% of the ingredient to be tested with the reference diet – this diet is the test diet; and
- determine apparent total and nutrient digestibility of the test diet.

The apparent total and/or nutrient digestibility of the ingredient is estimated using the following equation [assuming that the test diet (TD) has been made up in a ratio of 8:2 of the reference diet (RD): the ingredient].

Apparent total digestibility of the ingredient

$$= \frac{100}{20}\left(\text{dry matter digestibility of TD} - \frac{80}{100}\text{ dry matter digestibility of RD}\right)$$

Apparent nutrient digestibility of the ingredient

$$= \frac{100}{200}\left(\text{nutrient digestibility of TD} - \frac{80}{100}\text{ nutrient digestibility of RD}\right)$$

This procedure was first introduced by Cho *et al.* (1974) for estimation of the digestibility of an ingredient, and has been used widely; it had become standard practice to use 7:3 ratio of reference diet to ingredient. More recently however, De Silva *et al.* (1990) have shown that the more desirable mix of the reference diet to ingredient would be 80–85% to 20–15%.

4.6.2 Markers

It is necessary to consider the credibility of different markers used in digestibility studies. It is preferable not to use any external markers, even if such markers, to the best of our knowledge, are known to conform with the basic criteria as detailed earlier.

Among the endogenous markers, there is no agreement as to what is the best. Endogenous markers used by different workers and a summary of their findings are given in Table 4.6.

With respect to external markers, the most commonly used marker, without doubt, in fish as well as in other animals is Cr_2O_3. So far, no one has demonstrated that Cr_2O_3, or for that matter any of the other markers used, affects the digestive physiology of fish when fed in small quantities

Table 4.6 Different types of internal (endogenous) and external markers used in digestibility studies and authorities for chemical determination (cd) of some of the more widely used components

Marker (Abbreviation)	*Used by*	*cd*
Internal		
Hydrolysis-resistant ash (HRA) or acid-insoluble ash (AIA)	1, 3, 8, 9	8, 14, 18
Hydrolysis-resistant organic matter (HROM)	4, 5, 8, 9	5, 8
Crude fibre (CF)	4, 8, 9, 15	8, 14
Ash	9	9
Cellulose	4, 17	4, 7, 8
Silica	12	11
External		
Chromic oxide	Numerous, e.g. 2, 6	10
^{32}P (insoluble ammonium molybdate)	12	
Titanium (IV) oxide	13	
Celite (as a supplementation to AIA)	1	
Polythene	16	

1, Atkinson *et al.* (1984); 2, Austreng (1978); 3, Bowen (1981); 4 and 5, Buddington (1979, 1980); 6, Cho *et al.* (1974); 7, Crampton and Maynard (1938); 8, De Silva and Perera (1983); 9, De Silva *et al.* (1984); 10, Furukawa and Tsukahara (1966); 11, Hickling (1966); 12, Hirao *et al.* (1960); 13, Lied *et al.* (1979); 14, Tacon and Ferns (1979); 15, Tacon *et al.* (1983); 16, Tacon and Rodrigues (1984); 17, Van Keulen and Young (1977); 18, Van Dyke and Sutton (1979).
Source: De Silva (1989).

(<1% by dry weight of the diet) to fish or to be toxic. However, Bowen (1978) indicated that Cr_2O_3 may move along the gut at a different rate from the rest of the food material because of its higher density, and that such a differential rate of movement may cause errors in digestibility estimations. The differential rate of movement of food material dependent on density has been confirmed independently by De Silva and Owoyemi (1983). Nevertheless, there has been no proof forthcoming that such differences bring about significant errors in digestibility estimations.

With respect to prawns and shrimps, however, Leavitt (1985) has raised serious doubts about the use of Cr_2O_3 as a marker, based on almost the same reasoning as that of Bowen (1978) and De Silva and Owoyemi (1983). Leavitt (1985) recommends the gravimetric method for crustaceans. The problems with using this method have been already discussed with respect to fish. This is clearly an area which needs further research.

A few comparisons have been made between external and internal markers, using the same reference diet (De Silva, 1985a). There is no convincing evidence to show that the digestibility estimations based on external markers were erroneous and/or significantly different from those based on endogenous markers; the differences were slight, and could be attributed to analytical errors as well as errors in methodology.

Pandian and Marian (1985) suggested that nitrogen content of food could be used as an index of absorption efficiency in fishes. This concept is yet to be tested on a wider scale and, if shown to hold true, will certainly provide an easier method for digestibility evaluations, particularly in wild stocks and under culture situations.

4.6.3 Factors influencing digestibility

As for any metabolic process, digestibility is influenced by both biological and environmental factors. However, unlike the influence of biological and environmental factors on metabolic activities such as respiration, that on digestibility is less well defined and little understood. This is primarily because of the dearth of research on aspects of digestibility. At times the information available is contradictory. The available information is summarized in Table 4.7.

Apart from what is given in Table 4.7, there are obvious changes in digestibility associated with changes in the feeding habits and gut morphology occurring during the life cycle. The best example is seen in grass carp (*Ctenopharyngodon idella*), a commonly cultured fish in Asia and Europe. In the larval and early fry stages it feeds predominantly on zooplankton, becoming transformed into a macrophyte feeder in the early fingerling stage.

Digestibility varies from day to day, even when all environmental factors are kept unchanged and when fed on the same diet (De Silva and Perera, 1983). This variation was not found to be correlated to consumption.

Table 4.7 Some factors known to affect or investigated for their influence on the apparent dry matter or nutrient digestibility

Factor	*Species*	*Reference*
Feeding level, meal size	*Clarias gariepinus*	Henken *et al.* (1985)
	Cyprinus carpio	von Gongnet *et al.* (1987)
	Salmo gairdneri	Windell *et al.* (1978)
Size, age, density	*S. gairdneri*	Hastings (1969); Windell *et al.* (1978)
Dietary components – protein, lipid, fibre, etc.	*S. gairdneri*	Rychly and Spannhof (1979); Cho *et al.* (1976)
	Oreochromis niloticus	De Silva and Perera (1984)
	S. gairdneri	Beamish and Thomas (1984)
	S. gairdneri	Hanley (1987)
	O. niloticus	De Silva and Perera (1984)
Type of nutrient (e.g. protein)	*S. gairdneri*	Nose and Toyama (1966)
Physical state of diet	*S. gairdneri*	Bergot and Breque (1983)
Protein–energy ratio	*C. carpio*	von Gongnet *et al.* (1987)
Temperature, salinity	*S. gairdneri*	Windell *et al.* (1978)
	Colisa fasciatus	Pandey and Singh (1980)
	O. niloticus	De Silva and Perera (1984)

Source: De Silva (1989).

4.6.4 Digestibility estimations

It has been pointed out earlier that digestibility estimation based on total faecal matter collection is not recommended unless an automatic continuous faecal matter collection device is employed. Therefore this section essentially deals with digestibility determination based on markers.

The most important prerequisite is the collection of a representative faecal sample which ensures as far as possible that the nutrient leaching from the sample is minimized. Ideally, in digestibility estimation experiments, faecal trapping devices are incorporated into the experimental system. Such devices ensure that leaching of nutrients from the faeces and mixing of food particles with the faeces are minimized, and that reingestion of faecal matter is completely avoided. The most commonly used device is one that was developed by Cho *et al.* (1982); modified and simplified versions are very much in use (Figure 4.13).

Normal experimental tanks without such devices can be used as long as it is ensured that food particles are not mixed with faecal matter, fish are not permitted to reingest faeces voided, faecal samples are collected at predetermined intervals after a feeding and when collecting faecal samples care is taken to minimize breaking of 'faecal strands' as breaking can enhance leaching of nutrients.

It is also important to utilize fish of approximately the same size and age and in good condition. The experimental fish should be acclimatized to the test diet for at least 7–10 days and the processes of cleaning the tanks after feeding and 'mock' collection of faeces carried out during the acclimatization period. This will acclimatize the experimental fish to a routine which is expected to minimize any influences the experiment might impose on the digestive physiology (such as excitement) on the fish.

Generally, it is desirable to use 8–10 fish (of course this could vary with the size of your experimental fish as well as the tank size), provided with 1.0–1.5 l of water per individual. A single faecal collection is often insufficient to perform all the analysis, and samples collected over at least 7 days are recommended. Several such 'pooled' samples, each collected over the same number of days, are needed. All samples should be collected at a predetermined time after each feeding, or collected overnight. The time of collection of each day's sample and pooling of samples need to be consistent. Some workers 'strip' faecal matter from the rectal region. Advantages and disadvantages of the different methods of faecal collection are numerous and often a matter of opinion. They are dealt with by Austreng (1981), De Silva *et al.* (1990) and Hajen *et al.* (1993).

Faecal material is then preferably freeze dried, although oven drying is adequate, and the appropriate chemical analysis carried out to determine the relative concentration of the marker(s) and the nutrients.

CHAPTER 5

Larval nutrition and growth

5.1 INTRODUCTION

Nutrient requirements of all animals vary throughout their life-cycle. The changes that occur in the morphology and physiology of aquatic animals between hatching and maturity lead to a number of important variations in feeding and nutritional requirements through the larval, fingerling and adult stages. These variations occur in the morphology of the digestive organs, the digestive processes, the nutrient requirements and the feeding behaviour.

Five periods in the lifespan of a fish have been identified (Balon, 1975):

Embryonic	From fertilization to first exogenous nutrition;
Larval	When active swimming and feeding is taking place;
Juvenile	Beginning when the fins are fully developed, the scales have appeared and most organs have been formed and ending with the beginning of gametogenesis;
Adult	Beginning with the onset of gametogenesis;
Senescent	Following the last spawning event. This stage is most obvious in those salmonids which spawn only once and may not occur in some species.

The information contained in this chapter relates to the larval period of development. Prior to this, the animal's nutrition has been provided by the yolk sac and oil globules, and subsequent to this the organs are clearly developed and function much as those in an adult fish. The transition from an endogenous to an exogenous food supply which marks the onset of the larval stage is one of the most critical phases of the life-cycle and is the period when much of the mortality of hatchery-reared stock occurs.

5.2 MORPHOLOGY

The morphology of the digestive tract of adult fish has been discussed extensively in the previous chapter. However, the morphology of adult digestive tracts varies markedly not only between species, but also throughout development. Clear examples of these morphological changes are the increase in mouth size and the differentiation of the digestive tract.

Changes in mouth size are extremely important in considering larval nutrition as they affect the capacity of the fish to ingest food organisms. Larval fish provided with food particles that are too large will be unable to consume them and starvation is inevitable. Conversely, if the particles are too small the energy required to gather sufficient quantities of food for their needs will be too great and will result in poor growth. Therefore, one of the most important features of larval nutrition is ensuring that the particle size provided is correlated correctly to mouth size.

Considerable variation in larval mouth size occurs both between species and throughout the larval stage. Larvae of the Murray cod (*Maccullochella peelii*) have relatively large mouths and can easily consume newly hatched nauplii of *Artemia*. However, larvae of a flounder (*Rhombosolea tapirina*) have extremely small mouths and cannot accept *Artemia* nauplii or even nauplii of rotifers which are smaller (95–350 μm in cross-section). For this species it may be more appropriate to provide ciliate protozoans or mollusc velligers, both of which may be as small as 35 μm in cross-section. It is also important to increase the size of the food particle with changes in mouth size as the larvae grow. Studies of turbot (*Scophthalmus maximus*) have shown that early stage larvae prefer smaller copepod nauplii to bigger rotifers, although at a later stage adult copepods are required in the diet of the turbot larvae to allow them to metamorphose to juvenile fish (Nellen, 1986).

The morphology of the digestive tract is generally very simple in swim-up larvae. Teeth are often absent. The gut is relatively short, about half the standard body length, and the epithelial cells lining the digestive tract show no regional differentiation. The majority of these epithelial cells are absorptive enterocytes: high narrow cells with many microvilli on their luminal surface. Secretory cells are present in only small numbers.

After first exogenous feeding, a number of changes occur. Mucosal folds gradually develop and the intestine develops regional differentiation. Absorption of nutrients in these animals is characterized by pinocytosis of macromolecules (e.g. protein and lipid) in which the surface membrane of the epithelial cells engulfs the molecule. Material absorbed in this way is subsequently broken down intracellularly. Later in larval development, teeth appear on the jaws, the stomach and pyloric caecae develop in those species which have them and the intestine lengthens relative to body length and begins to coil (Stroband and Dabrowski, 1981). With the development of the mucosa, greater amounts of digestive enzymes are produced, which facilitate digestion within the gastrointestinal tract.

The significance of these developmental changes relate to the feeding frequency and prey size for larval animals. Reference has already been made of the requirement to match prey size with mouth size. In addition, the lack of teeth on the jaw and the absence of a stomach mean that the food items must be easily captured and digested independently of these mechanisms. The short relative gut length of younger fish means that a limited number of food items can be taken at once, and these items are unlikely to be retained for a lengthy period of time in the gut. As a result their digestion must be rapid and feeding must occur often to ensure sufficient provision of nutrients.

5.3 DIGESTIVE PROCESSES

The development of the digestive processes of larval aquatic animals goes hand in hand with increasing morphological complexity. The fact that the majority of the epithelial cells of the digestive tract are absorptive in swim-up larvae indicates that the secretion of digestive enzymes is likely to be limited early in development. This is reflected by the activities of some, but not all, digestive enzymes. Subsequent development of gastrointestinal complexity is correlated with changes in the activities of digestive enzymes. Figure 5.1 (Baragi and Lovell, 1986) shows the activities of a number of digestive enzymes in the gut of striped bass (*Morone saxatilis*) fed three different diets. Four of the enzymes, trypsin, chymotrypsin, carboxypeptidase and pepsin, are protein-degrading enzymes, and their activities can be correlated with developmental changes in the gastrointestinal tract associated with the change from intracellular to intragastrointestinal digestion of protein. Trypsin increases through larval development to age 12 days, whereupon it decreases to day 16. However, by day 25 it has increased again. Trypsin is most active at neutral pH. Pepsin, active at low pH and present only in the stomach, increases from day 16. This increase is probably due to the development of the stomach and the glands that secrete the enzyme and acid to produce the low-pH environment. Chymotrypsin shows no great variation in activity over development and carboxypeptidase shows a decline to day 16, but like trypsin carboxypeptidase activities increase at day 25. Amylase activity also remains largely unchanged in striped bass through larval development.

Similar results were obtained by Mahr *et al.* (1983), although over a longer time-frame. These workers found that the stomach of *Coregonus* spp. begins to differentiate after 23 days post hatching, and the first filling was observed around the 50th day. At this time the pH of the stomach fluid was close to neutral. The morphological development of the stomach was not completed before the 80th day, and efficient acidification of the ingested food was reached at about the 100th day.

In a study of the digestive processes of the roach (*Rutilus rutilus*), a stomachless fish, Hofer and Nasiruddin (1985) found that relative gut

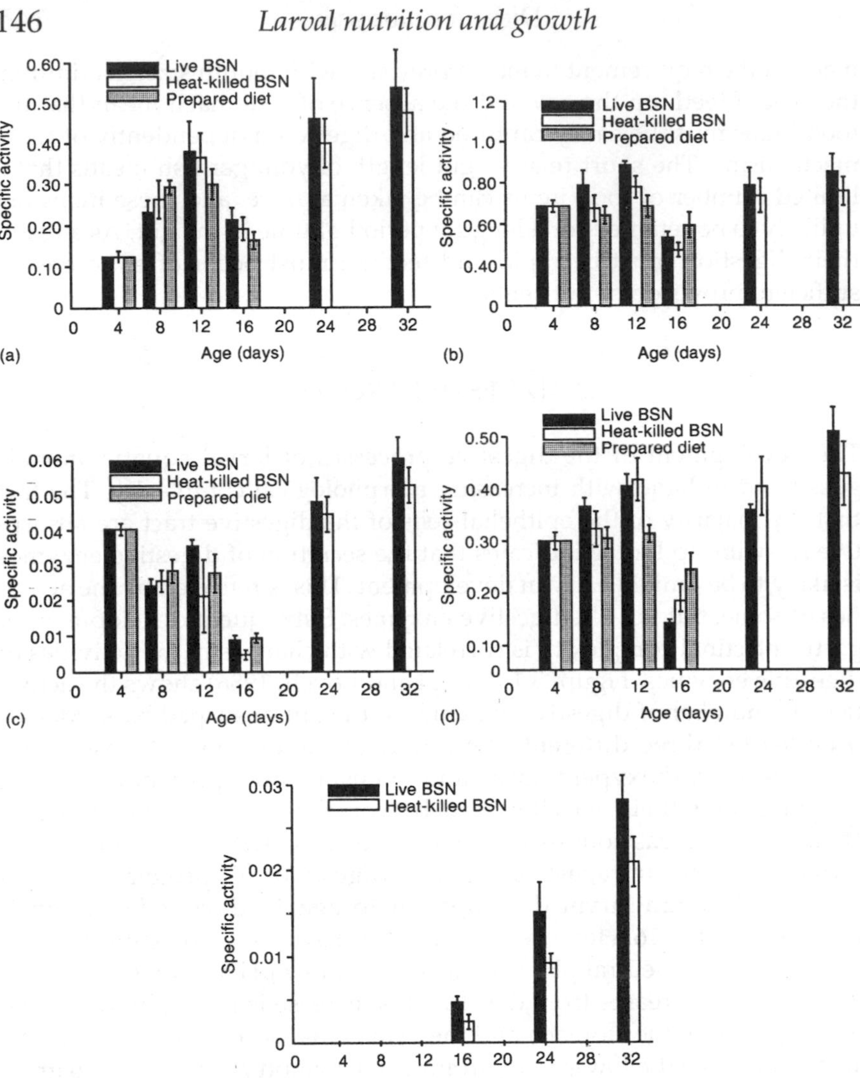

Figure 5.1 Specific activities of digestive enzymes in striped bass larvae fed live brine shrimp nauplii (BSN), heat-killed BSN or a prepared diet. Activities are expressed as micromoles of substrate hydrolysed per minute per milligram of protein at 30°C for (a) trypsin (substrate: tosyl-L-arginine methyl ester), (b) chymotrypsin (*N*-benzoyl-L-tyrosine ethyl ester) and (c) carboxypeptidase A (*N*-benzoylglycl-L-phenylalanine). Specific activity of α-amylase (d) is expressed as micromoles of maltose produced per minute per milligram of protein at 30°C, and specific activity of pepsin (e) is expressed as change in absorbance per minute per milligram of protein at 30°C. Values are means ± s.d. of four determinations. Fish fed the prepared diet died soon after day 16 and that treatment was discontinued. There was no effect of diet ($P<0.05$) on specific activity of any enzyme at any time period. Pepsin activity, which was not found until day 16, was higher ($P<0.10$) in fish fed live BSN at day 16. (From Baragi and Lovell, 1986.)

length, gut passage rate, tryptic activity and the ability to reabsorb digestive enzymes in the hind gut all increased with age from larvae to adults.

More recently Walford and Lam (1993) made a detailed study of the development of the digestive tract and the proteolytic activity in the sea bass or barramundi (*Lates calcarifer*). In barramundi, morphological transformation of the digestive tract is well advanced in 14-day-old larvae, and marked changes in pH of the presumptive stomach of early larvae (day 10 onwards) or in the stomach (day 20 onwards), was evident; by day 20 the pH ranged from 2 to 5 (mean of 3.5), suggesting that pepsinogenic activity may also commence by this time. Studies of Walford and Lam (1993) also confirmed the observations by Lauff and Hofer (1984), amongst others, of the importance of exogenous proteases from live prey to proteolytic activity in larvae. These enzymes act not only to digest the food, but also to activate the endogenous enzyme activities by cleaving the inactive zymogen forms (Chapter 4). In the case of barramundi larvae, it was concluded that the exogenous proteolytic enzymes (in this case from rotifers) and endogenous trypsin secretion induced by the ingested live feed would be sufficient to cause rapid breakdown of rotifers. However, because of the short larval gut length and therefore the short gut retention time of the food, the extracellular digestion based on tryptic activity may be insufficient to achieve complete hydrolysis of protein. The authors suggest that in barramundi larvae this is compensated for, to a certain degree, by the high level of pinocytotic activity in the rectal cells. These cells tend to absorb protein macromolecules, which are subsequently digested intracellularly, a phenomenon which has also been reported in the pond smelt (*Hypomesus transpacificus*) by Y. Watanabe (1984).

All these studies indicate that the digestive mechanisms are poorly developed early in larval life and any diet provided must take account of this deficiency.

5.4 FEEDING BEHAVIOUR AND ITS ROLE IN LARVAL NUTRITION

Despite the clear importance of nutrition in larval survival, growth and development, relatively little is known about the absolute nutrient requirements of the various larval aquatic animals. This is the result of a number of factors, but foremost among these must be the difficulty in working with such small animals and their sensitivity to handling, both of which make measuring and weighing larvae problematic. Difficulty experienced in inducing larvae to take artificial diets is also a major problem in these studies. Some examples of species whose larvae are not known to take an artificial diet include turbot (*Scophthalmus maximus*), milkfish (*Chanos chanos*), red drum (*Sciaenops ocellatus*), striped bass

(*Morone spp.*), black bass (*Micropterus* spp.), barramundi (*Lates calcarifer*), Murray cod (*Maccullochella peelii*), golden perch (*Macquaria ambigua*) and silver perch (*Bidyanus bidyanus*). Of other cultured species that will take an artificial diet, it is necessary to supplement such diets with live zooplankton. Species in this group include yellow tail (*Seriola quinqueradiata*), gilthead sea bream (*Sparus aurata*) and puffer fish (*Fugu rubripes*) (Wilson, 1991). In the case of puffer fish a complete diet of live food is not necessarily optimal, however, since provision of microparticulate diets in addition to live food enhances both growth rate survival rate (Kanazawa, 1991a).

It seems that a majority of species that require live food during their larval stage are induced to feed by the movement of the prey. The eyes generally develop early in aquatic animal larvae, and vision seems to be important in capturing prey (Sbikin, 1974). This capacity varies between species, and some fish, such as the sturgeon (*Acipenser* sp.), are capable of catching zooplankton without any light (Sbikin, 1974). There is insufficient information to determine whether or not early development of acute vision, such as that possessed by the sturgeon, is related to the ease with which fish larvae can be induced to feed on artificial diets. By comparison, in most cyprinid species that are cultured in the tropics, such as the Chinese carp, feeding in larvae is known to be by passive filtration.

There is of course considerable advantage in having the capacity to culture larvae on artificial diets since it removes the need to engage in secondary culture of a food organism, and it allows ready manipulation of the contents of the diet. A study by Dabrowski *et al.* (1987) showing that amino acids in live foods are catabolized at a lower rate and therefore used to a greater extent for protein synthesis than amino acids from dry diets fed to larval sturgeon indicates that development of artificial diets for larvae may be an extremely difficult process.

5.5 NUTRIENT REQUIREMENTS OF LARVAE

As already stated, the inability to induce larvae of many species to feed on artificial food has limited the ability to undertake experiments to determine optimum nutrient levels for these species. Most studies investigate the effect of enriching live food organisms with various nutrients, usually fatty acids (Lavens *et al.*, 1991). In those studies that have been conducted, some common features have emerged. Requirements for protein in larval fish are greater than in adult fish and the requirements for essential fatty acids are also greater. In addition, Wilson (1991) recommends that diets given to larval fish should also be overfortified with vitamins.

These generalizations regarding protein requirements hold true for ayu (*Plecoglossus altivelis*) (Kanazawa, 1991b), catfish (Wilson, 1991), gilthead sea bream (Kissil, 1991), salmonids (Cho and Cowey, 1991) and tilapia (Luquet, 1991). The requirement for increases in essential fatty acids is based on the above-mentioned studies on ayu, catfish and salmonids and is likely to be the case for other species which have not yet been studied.

From the above information it is apparent that larval fish require diets having a higher protein content and that provision of sufficient amounts of essential fatty acids is of great importance. It is in this context that an increasing number of studies have been initiated with a view to increasing the nutrient quality of live foods, particularly in respect of hatchery rearing of finfish species, which are nurtured on a single type of live food at any one time.

The live foods that have been most intensely investigated in respect of their nutritive suitability are brine shrimp and rotifers. In brine shrimp, *Artemia*, the main factor affecting its quality as a food source for marine larval organisms is its content of the essential fatty acid eicosapentanoic acid [C20:5(*n*-3), EPA]. Another essential fatty acid, docosahexaenoic acid, C22:6(*n*-3) (or DHA), is also low in *Artemia*. Simple methods have been developed to incorporate particulate products into the brine shrimp nauplii prior to offering it as prey for fish larvae by having the nauplii consume particles of a desired composition. This bioencapsulation is now known as *Artemia* enrichment or boosting, and Leger *et al.* (1986) dealt in detail with the different enrichment methods. T. Watanabe *et al.* (1983) reviewed the nutritional value of live food organisms used in Japan for mass propagation of fish. The nutritive value of rotifers, also deficient in *n*-3 fatty acids, is made suitable by culturing in a suitable medium such as *w*-yeast, by feeding with a mixture of homogenized lipids and baker's yeast or a marine alga, *Chlorella* spp, all of which are rich *n*-3 polyunsaturated fatty acids. Increasing the essential fatty acid content of live foods which are then fed to larvae has been shown to be beneficial for turbot (Guillaume *et al.*, 1991), gilthead sea bream and sturgeon (Hung, 1991).

While it is generally considered that EPA and DHA are important fatty acids in the nutrition of larval fish, the particular fatty acid required varies between species. Larvae of mahimahi (*Coryphaena hippurus*) disproportionately conserve DHA when starved and show greater stress resistance when fed *Artemia* enriched with DHA (Ako *et al.*, 1991). Larval sole (*Solea solea*), however, are able to grow and survive at high rates when fed a diet almost completely deficient in DHA but which contains EPA (Howell and Tzoumas, 1991).

The problems associated with nutrient quality of live feeds do not necessarily occur when rearing of larvae is done in outdoor ponds (or equivalent) in which planktonic growth is stimulated by fertilization, as in the case of rearing of Chinese and Indian carps or freshwater fish in Australia. In such conditions the availability of a range of suitable natural food organisms enables the cultured organism to obtain all its nutrient requirements without supplementation.

Dabrowski (1984) reviewed the feeding of fish larvae, and pointed out that fish larvae are susceptible to dietary deficiency in more spectacular ways than juvenile or adult fish. He concluded that in the formulation of dry compound diets more account should be taken of zooplankton composition.

Piggott and Tucker (1985), on the other hand, considered that a major portion of larval feed development should be involved with more precise control of nine factors:

1. nutritional balance of the formulation;
2. retention of nutritional components;
3. homogeneity of particles;
4. particle size and distribution;
5. density of particles;
6. water stability;
7. water solubility;
8. storage stability; and
9. packaging requirements.

It is apparent that the majority of these parameters are physical in nature and reflect the importance of matching the format of the feed to the morphology in larval fish nutrition.

Apart from providing a balanced diet, other problems related to weaning larvae to an exogenous food supply are pivotal to fish culture. Weaning is not necessarily dependent only on the provision of a nutritionally balanced diet of appropriate particle size, and other aspects are dealt with later in this book.

CHAPTER 6

Broodstock nutrition

6.1 INTRODUCTION

In the literature relating to reproduction and population dynamics of fish, it is often considered that the nutritional status of the animals in question affects their reproductive potential. Without going into detail of reproduction, which is beyond the scope of this book, there are a number of aspects of reproduction which may be affected by nutritional status. These are:

- the time to first maturity;
- the number of eggs produced (fecundity);
- egg size; and
- egg quality as measured by chemical composition, hatchability and larval survivorship.

Although there are relatively few studies of the effects of nutritional status on reproduction, there are a limited number of data on the effect of nutrition on all of these parameters. In this chapter we will consider the effect of nutritional status on the various aspects of reproduction and compare feeding and nutritional strategies to determine their effects on overall production levels.

6.2 ENERGY PARTITIONING FOR REPRODUCTION

You will recall that energy is partitioned by an animal between each of the various physiological processes involved in maintenance, growth and reproduction. The end result of this partitioning is that it is not possible for energy to be expended for more than one purpose, i.e. energy that is used for growth cannot also be used for reproduction. In general, the maintenance requirements of an animal are met first and then excess energy is divided between growth and reproduction. The relative partitioning of energy between growth and reproduction varies both between species and between strains of an individual species, and further generalizations are difficult to draw.

Reznick (1983) examined the trade-off between growth and reproduction in guppies, *Poecilia reticulata*. Two groups of guppies, non-reproducing females and reproducing females, each fed a natural diet, were compared to determine the amount of energy allocated to protein (somatic growth), fat (energy storage), and reproduction (gonadal material). In this experiment, those animals which did not expend energy in reproductive processes stored the excess energy as fat rather than increasing somatic growth or body protein. In a second experiment, Reznick (1983) also compared the pattern of allocation of nutrients to gonadal and somatic growth and energy stores in two strains of *P. reticulata* that used different amounts of energy for reproduction. In that experiment, the increased energy spent on reproduction was at the expense of somatic growth. In a study of the allocation of energy in roach, *Rutilus rutilus*, it was found that energy devoted to reproduction was able to be redirected from some of that used for locomotion (Koch and Wieser, 1983). While no clear pattern is apparent from these studies, it is clear that the energetic cost of reproduction is met by diversion of energy at the expense of another activity.

The total amount of energy available for utilization in the various physiological processes has been found to affect the size, quality and number of eggs produced. In one of the earliest studies of its type, Scott (1962) described a relationship in which various starvation regimes caused regression of the gonads in rainbow trout, *Oncorhynchus mykiss*. In that study, Scott based his estimation of fecundity upon histological examination of the gonads. In a more recent study, Springate *et al.* (1985) have determined the effect on rainbow trout of feeding what they termed either a full ration (0.7% of fish body weight per day) or half ration (0.35%) on the total and relative (number of eggs produced per kg of fish) fecundities of rainbow trout broodstock and on egg size. The data obtained from that study (Figure 6.1) indicate that the number of eggs produced by a fish on 0.7% per day is much greater than that produced by fish on 0.35% per day. In addition, the diameter of the eggs was significantly greater in those fish on 0.7% per day. However, it can be seen from Figure 6.1c that the relative fecundity of the fish on 0.35% per day was greater than that of fish on 0.7% per day. This occurs because of the interaction of two features. Firstly, smaller egg size means that a given volume of gonad will actually hold more eggs. The second is that broodstock on reduced rations grow less throughout the experimental period (6 months) and the proportion of the body present as gonad is greater in the smaller fish. That study is confounded by the fact that the animals fed at 0.35% per day were smaller at spawning than those fed twice as much. However, there is evidence that total egg volume is significantly greater in fish fed a lower ration when the influence of fish size is removed (Bromage and Cumaranatunga, 1988). The other effects of reducing rations for rainbow trout are modest reductions in the

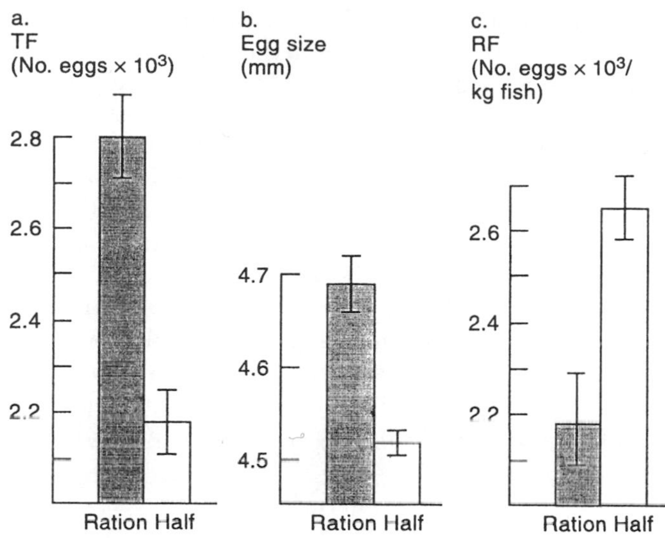

Figure 6.1 Effects of feeding either a full ration (0.7% of fish body weight per day) or half ration (0.35%) on the total (a) and relative (c) fecundities of rainbow trout broodstock and on egg size (b). Each histogram is the mean ± s.e. TF, total fecundity; RF, relative fecundity. (From Springate *et al.*, 1985.)

numbers of fish reaching maturity, and a delay of 2–3 weeks in the time of onset of spawning (Bromage and Cumaranatunga, 1988). Therefore it is apparent that reducing rations to rainbow trout, and possibly generally, results in reduced egg size but greater relative fecundity.

Springate and Bromage (1985) had previously found that, under the generally favourable conditions found in commercial hatcheries, egg size has no direct implications as far as overall egg quality (success in fertilization and hatching) and fry survival are concerned. Thus, for rainbow trout at least, there are competing interests in producing the largest or the maximum number of eggs from broodstock, and that by reducing feeding rates it is possible to produce a greater number of eggs of similar quality. The culturist must be careful, however, to ensure that rations are sufficient to allow growth and are not reduced to such an extent that the broodstock are only able to satisfy their maintenance requirements.

6.3 PROTEIN REQUIREMENTS OF BROODSTOCK

The need for a correct protein level in the diet for reproduction in higher vertebrates has been well documented. Detailed studies of fish are fewer and occasionally contain conflicting information. However, the literature is generally in agreement with the fact that there is an optimal protein level for reproductive success and that this protein level is related to the optimal level for growth of the species in question.

In a study by De Silva and Radampola (1990) on tilapia (*Oreochromis niloticus*), the effect of feeding diets varying in the level of protein on male and female growth rates and female fecundity was investigated. The results of this study can be summarized as follows: the optimal protein level for both male and female growth was 30% (Figure 6.2) and the greatest percentage of female spawning occurred amongst those animals fed 25% or 30% protein (Figure 6.3). There was a positive correlation between the size at first spawning and the protein content, with those receiving less protein spawning at a smaller size (Table 6.1), although those fed 20% protein spawned less frequently than those fed 25% or 30% (Table 6.1). As well as this, those fish receiving the lower protein had a higher relative fecundity (Table 6.2) and spawned a greater

Table 6.1 The mean (range) interspawning interval (days) and mean weight (g) at first spawning in *O. niloticus* reared under different dietary regimes

Protein	*Mean interspawning interval (days)*	*Mean weight (g)*
20	74 (34–153)	17.53 (10.5–25)
25	48 (17–66)	22.46 (12.3–48)
30	67	23.32 (15.4–36)
35	220	27.43 (12.4–61.7)

Source: Adapted from De Silva and Radampola (1990).

Table 6.2 The mean number of eggs per 100-g female in *O. niloticus* maintained under different dietary regimes (where available the range is given in parentheses)

Protein (%)	*No. of eggs*
20	608 (288–1098)
25	321
30	535 (337–734)
35	434 (228–628)

Source: Adapted from De Silva and Radampola (1990).

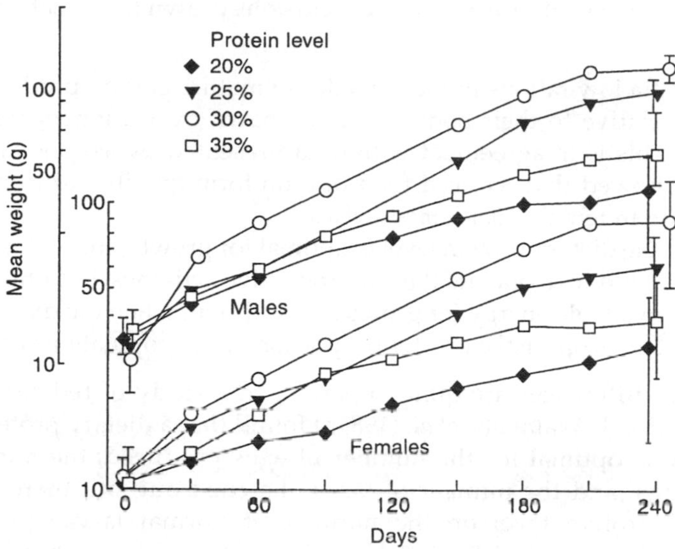

Figure 6.2 The mean weight of *O. niloticus* males and females maintained on isoenergetic diets of different protein content (vertical lines indicate s.d.). (From De Silva and Radampola, 1990.)

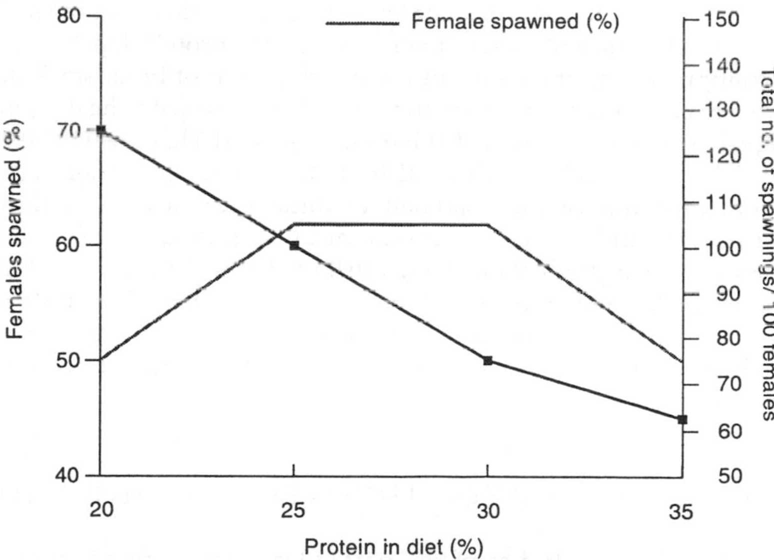

Figure 6.3 The percentage of females spawned and the total number of spawns per 100 females of *O. niloticus* in relation to the dietary protein level over the experimental period of 8 months (solid line indicates total number of spawnings/100 females). (From De Silva and Radampola, 1990.)

number of times (Figure 6.3). The conclusions drawn from that study are as follows:

- In tilapia low rations induce females to mature earlier, produce more eggs relative to their body size and spawn over a longer period of time. This is in agreement with data presented by Mironova (1978), who showed that reduced food of a uniform quality induced reproduction in female *Tilapia mossambica*.
- When the dietary protein level is optimal for growth, individual tilapia do not tend to accelerate their reproductive activities, but utilize most of the available energy for growth. In addition, when the dietary level of protein is optimal a greater proportion of the population spawns.

Species differences are quite apparent. In a study of red sea bream, *Pagrus major*, T. Watanabe *et al*. (1984a) found that a dietary protein level of 45% was optimal for the number of eggs produced, the number of viable eggs, and the number of larvae hatched out, but there was no effect of protein level on the number of normal larvae produced. T. Watanabe *et al*. (1984b) found an interesting relationship between dietary protein quality and reproductive success. By replacing white fish meal with cuttlefish meal, they were able to markedly increase the percentage viability of eggs, the number of larvae hatched out and the number of normal larvae. They were, however, unable to define the nutritional difference between cuttlefish meal and white fish meal. This is of particular interest as the quality of dietary protein has been, and will continue to be, of concern in the development of broodstock diets. Efforts are currently being made to replace fish meal with cheaper, more readily available grain meals. Cumaranatunga and Thabrew (1989) substituted legume meal for fish meal in diets of tilapia, *O. niloticus*, and found that fish fed fish meal had better ovarian growth and significantly larger oocytes, indicating that legume meal is an inadequate source of nutrients for egg production. They attributed this difference to higher levels of vitellogenic proteins and/or lipids in fish meal. This study suggests that diets will have to be carefully evaluated for the effects on reproduction if their use for broodstock is to be undertaken with confidence.

6.4 EFFECT OF DIETARY QUALITY ON REPRODUCTIVE OUTPUT

There are very few studies of the particular nutrient requirements of aquatic animals for gonadal development, but those that have been undertaken indicate great species variability. The questions addressed have often reflected the interests of the investigators and so general conclusions are extremely difficult to draw. Most work has concentrated on essential fatty acid requirements and fat-soluble vitamin requirements, and it has generally been assumed that the amino acid requirements of broodstock are similar to those for optimal growth. In a study that continued over 8 months, T. Watanabe *et al.* (1984b) found that low-protein, phosphorus-deficient and EFA-deficient diets produced eggs significantly low in hatchability, with most of the hatched larvae showing deformity. T. Watanabe *et al.* (1984c) considered that the most significant nutrient was n-3 PUFAs, which were found to be high in eggs of broodstock given diets containing high levels of n-3 PUFAs and low in eggs of broodstock given n-3 PUFA-deficient diets. In that study, however, those authors were unable to define the relationship between quality of eggs and their fatty acid distribution.

Mourente and Odriozola (1990) investigated the effect of broodstock diets on lipid classes and their fatty acid composition in the eggs and larvae of gilthead sea bream, *Sparus aurata*. The fish were given two diets of the same total lipid content but differing with respect to their fatty acid compositions. The fatty acid composition of eggs from the different groups were significantly different and were also reflected in differences in the composition of the larvae. However, the major proportion of the fatty acids in the larvae were 16:0, 18:1(n-9) and 22:6(n-3) and were independent of the diet. The authors drew no conclusions regarding improved survivorship of larvae fed either diet, but pointed out that high levels of 18:2(n-6) have been related to body deformation and inability to inflate air bladders in larval red sea bream.

Studies of rainbow trout have also indicated an important role of particular nutrients in the reproductive success of broodstock. Takeuchi *et al.* (1981) found that fish fed on diets without supplemental trace elements produced significantly lower percentages of both eyed and viable eggs than fish fed a sufficient diet. The contents of manganese, zinc and iron in the eggs of fish fed the diet without supplemental trace elements were also found to be significantly lower. T. Watanabe *et al.* (1984d) found that feeding broodstock rainbow trout diets deficient in EFA resulted in low growth rates, low rates of eyed eggs and low hatchability.

In the studies described, the diets given to the broodstock were designed to be deficient in either essential fatty acids, minerals or vitamins. In a hatchery situation, where animals are fed a commercially prepared and balanced diet suitable for growth, it is unlikely that broodstock would be faced with such deficiencies. Such profound changes in diet are unlikely to be experienced by farmed brood fish and therefore the effects observed are of little significance to the overall determination of diet and hence egg quality.

6.5 EFFECTS OF SPECIALIZED DIETS PROVIDED TO BROODSTOCK IMMEDIATELY PRIOR TO SPAWNING

The previously described studies were investigating the effects of nutrition on reproduction in long-term (greater than 1 year) experiments.

Some interesting short-term effects of broodstock have been described in red sea bream. These are particularly related to the fatty acid content and fat-soluble vitamin content of the eggs. In a comprehensive investigation, T. Watanabe *et al.* (1984e, 1985a, b) found that specialized diets given immediately prior to or during spawning of red sea bream affected the composition of the eggs. Pigments such as β-carotene, canthaxanthin or astaxanthin resulted in marked improvement in the percentage of buoyant eggs. On the other hand, feeding corn oil resulted in reduced viability of eggs. Both effects carried through to the number of abnormal larvae developed. Similarly, fatty acids and vitamin E, but not cholesterol, fed immediately prior to or during spawning resulted in increased levels of those compounds in the eggs.

A summary of particular requirements of broodstock of a number of species is contained in Table 6.3. This table reflects the paucity of relevant information regarding broodstock requirements. Clearly, there is a considerable need for further research in this area.

Table 6.3 Summary of the known nutritional requirements of broodstock

Species	*Requirement*
Ayu (*Plecoglossus altevilis*)	Vitamin E increases spawning success, egg survival and hatchability and larval survival. Requirement in broodstock is 34 mg/kg diet. Phosphorus increases spawning success. 20: 5 (*n*-3) and 18:3 (*n*-3) fatty acids are probably required
Carp (*Cyprinus carpio*)	Vitamin E increases gonadosomatic index and is required for vitellogenesis and for proper maintenance of *n*-6 fatty acids oocytes. *n*-3 fatty acids are probably required
Rainbow trout (*Oncorhynchus mykiss*)	Based on transfer of nutrients to eggs, requirements are 10 000–20 000 IU/kg diet for vitamin A, 100 mg/kg diet for vitamin E although lower for vitamin D. Low-protein/high-energy diets are as good as high-protein/low-energy diets for broodstock development. Trace minerals, particularly manganese, are required, and 20:4 (*n*-6) and 18:2 (*n*-3) fatty acids are EFAs for broodstock. EFAs are also required for high-quality sperm
Red sea bream (*Pagrus australis*)	Vitamin E is required and probably has the same effect as for ayu. *n*-3 fatty acids are required for buoyant (viable) eggs. β-Carotene and other carotenoids are important for egg viability. An unknown component of cuttlefish meal enhances spawning success

Source: Adapted from Kanazawa, (1988).

CHAPTER 7

Diet preparation

7.1 INTRODUCTION

Most aquaculture practices, particularly for finfish and shrimp, are either intensive or semi-intensive. Semi-intensive and intensive practices differ from extensive practices in that an external feed input is required for the well-being of the stock. Such a feed input constitutes one of the major recurring costs of the operation, its share often increasing in proportion to the increasing intensity of the culture operation. Therefore, it is crucial that the principles underlying diet preparation are well understood.

The science of fish nutrition has advanced over the last 2–3 decades, primarily in response to developments in commercial aquaculture. We are gradually approaching a phase, particularly in respect of species designated for 'luxury' markets, where the texture, colour of the flesh and so on dictate marketability, and hence the success and viability of operations. In the early investigation into fish nutrition, biologists approached the problem by researching the effect of natural food on cultured organisms. This, by the way, is an area of research which is still on-going and of great importance to warmwater aquaculture, and is an area in which our understanding is still very limited. One of the first studies pertaining to fish nutrition was that of Embody and Gordon (1924), who evaluated the proximate composition of various insects eaten by trout and found it to be approximately 49% protein, 15–16% lipids, 8% fibre and 10% ash. It is not surprising that these values closely resemble the composition of commercial trout diets used today. The subsequent stage of development was when nutritionists attempted to substitute on a nutrient basis other material in feed formulations, which also gradually led to our understanding of the nutrient requirements of cultured organisms, and to diet development.

7.2 AIMS OF DIET PREPARATION

Proper nutrition is one of the most important factors influencing the ability of cultured organisms (for that matter any living organism) to attain the genetic potential for growth, reproduction and longevity. The nutrient requirements vary between species, and within species between the different stages of its life-cycle (Chapter 3), and is also perhaps the most intensively investigated aspect of nutritional research. Of the species cultured commonly, the nutritional requirements of salmonids, channel catfish and common carp are the best understood (NRC, 1981, 1983). The available information on other cultured species has also been recently reviewed (Wilson, 1991).

We wish to redraw your attention to aspects of nutrient requirement of finfish species, in this instance in relation to practical diet preparation. Often the gross nutrient requirements of protein, lipid and carbohydrates of closely related species or species groups fall within a narrow range. Similarly, the differences in the requirements of most of the micronutrients such as the amino acids, vitamins, minerals and fatty acids are rather marginal between cultured species. For example, the same 10 essential amino acids are known to be required by all cultured finfish species hitherto investigated, and the quantitative differences in such requirements between species are mostly marginal (Tables 7.1 to 7.4). Luquet (1989) also pointed out the rather close agreement between amino acid requirements for coldwater fish (rainbow trout) and those of warmwater fish (channel catfish), when expressed in absolute terms and not as the percentage of the protein content. Accordingly, for example, the methionine, lysine and arginine requirements are respectively 20, 44 and 30 mg/100g/day (Cowey and Luquet, 1983). Keembiyahetty and Gatlin (1992) in their study on lysine requirements of juvenile hybrid bass concluded that apparent differences in requirement values reported for other species could be due to variations in dietary protein level, protein source and lysine availability as well as the difference in feeding rate. They went on to conclude that when all these factors are taken into consideration the lysine requirement of hybrid striped bass is more like the requirement values reported for other species. This implies and reinforces the fact that the quantitative differences in respect of the requirements reported for other amino acids for different species are in effect not real differences but apparent ones, manifested as a result of differences in other nutritionally related factors. Luquet (1989) suggested that extensive research on the determination of quantitative amino acid requirements does not seem to be a priority as indirect approaches provide a rather accurate estimate of requirements. We tend to agree with these ideas and believe these can be extended to practical diet formulation also.

The main objective and/or aim of diet formulation and preparation is to utilize the knowledge of nutrient requirements, locally available feed ingredients and digestive capacity of the organism for the development of a nutritionally balanced mixture of feedstuffs which will be eaten in adequate amounts to provide optimum production of the cultured organism at an acceptable cost. In practice, however, apart from the nutritional value, cost and availability of ingredients, other considerations have to be taken into account in diet preparation. Among such considerations are

Table 7.1 Amino acid requirements of some species of warmwater fish[a]

	Channel catfish	*Common carp*	*Japanese eel*	*Tilapia nilotica*
Arginine	4.3 (1.0)	4.3 (1.6)	4.5 (1.7)	4.2 (1.18)
Histidine	1.5 (0.4)	2.1 (0.8)	2.1 (0.8)	1.72 (0.48)
Isoleucine	2.6 (0.6)	2.5 (0.9)	4.0 (1.5)	3.11 (0.87)
Leucine	3.5 (0.8)	3.3 (1.3)	5.3 (2.0)	3.39 (0.95)
Valine	3.0 (0.7)	3.6 (1.4)	4.0 (1.5)	5.12 (0.78)
Lysine	5.1 (1.2)[b]	5.7 (2.2)	5.3 (2.0)	5.12 (1.43)
Phenylalanine	5.0 (1.2)[c]	6.5 (2.5)[e]	5.8 (2.2)[f]	3.75 (1.05)[h]
Methionine	2.3 (0.6)[d]	3.1 (1.2)[d]	3.2 (1.2)[d]	2.68 (0.75)[i]
Threonine	2.0 (0.5)	3.9 (1.5)	4.0 (1.5)	3.75 (1.05)
Tryptophan	0.5 (0.12)	0.8 (0.3)	1.1 (0.4)[g]	1.00 (0.28)
Crude protein in diet (%)	24	38.5	38	28

[a] Based on Wilson and Halver (1986). Values for tilapia from Santiago and Lovell (1988). Requirements expressed as percentage of dietary protein. Values in parentheses indicate requirements as percentage dry diet.
[b] Other values reported: 5.0 (1.5), total protein in the diet 30%.
[c] Diet contained 0.3% tyrosine, with 0.6% tyrosine in the diet; phenylalanine requirement was 2.0% of the protein.
[d] In the absence of cystine.
[e] In the absence of tyrosine, with 1% tyrosine in the diet; phenylalanine requirement was 3.4% of the protein.
[f] In the absence of tyrosine, with 2% tyrosine in the diet; phenylalanine requirement was 3.2% of the protein.
[g] Other values reported: 0.3 (0.1), total protein in the diet 12%.
[h] Tyrosine 1.79% of dietary protein.
[i] Cystine 0.54% of dietary protein.
Source: Lall (1991).

Table 7.2 Essential fatty acid requirements of certain warmwater fish species

Species	*Requirement*
Ayu	1% 18:3(*n*-3) or 1% 20:5(*n*-3)
Channel catfish	1% 18:3 or 0.5–0.75% 20:5(*n*-3) and 22:6(*n*-3)
Common carp	1% 18:2(*n*-6) and 1% 18:3(*n*-3)
Japanese eel	0.5% 18:2(*n*-6) and 0.5% 18:3(*n*-3)
Red sea bream	0.5% 20:5(*n*-3) and 22:6(*n*-3) or 0.5% 20:5(*n*-3)
Sea bass	1% 20:5(*n*-3) and 22:6(*n*-3)
Tilapia nilotica	0.5% 18:2(*n*-6)
Tilapia zillii	1% 18:2(*n*-6) or 1% 20:4(*n*–6)
Yellowtail	2% 20:5(*n*-3) and 22:6(*n*-3)

Source: Lall (1991).

Table 7.3 A summary of published vitamin requirements for growth of channel catfish, common carp and rainbow trout[a, b]

Vitamin	*Channel catfish*	*Common carp*	*Rainbow trout*
Vitamin A (IU)	5500	1000–20 000	2000–15 000
Vitamin D (IU)	500–4000	NR	2400
Vitamin E	50–100	80–300	30–50
Vitamin K	10	NR	10
Thiamin	1–20	NR	1–12
Riboflavin	9–20	4–10	3–30
Pyridoxine	3–20	4	1–15
Pantothenic acid	10–50	25	10–50
Niacin	14	29	1–150
Folic acid	NR or 5	NR	5–10
Vitamin B_{12}	0.02	NR	0.02
Choline	400	500–4000	50–3000
Inositol	NR	200–440	200–500
Vitamin	NR or 100	R	100–500

[a] mg/kg of diet unless specified.
[b] Values summarized from the published literature.
NR, not required.
R, required.
Source: Lall (1991).

Table 7.4 Mineral requirements of certain finfish

Mineral	*Rainbow trout*	*Channel catfish*	*Common carp*	*Japanese eel*
Calcium (%)	<0.1	<0.1	<0.1	0.27
Phosphorus[a] (%)	0.7	0.4	0.7	0.3
Magnesium (%)	0.05	0.04	0.05	0.04
Iron (mg/kg)	R	30	–	170
Copper (mg/kg)	3	5	3	–
Manganese (mg/kg)	13	2.4	13	–
Zinc (mg/kg)	15–30	20	15–30	–
Iodine (μg/kg)	R	–	–	–
Selenium (mg/kg)	0.15–0.38	0.25	R	R

[a] Inorganic phosphorus.
R, required.
Source: Lall (1991).

pelletability of the resulting diet, anti-nutritional factors in the ingredients and diet acceptability or palatability. Within this broad area, however, diet preparation may differ in detail, primarily being determined by the nature of the operation, species cultured, stage of the life-cycle cultured, whether a hatchery operation or grow-out and intensity of culture.

Diets are formulated and prepared to serve one of three functions:

- Diets which are expected to provide the organism with all of its energy requirements, gross major nutrient requirements as well as micronutrient requirements. These diets are used in intensive culture.

- Practical diets to supplement the natural food sources such as phyto- and zooplankton in the culture system. These diets are used in semi-intensive culture, and do not necessarily have to provide all the essential nutrient requirements. Such feeds are also commonly known as supplemental feeds.
- Semipurified diets and purified diets are used in experimental work for quantification of nutrient requirements. As the name implies in these diets, purified ingredients are used. For example, common protein sources in semipurified diets are casein and/or gelatin, which have approximately 90% protein and a known amino acid composition, as opposed to fish meal, which is the primary protein source used in practical diets and contains proteins, lipids, ash, fibre, minerals, vitamins and unknown growth factors. Purified diets permit controlled experimentation on nutrient requirements.

7.3 FORMS OF DIETS

Diets supplied to aquatic organisms could vary in form. Possible diets include

- live food(s), generally required for the culture of most aquatic organisms in their larval phases;
- forage materials (e.g. grasses and macrophytes), which may be introduced into the culture system or made to grow in the culture system (e.g. in freshwater crayfish culture); and
- prepared diets, including a wide array of feeds, ranging from simple, on-farm-based mixtures of a few ingredients to microencapsulated diets.

The great bulk of diets used in aquaculture belong to the last category, and are compounded using a number of ingredients.

7.4 FEED INGREDIENTS

Animal and fish feed ingredients are for the most part by-products from the human food processing industry. Here, feed ingredients are produced following extraction of high-value food for human consumption from raw materials which are considered unsuitable for direct human consumption. A wide variety of ingredients are available for use in fish and crustacean feeds. New (1987) recognized 10 such categories (modified after Gohl 1981):

1. grasses
2. legumes

3. miscellaneous fodder plants
4. fruits and vegetables
5. root crops
6. cereals
7. oil-bearing seeds and oil cakes
8. animal products
9. miscellaneous feedstuffs
10. additives.

Further, New (1987) summarized the major characteristics of each group, and what follows is an extract of his summary, modified suitably. It should also be noted that most of the feedstuffs mentioned in the following section are often considered and referred to as non-conventional feedstuffs. Details on the type of ingredients used and the proximate composition, as well as the essential amino acid composition of the more important feedstuffs, are given in various manuals (Gohl, 1981; New, 1987; Tacon, 1987a). No attempt will be made to go into these details here. The following sections deal simply with some generalities of the different feedstuffs and/or ingredients.

7.4.1 Grasses

Grasses are normally utilized either fresh (as pasture) or in the form of hay or silage. Dried grass is also used in feeds for other livestock and is a potential minor ingredient in fish and shrimp feeds as a source of carotenoids. Being characteristically very high in fibre content, grasses are of limited value in fish feeds except for herbivorous fish.

7.4.2 Legumes

The leaves and stems of legumes are, like the grasses, widely used as fodder for terrestrial animals. A few (e.g. ipil-ipil and alfalfa) have successfully been used in feeds for aquaculture. Legume fodder is rich in

Table 7.5 Fatty acid composition of the triglyceridge fraction of some plant oils

	Per cent fatty acid						
	Soybean	*Rapeseed*	*Corn*	*Cottonseed*	*Sunflower*	*Peanut*	*Coconut*
Saturated	14.0	4.5	9.4	30.0	17.0	14.5	91.5
Monounsaturated	23.2	55.5	45.6	18.5	29.0	53.0	6.0
Polyunsaturated	62.8	39.5	45.0	51.5	52.0	27.5	2.5
Linoleic [18:2(*n*-6)]	54.5	29.5	45.0	51.5	52.0	27.5	2.5
Linolenic [18:3(*n*-3)]	8.3	10.0	–	–	–	–	–

Source: Lim and Akiyama (1992).

protein (20–50%) and minerals. Of the legumes it is the seeds of certain legumes that have potential value as aquaculture feed ingredients, in spite of the fact that many contain anti-nutritive factors when raw. Processing (heat treatment) usually renders them safe for use. Leguminous seeds are often rich in lysine but poor in methionine. Some examples of leguminous plants which have been found to be suitable for incorporation into fish diets are acacia, clover, lucerne, groundnut (peanut), gram, lentil, locust beans, chickpea, guar, ipil-ipil, lima beans, field peas, mung bean, cowpeas and soybean.

Soybean meal is one of the most commonly used legumes in finfish and crustacean feeds. Its relatively high protein content and ready availability has made it a potential ingredient for replacing the more expensive and less readily available fish meal as a protein source in diets. Akiyama (1988) and Lim and Akiyama (1992) reviewed soybean meal utilization by shrimp and full-fat soybean meal utilization by fish respectively. According to Lim and Akiyama (1992) full-fat soybean meal has one of the best amino acid profiles among vegetable proteins in meeting the essential amino acid requirements of fish, and is also a good source of linoleic and linolenic acids and phospholipids. A comparison of the different legumes in respect of the fatty acid and amino acid composition is given in Tables 7.5 and 7.6 respectively.

7.4.3 Miscellaneous fodder plants

The leaves and other aerial parts of many plants, other than those specifically grown for fodder, are classified as miscellaneous fodder. These plant parts are pretreated, by soaking in water and/or sun drying, and used as a leaf meal. While having local significance as aquaculture feed ingredients, the nutrient digestibility (though crude protein levels on a dry matter basis are often quite high) is low. The plants in this category which have proved to be useful in aquaculture feeds are listed by New (1987). Miscellaneous fodder plants have been investigated to substitute for fish meal in aquaculture diets. Ng and Wee (1989) reported that cassava leaf meal (*Manihot esculenta*) can be used to replace up to 20% of fish meal protein in compounded diets for *Oreochromis niloticus* without significant loss in performance. Similarly, Wee and Wang (1987) have reported that ipil-ipil leaf meal (*Leucaena leucocephala*) can be used to replace 25% of the fish meal protein in *O. niloticus* diets without ill-effects or loss in performance. Research findings indicate an increasing possibility of using various types of leaf meals to replace the fish meal component, to varying degrees, in compounded diets for tropical species. Unfortunately, however, such findings rarely get translated into commercial diets. It is not uncommon, however, in certain culture practices to use fodder plants directly as a feed. The best example is the use of morning glory (*Ipomea* spp.) as a fodder for silver barb (*Puntius*

Table 7.6 Essential amino acid composition of some plant seed meals

	Amino acid content (per cent protein)					
	Roasted full-fat soybean	*Dehulled solvent extracted soybean meal*	*Peanut meal*	*Cottonseed meal*	*Sunflower seed meal*	*Rapeseed meal*
International feed no.	5-04-597	5-04-612	5-03-650	5-01-621	5-03-871	5-04-739
Arginine	7.4	7.4	9.5	10.2	9.6	5.6
Histidine	2.7	2.5	2.0	2.7	2.7	2.7
Isoleucine	5.7	5.0	3.7	3.7	4.9	3.7
Leucine	6.8	7.5	5.6	5.7	8.3	6.8
Lysine	6.3	6.4	3.7	4.1	4.2	5.4
Methionine	1.4	1.4	0.9	1.4	2.5	1.9
(+ cystine)	2.8	2.9	2.4	3.3	4.1	2.7
Phenylalanine	5.5	4.9	4.2	5.9	5.1	3.8
(+ tyrosine)	8.7	8.3	7.4	7.9	8.1	6.0
Threonine	4.4	3.9	2.4	3.4	4.2	4.2
Tryptophan	1.4	1.4	1.0	1.4	1.3	1.2
Valine	5.3	5.1	3.9	4.6	5.6	4.8

Source: Lim and Akiyama (1992).

gonionotus) culture in Thailand. This species feeds selectively on the tender leaves, not eating the more fibrous stalks. This form of feeding has an added advantage in that it does not cause significant deterioration of the water quality.

7.4.4 Root crops

Root crops, being rich in carbohydrates, are an excellent source of energy for many classes of livestock. However, their value as ingredients for aquaculture feeds is limited. This is partly because of their high value for human food and partly because of the inability of most finfish species to digest carbohydrates. Root crops, with some exceptions, are very deficient in protein, calcium, phosphorus and vitamins. Waste from root crops can be utilized in small quantities in compound feeds, but generally many require heat treatment to destroy the toxins they contain. Some root crops have special value in aquaculture because of their ability to increase the water stability of diets, with potato and cassava starches commonly used as binders. Other plants in this category include yams, carrots, Jerusalem artichokes and dasheen (yaro).

7.4.5 Cereals

Cereals and cereal by-products, despite their high carbohydrate content, form an important component in aquaculture diets. Again the starch content helps to increase the water stability of the feed, particularly when heat is used in processing. Cereals also contribute significantly to the protein and lipid content of the diet. Though deficient in some amino acids (e.g. lysine), they can be used to balance high-protein animal and vegetable ingredients.

7.4.6 Oil-bearing seed by-products

Many plants are grown specifically for the oil which their seeds or fruits produce. Vast quantities of by-products from the vegetable oil industry are produced, and these are the staple ingredients of animal feedstuffs, being high in protein and low in carbohydrate. All are potential ingredients of aquaculture feeds. Oil seeds are generally higher in protein content than cereals, varying from 20% to 50% protein. However, oil seeds are deficient in the essential amino acid lysine, and some are also deficient in threonine. Examples of plants whose seed by-products are used in aquaculture are the leguminous plants soybean and groundnut, together with mustard, rape, sunflower, safflower, kapok, cotton, oil palm, linseed, poppy, sesame (gingelly) and para rubber (caoutchouc).

In considering the use of ingredients from this group, it is essential to understand the terminology used in describing oil-seed by-products because ingredients with apparently similar names have completely different analytical characteristics. The external coating of some seeds is not always completely removed before oil extraction, for example in the case of sunflower, groundnut and cotton seed. The material which remains after oil extraction is therefore referred to in several ways: decorticated (when the coating is removed before extraction), dehulled (without hull) or undecorticated (hull and coat intact). Some intermediary products between decorticated and undecorticated exist. Some tables of feed composition refer to these as 'with some hulls'. Generally, decorticated products are higher in protein and lower in fibre than undecorticated products.

The other major set of terms applied to this class of feeds refer to the method of oil extraction used. This also has important analytical consequences. Expeller seeds have the oil removed by mechanical process, while extracted seeds undergo oil removal by a highly efficient chemical process using solvents. Sometimes the term 'solvent extracted' is applied to this later product. The characteristics of these products are that the expeller residuals are much higher in oil content and lower in protein content than extracted products.

Two other terms are applied to oil-seed residues. These are cake and meal. Normally, if a product is referred to as a 'cake', it means it is an expeller residue. Similarly, a 'meal' normally refers to an extracted product. However, there can be some confusion here because the word 'meal' can also be used to refer to a ground or milled product. So, the words 'groundnut meal' might refer either to groundnut cake which had been ground into a meal or to extracted groundnut. When in doubt, the chemical analysis is the only criterion available to determine the extraction process used. Some illustrations of the important analytical differences designated by these terms are given in Table 7.7. Oil-seed hulls are also available by-products of the vegetable oil industry, but being

Table 7.7 Examples of the effect of processing on analytical characteristics of oil-seed proteins

Country and material	*Water*	*Lipid*	*Protein*	*Fibre*	*NFE*	*Ash*
Pakistan/cottonseed						
Decorticated	7.3	5.2	36.7	8.8	34.9	7.1
Undecorticated	6.5	8.9	21.5	24.5	32.5	6.1
USA/groundnut						
Expelled	10.8	7.3	45.1	6.8	24.7	5.3
Extracted	8.5	1.2	47.4	13.1	25.3	4.5

Note: Boxes denote the analytical component most affected by the processing differences between the alternatives in each pair of ingredients.
NFE, nitrogen-free extract.
Source: New (1987).

extremely high in indigestible fibre content are of little value for aquaculture feeds.

7.4.7 Animal products

Animal by-products are either of terrestrial, avian or marine animal origin. They constitute the most important (and often the most expensive) ingredients of aquaculture feeds. These feedstuffs are necessary to balance the amino acid and vitamin deficiencies in cereals and other plant products in complete diets. Animal by-products, particularly fish meal, appear to contain unidentified growth factors. Some examples are blood, feather meal, poultry by-products meal, fish meal, meat meal, raw fish, fish oils, fish silage, shrimp meal and meal by-products.

As with some of the plant by-products, certain animal by-products also show essential amino acid imbalances. Blood meal, meat and bone meal are all deficient in methionine, while hydrolysed feather meal is deficient in lysine. In addition, the sole use of certain animal by-products can result in dietary imbalances. Blood meal is very rich in leucine but contains only low levels of isoleucine. Leucine acts antagonistically to isoleucine. Accordingly, when high levels of blood meal are incorporated into a diet, the antagonism between leucine and isoleucine will result in the fish suffering from isoleucine deficiency.

By far the most suitable animal product for incorporation into fish diets is fish meal. Fish meal is almost always made from marine fish, and can be based on the by-catches or on a product of a specific fishery. In 1990 nearly 27.7% of the world catch of fish and shellfish was turned into fish meal and fish oil (Bimbo and Crowther, 1991). The bulk of fish meal production comes from the anchovy fishery off the Peruvian and Chilean coasts, a single-species fishery based on the Peruvian anchovy, *Engraulis ringens*. In the large-scale manufacture of fish meal, the oil is often extracted and the residue dried and crushed. The quality of fish meal is affected by the type of raw material, the nature of drying and the cooking temperature. Often antioxidants such as ethoxyquin (400–700 p.p.m.) are added to processed meal. The basic steps involved in fish meal manufacture are shown in Figure 7.1. For details readers should consult Windsor and Barlow (1981). The average proximate and amino acid composition of different types of fish meals is given in Tables 7.8 and 7.9. Also included in Table 7.9, for comparison, is information on fish silage and dried whole egg. It is important to note that the proximate and amino acid composition of the different types of fish meal differ in detail, but in general all types of fish meal conform closely to that of the whole egg. Marine protein sources have always been important components of aquaculture diets, although shortages of fish meal are stimulating research on methods of replacing them, either partially or completely, with other ingredients. Ingredients of marine origin are

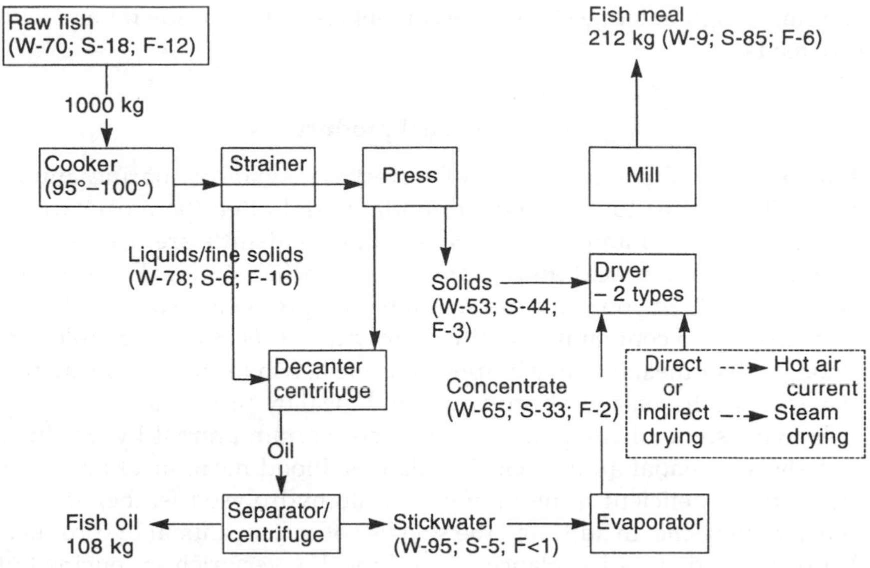

Figure 7.1 The steps involved in fish meal manufacture. Per cent water (W), solids (S) and fat (F) at each stage of processing, together with the average amount of fish meal and fish oil produced from 1000 kg of raw fish, are also indicated. The values are based on Windsor and Barlow (1981).

important sources of polyunsaturated fatty acids (PUFAs), particularly the important *n*-3 series.

Apart from fish meal, researchers continue to seek methods and/or techniques to improve the utilization of fishery by-products. One such technique is to produce fish silage. Fish silage is produced as a result of self-digestion, generally in an acidic medium with formic acid. The acidic medium prevents the multiplication of putrefying bacteria, but enhances the activity of endogenous, protein-digesting enzymes. The major problem with fish silage, however, is that its production is only economic in the wet form. In this form, however, silage is not a suitable feed because it can result in a lot of wastage and consequently result in an increase in the biological oxygen demand (BOD) of the culture medium. Perhaps ultrafiltration and/or reverse osmosis might enable the removal of the bulk of water and make the product more cost-effective, as well as enhance its wider usage as a feedstuff in aquaculture. In Norway fish silage is being used to some degree by being incorporated into soft pellets for use in salmonid culture.

Traditionally in Asian aquaculture, particularly in China and India, animal by-products such as silk work pupae have been, and are being, used as a feedstuff either singly or in compounded diets. In the early periods silkworm pupae were used directly in carp polyculture. Increasingly, silkworm pupae are used as an additional protein source in pelleted feeds in semi-intensive practices (Nandeesha *et al.*, 1990).

Table 7.8 Average proximate composition of selected animal by-products (all values are expressed as per cent by weight on an as-fed basis

Animal by-product	Average composition (per cent by weight)							
	Water	*CP*	*EE*	*CF*	*NFE*	*Ash*	*Calcium*	*Phosphorus*
Chicken (*Gallus domesticus*) eggs								
Whole egg (excluding shell), dried	4.0	46.5	41.6	0.0	4.3	3.6	0.20	0.74
Fish meal								
Anchovy (*Engraulis ringens*)	8.2	65.3	7.1	1.0	3.4	15.0	4.03	2.61
Herring (*Clupea harengus*)	7.9	72.7	8.5	0.8	–	10.1	2.04	1.42
Sardine/pilchard	8.5	65.0	6.7	1.0	3.5	15.3	4.44	2.72
Tuna (*Thunnus* spp.), mixed	7.0	59.0	6.9	0.8	4.4	21.9	7.86	4.21
Menhaden (*Brevoortia tyrannus*)	7.8	61.3	9.3	1.0	1.4	19.2	5.11	2.92
Red fish	8.0	57.0	8.0	1.0	–	26.0	7.70	3.80
White fish[a]	9.1	63.2	4.2	0.9	0.8	21.8	7.17	3.80
Freshwater (various species)[b]	9.0	66.7	9.1	1.0	–	14.9	5.40	2.90
Acid-preserved silages, fresh[c]								
Tilapia (*Oreochromis niloticus*), whole	71.9	15.6	4.2	–	–	5.0	–	–
Sprat (*Sprattus sprattus*), whole	74.3	16.7	6.4	–	–	2.7	–	–
Winter sprat (*S. sprattus*), whole[d]	65.7	15.6	13.9	–	–	3.3	–	–
Herring (*C. harengus*), whole	77.7	15.5	3.4	–	–	2.1	–	–
Herring (*C. harengus*), offal	68.1	14.5	16.3	–	–	2.4	–	–
Sandeels (*Ammodytes tobianus*), whole	77.7	15.4	3.4	–	–	2.4	–	–
White fish offal (excluding viscera)	78.9	15.0	0.5	–	–	4.2	–	–
Mackerel (*Scomber scombrus*), whole	70.2	16.9	12.0	–	–	2.1	–	–

[a] Includes various marine species such as Gadidae, Lophidae/Rajidae, all which have a low lipid content.
[b] Means of various freshwater species.
[c] Acid-preserved silages are produced by the external addition of mineral or organic acids to the macerated whole fish or wet fish by–products.
[d] Presented for a high-lipid sprat silage, after a 2-week storage period at 20°C with added ethoxyquin antioxidant.
CP, crude protein; EE, lipid or ether extract; CF, crude fibre; NFE, nitrogen-free extractives.
Source: Tacon (1988).

Table 7.9 Average essential amino acid (EAA) composition of selected animal by-products (all values are expressed as a percentage on an as-fed basis)

	Average EAA composition (per cent dry or wet meal)[a]											
Animal by-product	*Arg*	*Cyt*	*Met*	*Thr*	*Iso*	*Leu*	*Lys*	*Val*	*Tyr*	*Tryp*	*Phe*	*His*
Chicken (*G. domesticus*) eggs												
Whole egg (excluding shell), fresh	0.76	0.29	0.40	0.61	0.76	1.07	0.83	0.86	0.50	0.19	0.70	0.30
Whole egg (excluding shell), dried	2.94	1.09	1.48	2.26	2.87	4.03	3.10	3.30	1.91	0.73	2.59	1.10
Fish meal												
Anchovy (*E. ringens*)	3.67	0.61	1.94	2.78	2.99	4.98	5.08	3.52	2.17	0.76	2.63	1.52
Herring (*C. harengus*)	4.61	0.71	2.14	3.01	3.21	5.30	5.66	4.37	2.23	0.80	2.77	1.71
Sardine/pilchard	3.25	0.76	1.95	2.70	3.09	4.42	5.55	3.64	2.29	0.58	2.34	1.88
Tuna (*Thunnus* spp.), offal	3.42	0.44	1.46	2.31	2.41	3.81	4.04	2.80	1.72	0.56	2.16	1.78
Menhaden (*B. tyrannus*)	3.58	0.57	1.77	2.43	2.81	4.64	4.70	3.27	1.97	0.68	2.40	1.44
White fish	4.16	0.67	1.72	2.56	2.71	4.38	4.56	3.05	1.86	0.64	2.30	1.45
Red fish	4.10	0.40	1.80	2.60	3.50	4.90	6.60	3.30	–	0.60	2.50	1.30
Freshwater (various species)	4.62	0.47	1.92	3.25	3.27	4.87	5.89	3.50	–	0.62	2.92	2.03
Catfish (*I. punctatus*), offal	3.92	–	1.23	2.43	1.94	3.47	3.20	2.21	1.62	–	2.22	1.15
Catfish (*I. punctatus*), bone	2.75	–	0.72	1.19	0.99	1.59	1.70	1.21	0.64	–	0.97	0.58
Acid-preserved silages, per cent dry matter basis												
Catfish (*I. punctatus*), offal	5.40	–	1.49	2.86	2.51	4.38	4.66	3.33	2.30	–	2.74	1.44
Winter sprat (*S. sprattus*), whole[a]	2.69	1.27	1.27	2.16	1.90	3.54	3.86	2.65	–	0.25	2.60	1.22
Whiting (*M. merlangus*), whole, per cent total amino acids[b]	7.34	0.61	3.67	3.67	4.65	7.96	8.93	5.75	3.55	–	3.55	1.96

[a] Presented for silage after 8-weeks' storage at 20°C with added ethoxyquin.
[b] Composition for silage after 6 months' storage at 18–22°C and expressed as a percentage of total recovered amino acid.
Arg, arginine; Cyt, cystine; Met, methionine; Thr, threonine; Iso, isoleucine; Leu, leucine; Lys, lysine; Val, valine; Tyr, tyrosine; Tryp, tryptophan; Phe, phenylalanine; His, histidine.
Source: Modified after Tacon (1987b).

7.4.8 Miscellaneous feedstuffs

Many other ingredients have potential use in aquaculture feeds, however their value has not yet been fully assessed. Some of these ingredients are also referred to as unconventional or non-conventional feedstuffs though many, such as cane sugar molasses, are conventional ingredients in feedstuffs for other animals. This group of feedstuffs includes leaf protein concentrate, minerals, seaweed, by-products of the sugar and of fermentation industries, lipids, microbial proteins, algae, manures and celluloses.

7.4.9 Additives

An increasing diversity of additives is being used in animal feedstuffs, including synthetic amino acids, vitamins, binders, antioxidants, preservatives, prophylactic medicines, hormones and growth promoters. Most of these have very specific uses and may be non-nutritive.

(a) Amino acids (synthetic)

The major synthetic amino acids available for supplementation are L-lysine and DL-methionine. These are used as chemoattractants as well as to supplement deficiencies in a compounded feed.

(b) Vitamins

Individual vitamins or premixes of vitamins prepared for specific purposes are commercially available. The storage and mixing of vitamins and other trace substances require special care and facilities, and it is not recommended that farmers prepare their own or attempt to add individual synthetic vitamins to their feeds. Vitamins are used to supplement for deficiencies in the ingredients of compounded feeds. They have also been shown to provide additional protection against lipid oxidation, enhancing its storage life (Gatlin *et al.*, 1992).

(c) Binders

Binders are substances used in diets to improve their pelletability, enhance their durability, preserve their physical form during storage and to preserve the water stability. Commonly used binders in compounded feeds are carboxymethylcellulose (CMC), hemicelluloses, bentonites, agar, carageenin and collagen. Heat treatment during diet preparation results in carbohydrates gelatinizing, also providing good binding qualities to the resultant feed when high-carbohydrate ingredients are used.

(d) Antioxidants

Antioxidants are usually included in vitamin premixes or added to lipids to prevent or delay the onset of rancidity. Rancidity makes feeds unpalatable and generates toxic chemicals. Antioxidants can be naturally occurring substances, such as vitamin E, or synthetic chemicals. The commonly available commercial antioxidants, under a variety of trade names, are BHT (butylated hydroxytoluene), BHA (butylated hyroxyanisole) and ethoxyquin.

(e) Preservatives

Several substances may be added to feeds to control the rate of deterioration, particularly from fungal attack. These compounds are termed preservatives. Most are sodium or potassium salts of propionic, benzoic or sorbic acid.

(f) Chemoattractants

Chemoattractants are synthetic chemicals or natural ingredients containing chemicals, such as free amino acids, which elicit feeding responses. Generally, extractive compounds in muscles of molluscs and crustacea are believed to be principal flavour attractants. Attractants are a mix of chemicals comprising nitrogenous compounds including free amino acids, low molecular weight peptides, nucleotides and related compounds and organic bases.

(g) Hormones

Anabolic steroids have been commonly used in domestic animal feeds until relatively recently, when they were banned in most parts of the world. The concerns about hormone residues in products destined for the market place which led to the ban is also enforced on cultured fish in most countries. However, steroids are used in the culture of certain species, not with a view to enhancing growth but with a view to obtaining populations of a single sex, generally all-male populations of certain species. This is achieved by giving anabolic steroids in the diets during the fry stages. Use of growth-promoting hormones such as thyroid hormones is under investigation, but is yet to have wide commercial applicability.

(h) Antibiotics

Antibiotics are added to feeds to treat disease. However, routine use of antibiotics is done supposedly as a prophylactic measure in some culture practices, most notably in shrimp culture in South-East Asia. Routine use of antibiotics in feeds is not recommended, as it leads to resistant strains of bacteria.

(i) Carotenoid supplements

Xanthophylls and carotenoids are the most important classes of pigments for fish and crustaceans. Canthaxanthin and astaxanthin add colour to the flesh and eggs. High-value culture species such as salmon and red sea bream do not have the ability to convert xanthophylls to carotenoids, and therefore must receive these pigments in their diets. Carotenoid supplementation is provided to these species by adding natural material containing the specific pigments such as paprika, krill products and processing waste of shrimp and crab.

7.5 DIET FORMULATION

Diet formulation is not easy. It is a process in which the appropriate feed ingredients are selected and blended to produce a diet with the required quantities of essential nutrients. As indicated earlier, no single ingredient can be expected to meet all the nutrient requirement of a cultured organism. By selecting various ingredients in the correct amounts, a compounded ration which is nutritionally balanced, pelletable, palatable and easy to store and use may be formulated.

The basic information required for feed formulation is:

- nutrient requirements of the species cultivated;
- the feeding habits of the species;
- local availability, cost and nutrient composition of ingredients;
- ability of the cultured organism to utilize nutrients from various ingredients as well as the prepared diet ;
- expected feed consumption;
- feed additives needed; and
- type of feed processing desired.

Many factors need to be considered when formulating feeds for use in aquaculture. In aquaculture, as in any other form of husbandry, both nutrition and feed cost have to be taken into account. Feed cost is considered to be the highest operational cost in both intensive and semi-intensive aquaculture systems, and therefore special consideration needs to be given to this aspect in diet formulation. Supplying adequate nutrition for various aquaculture species involves the formulation of diets containing about 40 essential nutrients and the proper management of a multitude of factors relating to diet quality and intake. In essence, bioavailability of nutrients, diet acceptability (palatability), feed manufacture, storage methods and chemical contamination can have profound effects on the quality of the diet and hence performance and production of cultured organisms. Undoubtedly, formulation needs to take into consideration the nature of the culture practice. In intensive culture systems the formulated feed needs to provide all the nutrient (and energy) requirements of the organism cultured. For semi-intensive culture the formulated feed needs only provide 'supplementary nutrition' to the cultured organism.

Lall (1991) made some general observations which are important and of relevance to all feed formulation. These observations expand on several aforementioned considerations and are summarized below.

- It is important to ensure that feed formulae developed are nutritionally and economically sound. An 'economic diet' is expected to produce a kilogram of healthy fish at the least cost under normal growing conditions. It needs to be understood that the feed at the lowest price per kilogram of fish is not necessarily synonymous with least cost production. Linear programming is now widely used for formulating least-cost diets. In most instances feeds formulated without due consideration to nutritional aspects, purely by the computer, have not met with success. It is imperative that the information that nutritionists have gathered by experience, which cannot always be programmed into a computer, must be considered.
- In the development of economical diets the seasonal changes in the availability and composition of ingredients should be taken into account. Feed companies need to take advantage of availability of feed ingredients at economical prices at various times of the year, rather than base diets solely on one particular type of ingredient.
- When protein levels are stated it is expected that the protein will be of good quality, easily digestible and with an acceptable balance of amino acids. It is believed that fish feed to satisfy the energy requirements. Accordingly, the absolute amount of proteins, vitamins and minerals ingested will depend to a large extent on the energy intake. Therefore, the balance of energy to the nutrient supplies is more critical than the absolute level of specific nutrients. Accordingly, the dietary requirements, in particular the protein requirements, of many culturable species are being re-evaluated in this light (Kim *et al.*, 1991; El-Sayed and Teshima, 1992).
- Good quality feeds can be made only from good-quality ingredients. Fresh feed is more palatable than rancid or stale feed. Also, many vitamins tend to deteriorate with improper storage, as well as the length of storage. Nutritional deficiencies are encountered when feed is not properly mixed or when it is fed in a stressful environment.

7.5.1 Factors to be considered in diet formulation

As mentioned previously, many factors need to be considered in order to formulate a feed which will provide optimal growth performance of the cultured organism whilst also being cost-effective. These factors will be addressed in greater detail below.

(a) Nutrient requirements

It is of vital importance that a formulated feed meet the nutritional requirements of the cultured organism. As pointed out earlier, depending on the intensity of the culture, the feed may be expected to provide all of the nutrition required, or only some of it. In order to do this, it has been suggested that the species, strain, stage of development and health, as well as temperature and environmental conditions of the culture system, be taken into account. All these factors have been shown to affect the nutritional requirements of fish (Chapter 3).

The most notable differences in nutritional requirements are seen in the essential fatty acid and energy requirements. Consequently, it is necessary to vary the lipid and energy sources to accommodate these changes. As pointed out earlier, although the essential amino acid requirements tend to remain fairly constant, both within and between species, it is important that the protein sources selected meet these requirements. Likewise the requirements for micronutrients do not differ widely between species and may generally be met by standard commercial vitamin/mineral premix supplementations.

(b) Composition of ingredients

A knowledge of the nutrient composition and available energy of dietary ingredients is essential for their selection for use in diet formulation. The most comprehensive information on feed composition is provided in the United States–Canada tables of feed composition (NRC, 1983), and in Gohl (1981) for tropical feeds. The values given in such composition tables are average values. The compositions of feedstuffs are known to vary regionally, seasonally and also with soil fertility and type of processing and storage (Table 7.7). Therefore, it is desirable that each batch of feed ingredient is analysed for actual content prior to feed formulation. The most variable constituents tend to be the protein and essential amino acid concentrations.

Feedstuffs also need to be screened for enzyme inhibitors and other indigenous toxins, as well as for aflatoxins and other mycotoxins which are contaminants from mouldy grains.

(c) Digestibility and nutrient availability

If feed formulation is performed correctly, a knowledge of the digestibility of the individual nutrients of all the ingredients is essential. The process of estimating these is tedious and time-consuming (Chapter 4).

Table 7.10 Digestible energy values for finfish feeds as suggested by New (1987)

Nutrient	*GE[a] (Kj/g)*	*DE (estimated (Kj/g)*
Carbohydrate		
Non-legumes	17.15	12.55
Legumes	17.15	8.37
Proteins		
Plant	23.01	15.90
Animals	23.01	17.78
Fats	38.07	33.47

[a] GE = amount of heat released when a substance is completely oxidized in a bomb calorimeter at 25–30 atmospheres of oxygen. This is also referred to as physiological fuel value.
GE, gross energy; DE, digestible energy.

However, general digestibility values for most of the major ingredients for the widely cultured species are presently available and are often used in feed formulation.

Apart from nutrients, the digestible energy and metabolizable energy are also required. Estimation of digestible energy and metabolizable energy has shortcomings (Cho and Kaushik, 1990). However, they are valuable in determining the percentage of utilization of the feedstuff by the organism. For preliminary formulations New (1987) recommends the use of the digestible energy values given in Table 7.10.

Apart from straightforward digestibility considerations, the bioavailability of certain nutrients in different feedstuffs needs to be taken into consideration. The bioavailability of minerals differs significantly amongst feedstuffs in contrast to other nutrients. T. Watanabe *et al.* (1988) reviewed the bioavailability of minerals in fish meal to fish. These authors stated that the bioavailability of phosphorus in fish meal appears to be correlated to the presence of gastric juice in the stomach. This indicates that stomachless fish such as cyprinids need to be provided with their phosphorus requirement from another source. Similarly, the bioavailability of magnesium in white fish meal is known to be low.

(d) Other dietary components

Certain ingredients are added to the diets for physiological or economic reasons, and include binders, antioxidants, etc. (section 7.4.10). These are added in small quantities and often have no direct nutritional value and do not act as energy sources. However, when formulating appropriate diets, allowances need to be made for their inclusion.

(e) Dietary interactions

Although only emerging as important in respect of fish nutrition, study of the interactions of various nutrients in the diet is a very active area of research in domestic animal and human nutrition. Four main types of nutrient interactions are known in finfish. These are, micronutrient–macronutrient or other dietary component interactions, mineral–mineral interactions, vitamin–mineral interactions and vitamin–vitamin interactions. These interactions are influenced by a number of factors including diet composition, diet processing, species and age of the cultured organism and environmental factors.

The basic interactions of vitamins, minerals and diet composition in the diets of fish were reviewed by Hilton (1989). It is not intended to go into detail of what is known on nutrient interactions, however an example for each of the preceding interactions that are known are presented in order to familiarize you with the phenomenon.

Micronutrient–diet composition interactions

Factors such as protein quality and quantity, energy source and content have significant effects on the requirement and/or metabolism of most nutrients.

Thiamin (vitamin B_{12}) availability is known to be influenced by fat and protein content of diets of the same calorific value. Experiments conducted on trout, *Oncorhynchus mykiss*, at 15°C have indicated that when fed thiamin-deficient high carbohydrate diets, trout develop thiamin deficiency signs earlier and have higher mortalities than those fed a thiamin-deficient high-fat diet. This is indicative of a thiamin-sparing effect with high-fat diets. Similarly, metabolism of pyridoxine (vitamin B_6) is related to dietary protein or amino acid metabolism. It could be that fish meal and vegetable protein diets require different vitamin B_6 levels to be present.

Mineral–mineral interactions

A number of interactions between the essential minerals are known to exist. Magnesium requirement is dependent on the calcium and phosphorus content of the diet, while copper and zinc may be antagonists. However, the magnesium requirement is not known to increase when the dietary calcium or phosphorus is increased (Knox *et al.*, 1981). Hill and Matrone (1970) have suggested that as a result of this antagonism these two minerals may compete for binding sites on proteins involved in mineral absorption and/or the synthesis of metalloenzymes. Selenium appears to interact with sulphur, zinc, mercury and cadmium. Selenium is a component of the antioxidation enzyme glutathione peroxidase. Apart from this function it is believed that selenium has the ability to modify the toxicity of other heavy metals.

Vitamin–mineral interactions

Several vitamin–mineral interactions in fish have been reported. Fish obtain calcium from the surrounding medium via their gills and from the food via the intestinal mucosa, the former being the major route of uptake. Hilton (1989) summarized the reported interactions between vitamin D and calcium, and suggested that water chemistry, in particular, will dramatically affect vitamin D metabolism in fish and its interactions with calcium uptake and regulation. The interaction between ascorbic acid and minerals was reviewed by Hilton (1984). Two major interactions occur in fish; these are between ascorbic acid and iron and between ascorbic acid and copper.

Ascorbic acid (vitamin C) is involved in the metabolism of iron in fish. Vitamin C deficiency results in a reduction in serum iron levels and a redistribution of tissue iron stores in rainbow trout, as well as a reduction in both the haemoglobin level and haematocrit. Ascorbic acid–copper interaction(s) do not appear to be straightforward, in that ascorbic acid seem to have a differential effect on dietary copper depending upon the copper levels in water.

Vitamin–vitamin interactions

Vitamin–vitamin interactions are also common. There have been reports on the interaction of vitamin B_{12} and folic acid in the Indian major carp, *Labeo rohita*; when both these vitamins were deficient, the appearance of deficiency signs (anaemia) was accelerated.

Generally, the possible interactive effects are not taken into consideration when diets are formulated, primarily because very little is known of such effects. It is important, nevertheless, to be aware of the possibilities of the existence of such interactive effects between nutrients because they might help to explain some unforeseen symptoms or deficiency signs which are not immediately obvious from the formulations.

(f) Flavour quality

The influence of environmental factors on the organoleptic properties of marine fish has been known for some time. These factors include not only dissolved substances, the composition of the food chain, but also physical factors such as temperature, photoperiod and water cycle. At lower temperatures, fish tend to accumulate more unsaturated fatty acids in their tissues and organs. As the fatty acid composition is related to flavour and texture, alterations that occur during processing and storage will affect the organoleptic quality of the fish product. Therefore, maintenance of uniform environmental factors would help to obtain uniformity in the organoleptic qualities of cultured fish.

The type of culture will determine the degree to which the total environment (ocean, lakes, rivers and ponds) and the local environment (location of fish with respect to the environment) influence the organoleptic properties of the fish. Most work performed to date has investigated the relationship between flavour of cultured finfish and the total environment, rather than the influence of dietary quality and specific nutrients on flavour.

In ponds, fish are generally reared under near static water conditions. In such an environment, the organoleptic properties are likely to be influenced to a great extent. A common problem in pond culture in the tropics is the presence of 'off-flavour'. This is thought to be due to a compound known as geosmin. Geosmin is produced by a fungus of the group Actinomycetes and is a member of the blue–green algal genus *Oscillatoria* spp. These organisms grow on mud of high organic content, particularly on the interface between the reduced mud and the oxidized water layer above it.

Other possible sources of off-flavour are industrial wastes. The more important chemicals that impart off-flavour are phenols, tars and mineral oils. Some of these substances are capable of imparting off-flavour even when present at low concentrations. In common carp culture, *o*-chlorophenol and *p*-chlorophenol at concentrations of 0.015 and 0.06 mg/l respectively are known to impart off-flavour.

The organoleptic characteristics of common carp grown in ponds fertilized by liquid manure and those grown on diets of grain or high-protein pellets have been compared. Carp grown in the fertilized ponds were claimed to be superior in flavour and colour to those fed the prepared diets. However, fish raised in fertilized ponds had a flesh fat content of about 6%, while in those fed prepared diets the fat content ranged from 14% to 22%. It would therefore appear that, all other factors being equal, the compositional constituents in the flesh more strongly influenced the eating quality of the fish than did the flavour components contributed by the diets. The development of a soft-dry pellet for yellowtail culture in Japan is also claimed to have resulted in a product possessing different organoleptic properties from that based on trash fish feed. This fish is reported to have better consumer acceptability (T. Watanabe *et al.*, 1990), although no analyses are available to determine if it was composition of flavour provided by the diet that was responsible.

Recently, Johnsen and Dupree (1991) evaluated the influence of commonly used feed ingredients on the flavour quality of farm-raised channel catfish. In this study ingredients were replaced in semipractical diets, at levels used in commercial feeds. The findings of this study were that, for the farm-raised catfish industry, the practice of least-cost formulation of feeds may be followed without concern that flavour quality of fish will be affected adversely.

There is no doubt that diet influences the organoleptic properties. However, up until now nutritionists have been mostly concerned with the development of feeds that result in better growth and food conversion factors in an attempt to reduce the feed costs associated with commercial aquaculture. It is only now, in an era of increasing competition between producers, particularly in the luxury market, that dietary manipulation to improve organoleptic properties of the product is being considered. Perhaps the next decade or so will witness some major developments in this direction.

Generally, most off-flavours disappear when the animals are kept in clean, running water for a week or so.

7.5.2 Diet formulation

The nutritional considerations that should be taken into account in diet formulation are the energy content of the diet and the digestible/metabolizable energy to nutrients ratios, particularly the protein to energy ratio. These are followed by a calculation of the protein content and amino acid balances, selecting lipid type and level to satisfy the essential fatty acid and energy requirements and augmentation of vitamins and minerals.

The mathematical techniques used for feed formulations are simple, and are becoming easier with the availability of various software packages. Diets that contain few feedstuffs or in which levels of, say, protein, energy and minerals are fixed, may be formulated using simple algebra or simultaneous equations.

(a) Pearson squares

Formulation of diets with a few ingredients, and in which amino and fatty acid balances are not taken into consideration, is best and most simply achieved using Pearson squares. The complicated 'least-cost formulae' used in Pearson squares are based on series of simultaneous equations. The following simple formulations are based on examples cited by New (1987).

Assume that a nutritionist has been requested to formulate a diet of 26.0% protein and 8.0% lipid using locally available ingredients given below. From past experience the nutritionist decides that 10% fish meal needs to be incorporated, and hence the task is primarily to determine the proportion of ingredients that will make up the diet.

	Lipid (%)	Protein (%)
Fish meal	6.0	55.0
Groundnut cake	13.7	34.5
Soybean meal (fat extracted)	1.3	46.8
Rice bran	2.4	13.3
Maize meal	4.5	9.8

Find the proportions of the different ingredient combinations (in part) that would initially give a dietary protein level of 26% when mixed with 10% fish meal.

- Fish meal (10%) will contribute 5.5% protein to the diet. Therefore, the other 90% of the ingredients will have to make up to 20.5% of protein. If this portion of the formulation is treated separately, the non-fish meal portion of the diet must contain 20.5 × 100/90 = 22.8% protein.
- In order to provide 22.8% protein the amount of each of the four possible pairs of ingredient combinations which will supply this level of protein can be calculated by constructing a 'cross'. Note that because of the relatively low levels of protein in rice bran and maize meal the combinations that are likely to yield 22.8% protein will have to contain either groundnut cake or soybean meal.

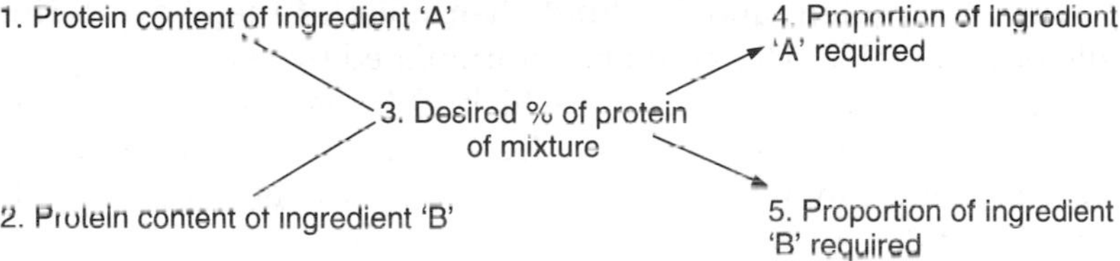

Substituting, for example, the values for groundnut cake and rice bran and subtracting 3 from 1 and 3 from 2 (ignore the signs):

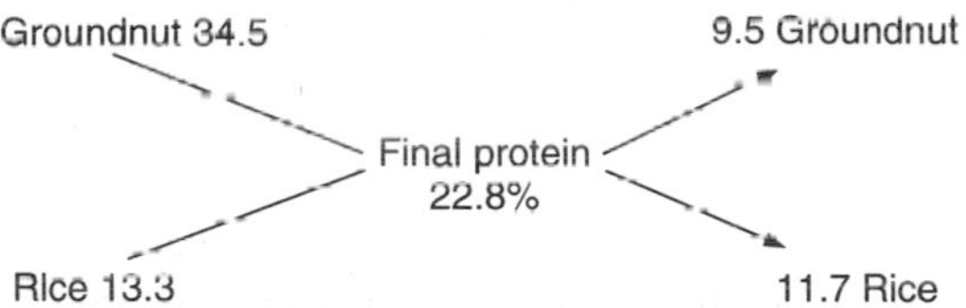

The actual amount of ingredient to be used is:

$$\text{Groundnut} = \frac{9.5}{11.7 + 9.5} \times 100 = 44.8\%$$

$$\text{Rice bran} = \frac{11.7}{9.5 + 11.7} \times 100 = 55.5\% \text{ (diet a)}$$

These ingredients, however, should constitute only 90% of the mix. Therefore the amount of groundnut cake and rice bran to be incorporated in the final mix is 40.3% (44.8 × 0.9) and 49.7% (55.2 × 0.9) respectively.

In the above manner three other combinations of ingredients can be obtained to make up a diet containing 10% fish meal and with an overall protein level of 26.0%.

These are:

47.3% groundnut cake	+ 42.6% maize	(diet b)
25.5% soybean	+ 64.5% rice bran	(diet c)
31.6% soybean	+ 58.4% maize	(diet d)

If you were to calculate the lipid level for the different combinations, for the four formulae, it would be

Fish meal + rice bran + groundnut cake	= 8.06% (diet a)
Fish meal + maize meal + groundnut cake	= 9.0% (diet b)
Fish meal + rice bran + soybean meal	= 2.48% (diet c)
Fish meal + maize meal + soybean meal	= 3.64% (diet d)

This shows that the lipid level of 8%, as specified in the original diet needed, is achieved by coincidence by the combination of fish meal, rice bran and groundnut cake. However, if the specification was to formulate a diet of 26% protein and 5% lipid then none of the above formulae would be acceptable. This could be reformulated by using the ingredient combinations (except fish meal, of which 10% has to be included in the final diet) by considering the ingredient mixes that were to yield 26% protein as a single ingredient and constructing suitable crosses similar to that used in balancing the protein level originally.

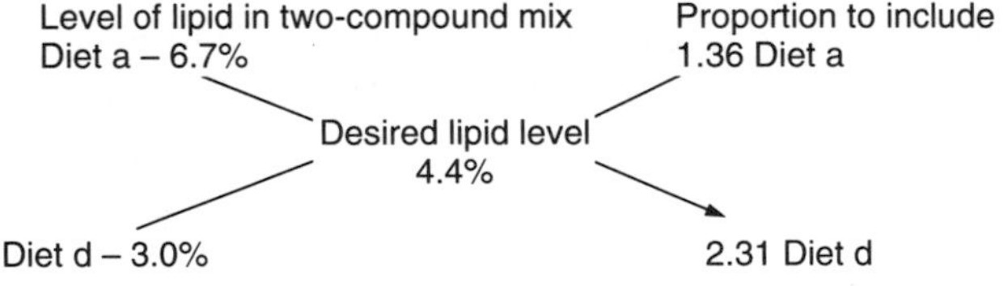

Thus the proportion of the ingredients in diet a to be used in the final formula would be

$$\frac{1.36}{1.36+2.31}\times 100 = 37.1\%$$

and similarly the proportion of ingredients in diet d to be used would be

$$\frac{2.31}{1.36+2.31}\times 100 = 62.9\%$$

Accordingly the final formulation would appear as

Fish meal 10% × 100%	= 10.00%
Rice bran 49.7% (from a) × 37.1%	= 18.4
Groundnut cake 40.3% (from a) × 37.1%	= 15.0
Maize meal 58.4% (from d) × 62.9%	= 36.7
Soybean meal 31.6% (from d) × 62.9%	= 19.9

The contribution of each of the ingredients to a diet of 26% and 5% lipid will then be

	Inclusion level (%)	Percentage contribution to Protein	Lipid
Fish meal	10.0	5.5	0.6
Rice bran	18.4	2.45	0.44
Groundnut cake	15.0	5.18	2.06
Maize meal	36.7	3.60	1.65
Soybean meal	19.9	9.31	0.26

The fact that amino acid and fatty acid requirements were not balanced for this diet does not imply that this is not always the case. In this instance we assumed that a correct balance will be provided by the ingredients utilized, and for most diets used in semi-intensive systems this assumption is acceptable. However, for the formulation of practical diets for intensive culture, such an assumption is not appropriate.

(b) Linear programming

Another mathematical technique available to nutritionists for selecting the best combination of feed ingredients to formulate diets at the least possible cost is linear programming. It suffices to mention that the information necessary for feed formulation using linear programming includes those mentioned earlier, and

- nutrient content and DE or ME of ingredients;
- unit price of feedstuffs including vitamin and mineral mixtures;
- any other additives to be used in the feed; and
- minimum and maximum restriction on the amounts of each ingredient in the feed.

Least-cost linear programming software for diet formulation is readily available, the price varying with the sophistication required. A commonly used spreadsheet such as Lotus 1-2-3 can also be utilized for formulating feeds, incorporating a smaller number of variables. It should be noted that least-cost feed formulation is not always practical for small-scale aquaculturists using on-farm feed manufacture facilities where the choice of ingredients available is limited.

7.6 FEED MANUFACTURE

Feed formulation is followed by manufacture. The technology will differ, at least in details, depending on the type of feed to be manufactured. The feed types in aquaculture can be schematically represented (Figure 7.2).

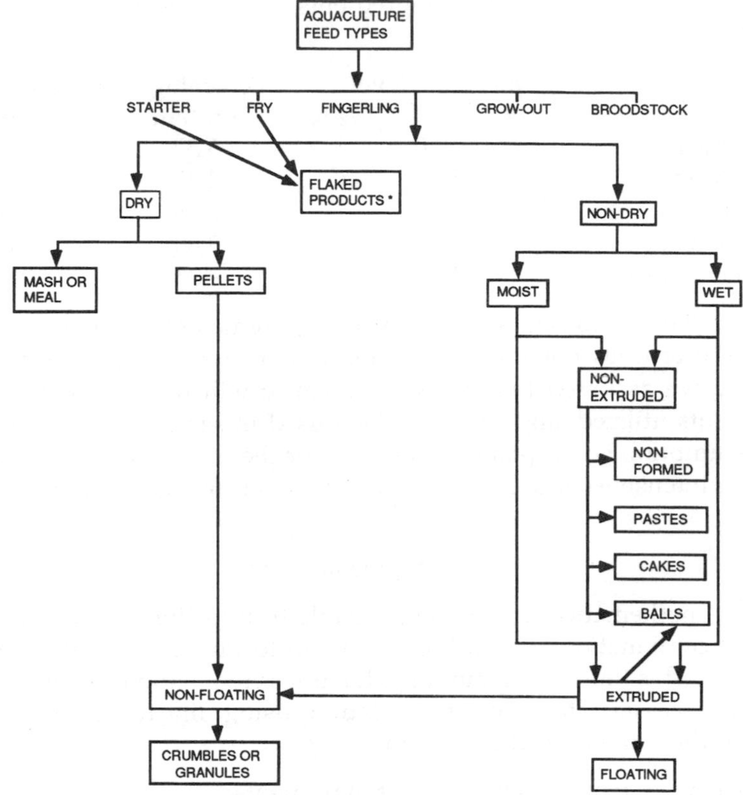

Figure 7.2 Feed types used in aquaculture.

7.6.1 Types of feeds

Feeds can be classified based on the stage of the life-cycle at which they are targeted. Accordingly, there are starter feeds, fry feeds, fingerling feeds, grow-out feeds and broodstock feeds. This does not necessarily imply that for the production of a cultured species all five types of feeds are required or used. Starter and fry feed may be the same, while grow-out and broodstock feed may also be the same. In addition, product quality feeds are used in many cultures to increase the quality of the final saleable product.

(a) Starter feeds

Starter feeds are given to first-feeding fry or larvae when their endogenous food supply (yolk) is exhausted or about to be exhausted. The transformation from an endogenous to an exogenous food supply is crucial to all aquatic organisms. It is the stage that excessive mortality occurs owing to the inability of larvae to adapt to an exogenous food supply (Chapter 5). Starter feeds should be nutritionally complete, easily digestible, and be of the appropriate particle size. Starter feeds differ in composition and type depending on the nutritional requirement and size of the organism at first feeding. Starter feeds are generally in the form of fine crumbles or flakes. In many cases, most notably shrimps and some cultured marine finfish, the first feeding is based on live foods rather than on formulated starter diets.

(b) Fry feeds

Fry is the term used for the unmetamorphosed young stage in the life-cycle of finfish. Fry feeds generally contain higher levels of protein because it is believed that the protein and energy requirements on a unit mass basis are much higher in the early stages of life. The highest relative weight gain (specific growth) is achieved in the fry stages, and it is important to ensure that full growth potential is realized during this stage of development for all culture systems. Fry feeds are generally in the form of flakes or crumbles.

(c) Fingerling feeds

The fingerling stage is defined as that between metamorphosis to a growth of about 10–20 g. Fingerling diets vary from crumbles to pellets, depending on the species cultured and their size. Fingerling diets also tend to contain less protein and energy than fry and starter diets.

(d) Grow-out feeds

During the grow-out stages, weight increase usually occurs at a uniform rate, decreasing slightly as the fish increases in weight. Accordingly, the nutritional requirement during grow-out is rather uniform. For grow-out diets, it is important to ensure that the protein in the feed is used mostly for growth and not for metabolic activity. During grow-out the biomass of the culture system increases considerably, and consequently the total quantity of feed needed is at its maximum. Therefore, the greatest feed cost-saving could be made at this stage of the culture.

(e) Broodstock feeds

During sexual maturation, somatic growth slows down and gonadal growth accelerates until spawning. Feed quality during this period is known to affect the quality of offspring. Therefore, broodstock feeds should be formulated to meet the nutritional needs of the reproducing fish. Broodstock nutrition (Chapter 6) is not that well known for most cultured species. Generally, formulations for broodstock feed simply contain higher levels of protein.

(f) Product quality feeds

These are feeds which are used to increase the market quality of the product. These feeds are used near to harvesting to enhance the consumer acceptability of the final product. An example of this is carotenoid supplementation for some species to enhance flesh colour. Product quality feeds are used mostly in respect of high-value species, and even then not too commonly. This is an area of diet formulation where there is scope for development.

7.6.2 Forms of feeds

As evident from Figure 7.2, feed types can take numerous forms. However, they basically fall in to one of two general forms: dry and non-dry (moist).

(a) Dry feeds

Dry feeds are generally made from dry ingredients or from mixtures of dry and moist ingredients. However, even though it may be implied by the name, these feeds are not entirely devoid of moisture, generally containing 6–10% water, depending on the environmental conditions. Dry feeds may be subdivided further. Feeds that are simple mixtures of dry ingredients are termed 'mashes' or 'meals'. Dry feeds that are compacted into a defined shape, generally by a mechanical means, are called pellets. Depending on the formulation and compacting techniques these diets may be floating or non-floating (sinking) in water.

In conventional steam pelleting a mixture of dry ground ingredients is forced together to form large stable particles by the application of heat, moisture and mechanical pressure. There is little, if any, cooking involved, and the chemical nature of the material remains relatively unchanged.

(b) Non-dry feeds

Non-dry feeds can be either wet or moist. Generally, wet feeds are those made from wet ingredients such as 'trash fish', slaughterhouse waste, undried forage, etc., and contain 45–70% moisture. Moist feeds, on the other hand, are made from mixtures of wet and dry raw materials, or from dry ingredients to which water is added. The moisture content of these feeds ranges from about 18% to 40%. However, the distinction between these two forms is not great. Non-dry feeds, both moist and wet, may be either extruded to form a pellet or non-extruded (non-formed), resulting in balls, cakes, etc. During the process of extrusion, the raw material is forced down a tapering shaft and through a die plate under pressure in an atmosphere of steam. This effectively exposes the feed to controlled conditions of high temperature, pressure and moisture. The extrusion process cooks the carbohydrates, primarily causing the starch granules to gelatinize, and increasing the binding quality upon cooling.

By altering the extrusion process, pellets of variable floating and/or sinking rates may be produced. This is achieved by trapping air instead of water in the pellet as it leaves the high-pressure chamber. These air pockets are then stabilized by the rapidly cooling gelatinized starch.

Extruded diets are known to be better utilized. In certain instances extrusion of ingredients with high starch content, prior to pelleting, is reported to have had an influence on the utilization of the diet (Pfeffer *et al.*, 1991).

7.6.3 Types of machinery

New (1987) summarized the types of machinery/equipment needed for the production of the various types of feeds for the aquaculture industry, providing adequate descriptions of each. The basic apparatus required are: grinders, mixers, elevators and conveyors, mincer/extruders, cooker/extruders, cooler/dryers, fat sprayers, steam boilers. The engineering and design of feedmills vary, in detail in design as well as in the production process, and these aspects are beyond the scope of this book.

7.6.4 Manufacture

Feed manufacture offers a challenge in that the processes involved are expected to stabilize and maintain the nutrient components under the conditions encountered during shipping, storage and feeding. Fish feed manufacture on an industrial scale is complex and it is beyond the scope of this book to consider it in detail. However, it is important that an overview of the steps and processes in diet manufacture is considered briefly.

Basic steps in diet manufacture are schematically represented in Figure 7.3. These steps are generally termed grinding, mixing and pelleting, and are covered in reasonable detail below.

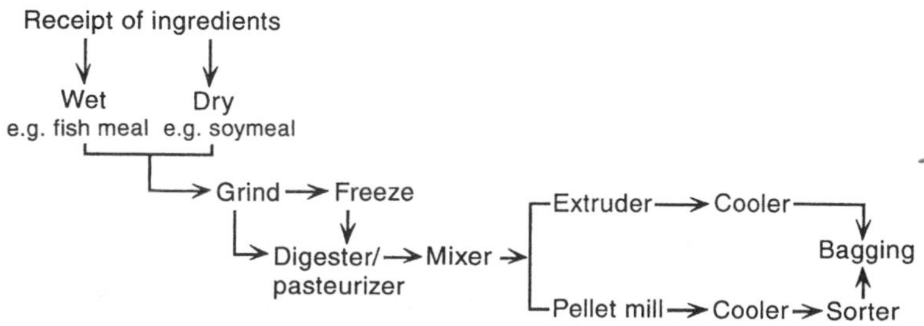

Figure 7.3 The basic steps involved in aquafeed manufacture.

(a) Grinding

Grinding reduces particle size and increases the surface area of ingredients, thereby facilitating mixing, pelleting and digestibility. The types of grinder mills used may vary. The most commonly used mills for grinding are plate mills and hammermills. In a plate mill the feed is sheared between two roughened plates, one or both of which are 'rotating'. Generally, plate mills are not suitable for aquafeeds as they are incapable of grinding the particles finely. In a hammermill the grinding chamber consists of series of non-moving or swinging hammers attached to a rotor. The hammer breaks up the incoming material, which is often forced through a steel screen (Figure 7.4). The steel screens are available with different hole sizes depending on the desired particle size.

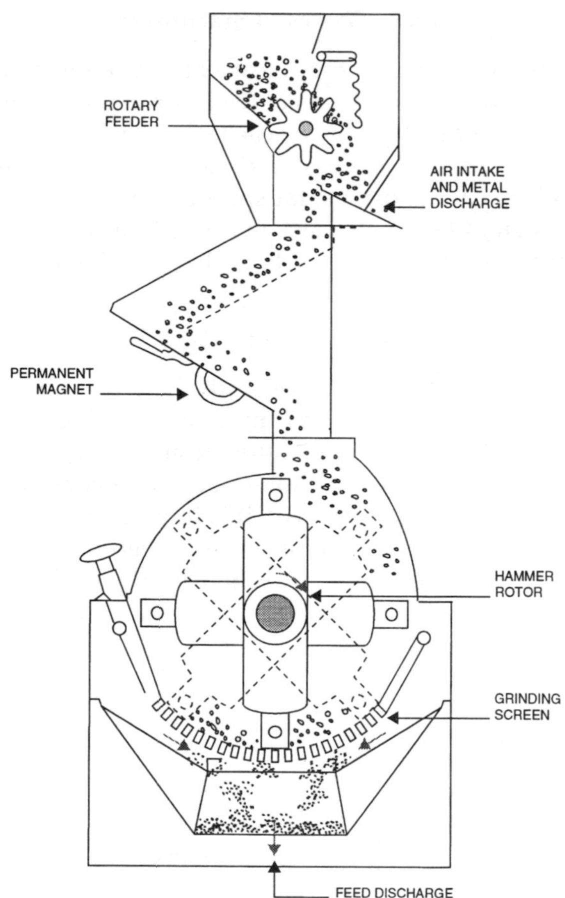

Figure 7.4 A cross-sectional view of a hammermill. (From New, 1987.)

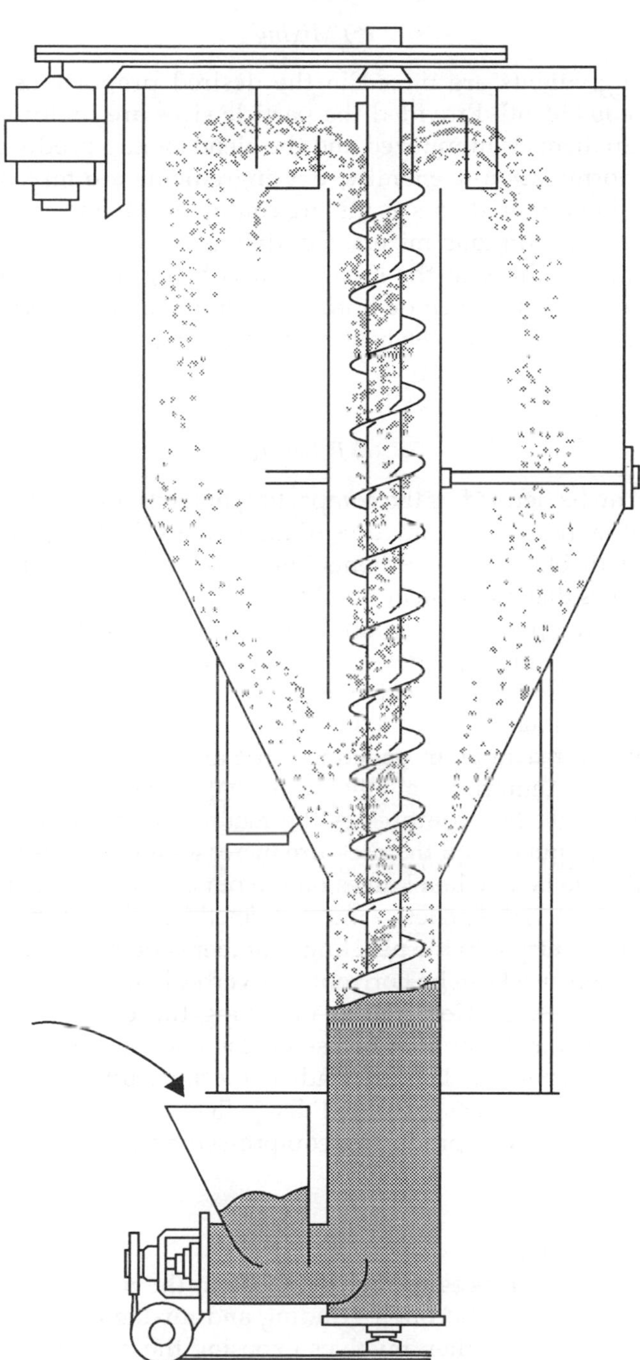

Figure 7.5 A vertical mixer used in feedmills.

(b) Mixing

Ground ingredients are mixed in the desired proportions to form a homogeneous blend. Provided the particle sizes are uniform, segregation of ingredients is minimized and the blend should produce pellets of a similar formulation. Generally, dry ingredients are mixed first, followed by liquid ingredients as mixing continues. Mixing can be done in batches or in continuous mixers. Continuous mixers are such that the material moves through the mixer as it is being mixed. The types of mixers used vary and include horizontal ribbon mixers, vertical mixers and turbine mixers. A sketch of a typical vertical mixer is shown in Figure 7.5.

(c) Pelleting

Pelleting can be defined as the compacting of feeds formed by extruding individual ingredients or mixtures of ingredients. Pelleting converts the homogeneous blend of ingredients into durable forms having physical characteristics that make them suitable for feeding. The arrangement of a typical pelleting plant will essentially consist of a supply bin, pellet mill, cooler, pellet crumbler, sifter and collector(s) (Figure 7.6). Maximum compaction is needed for good pellet quality. However, compaction and capacity are antagonistic and hence an economic balance has to be reached in the machine. In pelleting a die and roller assembly is used. The mash is fed uniformly into the die or shaft and comes in contact with the roller, which then compresses the material into the die holes. The process is continued until the pressure increases and moves the material through the die hole, when knives on the outside of the die cut or break off plugs of material – pellets – into a defined length (Figure 7.7). Steam at the point of entry into the pelleting chamber should be as dry as possible. The coolers used can be horizontal or vertical, depending on the type of plant (Figure 7.8). Coolers aid in drying the pellets, and in some instances a dry air stream may be used to facilitate the process.

Three basic types of pellets are made in the aquaculture industry, these being compressed pellets, extruded dry pellets and semimoist extruded pellets. The process of producing compressed and semimoist pellets is detailed below.

Compressed pellets

As in all pelleting processes the first steps in the manufacture of compressed pellets is the thorough grinding and mixing of the ingredients. Compressed pelleting then involves exposing the mixture to steam for 5–20 s, obtaining 85°C and 16% moisture, followed by forcing the mix through a metal die, and is summarized overleaf.

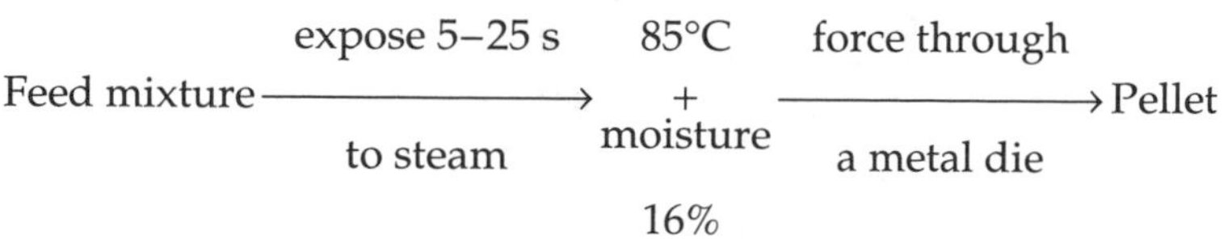

1. Pellet mill mash supply bin
2. Pellet mill
3. Vertical cooler (could be horizontal)
4. Pellet crumbler
5. Bucket elevator
6. Scalper
7. Cyclone collector
8. Fines from cyclone returned either to mash supply bin or to pellet mill feeder
9. Overs top scalping screen flowed either to pellet bin or to crumbler for re-crumbling
10. Overs bottom scalping screen flowed either to crumble bin or to mash bin or pellet mill feeder for re-pelleting
11. Fines returned either to mash supply bin or to pellet mill feeder for re-pelleting

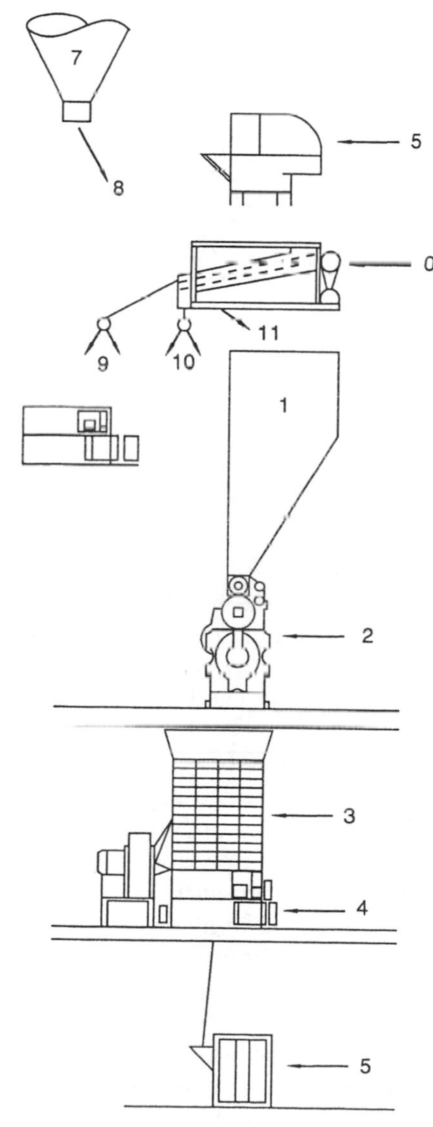

Figure 7.6 The arrangement of a typical pelleting plant.

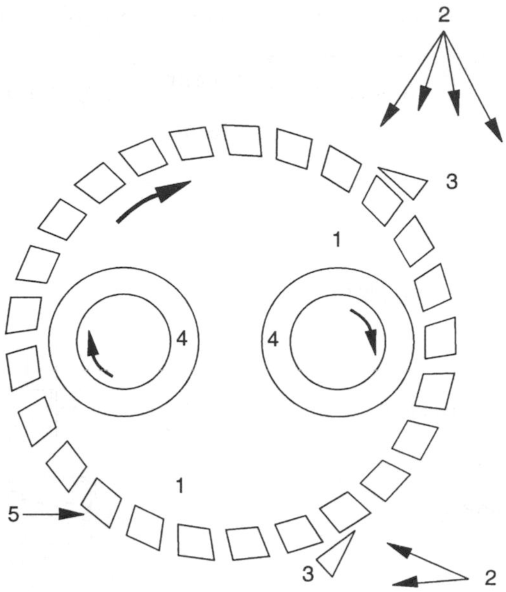

Figure 7.7 A typical die and roller assembly used in pelleting. 1, Meal or mash; 2, pellets, 3, blades; 4, rollers, 5, die.

The combination of heat, moisture and pressure compresses the mixture into a compressed pellet in which the starch is gelatinized. This method is also known as steam pelleting. Pellet quality is influenced by the fat level, moisture and humidity. Very low (< 2%) or high (>10%) fat levels are not desirable. Low fat levels make the pellet unduly hard, whereas high fat levels make pelleting difficult. Excessive moisture results in soft pellets, and insufficient moisture results in crumbly pellets.

Extruded dry pellets

Formation of extruded dry pellets involves the use of different physical conditions and dies to those employed for compressed pellets and results in a very different product. Here the temperature is increased to about 125–150°C in a pressurized conditioning chamber (20 s) and the moisture is increased to about 20–24%, enhancing gelatinization of starch. This results in the mixture being made into a dough-like consistency, which is then forced through a die at high pressure.

As the pellet leaves the die, the fall in pressure causes the trapped water (which is in liquid form because of the high pressure) to evaporate, and the gelatinized material expands, forming air pockets. When cooled, the density is generally about 0.25–0.3 g/cm^3, so that the pellets float or sink only slowly. By adjusting the ingredient combination and the cooking conditions, floating or sinking pellets can be produced.

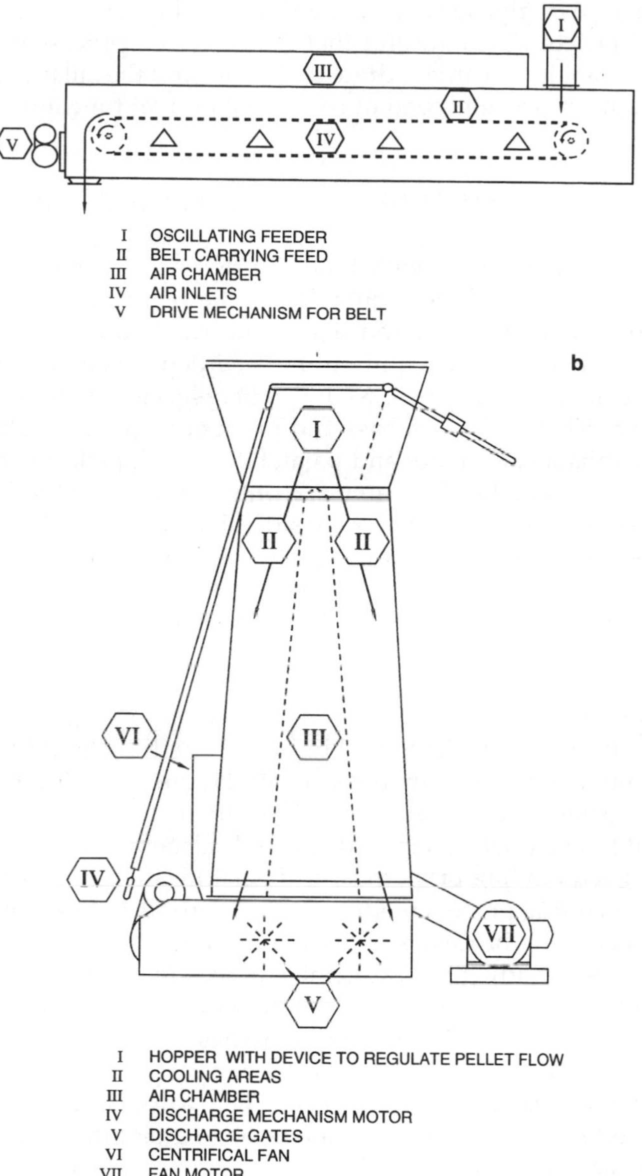

Figure 7.8 A horizontal (a) and a vertical (b) cooler used in feed plants. (From New, 1987.)

Hilton *et al.* (1981) made a detailed study on the effect of extrusion processing and steam pelleting diets on pellet durability and the physiological response of rainbow trout to the two types of pellets. However, the study of Hilton *et al.* (1981) indicated that liver–body weight ratio and percentage liver glycogen were significantly higher in trout reared on extruded pellets. This indicates that the extrusion process may increase the bioavailability of carbohydrate in the diet and the enlarged livers and increased liver glycogen content could impair liver function.

7.7 WATER STABILITY OF FEEDS (PELLETS)

Apart from being able to withstand the rigours of handling and transportation, commercial feeds should also be relatively stable in water, minimizing disintegration and loss of nutrients due to leaching. The degree of water stability required in a feed depends on the species cultured and its feeding habits. Many finfish species such as salmonids, channel catfish, tilapias, sea bass and grouper respond quickly to a diet that is of suitable pellet size and palatability. Such pellets need to retain their physical stability in water for only a few minutes. In contrast, shrimp feeds are left uneaten for considerable periods of times, therefore requiring greater integrity in the aquatic environment.

Water stability of pellets is influenced by a number of factors, foremost amongst which are diet composition, the manufacturing process and nature of the binders used.

- Composition of diets. The proportion of ingredients which are difficult to grind or have no binding properties should be kept to a minimum (e.g. rice bran, bone meal). Hydroscopic ingredients such as salt, sugar and molasses absorb water, making the feed moist and crumbly even before being dispersed. Generally, starchy products have good binding properties, and gelatinization of the starch in the manufacturing process renders the final product more stable.
- Manufacturing processes. Grinding is common to all manufacturing processes. Grinding increases the surface area of a feed and thereby permits more space for steam condensation during the conditioning process, resulting in harder and more desirable pellets. Hastings and Higgs (1980) studied the effect of processing parameters on the water stability of a standard catfish feed. Their work clearly demonstrated that certain processing parameters have a significant influence on water stability of pellets (Table 7.11).
- Binders. According to Stivers (1970) there are at least three actions by which binders increase the hardness and water durability of pellets. Binders reduce the void space in the mix and thus provide a more compact and durable pellet. Some possess adhesive action and by their ability to stick particles together provide durability to the pellet. Others are thought to undergo changes during the pelleting process and exert a chemical action which changes the nature of the feed

Table 7.11 Effect of processing parameters on water stability of a standard catfish feed

Processing parameter	*Water stability (per cent DM retained at 10 min in running water)*
Unground, no steam, thin die	21.5
Unground, no steam, thick die	23.3
Unground, added steam, thin die	32.3
Unground, added steam, thick die	78.9
Ground, no steam, thin die	65.8
Ground, no steam, thick die	74.5
Ground, added steam, thin die	84.9
Ground, added steam, thick die	88.0

Source: Hastings and Higgs (1980).

mixture, hence increasing pellet durability. Several studies have been carried out to evaluate different types of natural, modified or synthetic substances used as binding agents for aquafeeds. These have been reviewed by Huang (1989). As shrimp feeds require greater binding properties, most research on binders has been done on shrimp feeds. Results of a study by Dominy and Lim (1991) investigating 17 different binders are summarized in Table 7.12.

7.8 FEED STORAGE

A manufactured diet requires storage at least at the place of manufacture and on the farm. Feeds are composed of perishable biological material which deteriorates with storage. Therefore it is always desirable to minimize storage time.

Deteriorative effects during storage are caused by:

- oxidative damage;
- microbial damage;
- insect and or rodent damage/infestation; and
- other chemical changes during storage.

The causes are detailed below.

7.8.1 Moisture and heat

These processes of feed deterioration are accelerated by a variety of storage conditions. Of these, temperature and humidity can be singled out as the most important environmental factors that govern storage or shelf life. These factors affect the moisture content of the diet, the rate at

Table 7.12 Effect of different binders on the water stability of shrimp pellets after 8 h in seawater

Binder	*Type of binder*	*Dry matter retained (%)*	*Amount used (%)*	*Cost (US$ per tonne)*
EX-5819	Xanthan and locust bean gum	88.0^{a}	1.0	156.9
EX-5820	Xanthan and locust bean gum	87.5^{ab}	0.43	61.4
RE-9556/57	Carrageenin mix	89.8^{bc}	0.5	37.4
Gampro	Wheat gluten	86.0^{cd}	4.0	60.0
Aquabind	Ethylene/vinyl acetate copolymer	85.7^{cd}	4.0	72.0
Gampro-Plus	Modified wheat gluten	85.5^{d}	4.0	72.0
Pel-Plus 100	Mineral	85.5^{d}	3.0	23.0
Pel-Plus 250A	Mineral	85.3^{d}	4.0	13.7
Ameri-Bond 2000	Lignin sulphonate	85.3^{d}	2.0	3.7
BASFIN	Urea formaldehyde	85.2^{d}	1.0	13.2
Nutriflex 40 Mega	Collagen protein	85.0^{de}	0.25	4.7
RE 9556	Carrageenan mix	85.0^{de}	0.5	37.4
Aqua-Firm 2A	Urea formaldehyde	84.1^{ef}	1.0	17.0
D-357	Modified lignin sulphonate	83.9^{ef}	2.0	NA
AP-520	Plasma protein	83.3^{fg}	4.0	108.0
Nutri-Binder	Modified sorghum	83.0^{fg}	5.0	13.3
Aqua-Firm 1A	Urea formaldehyde	82.5^{g}	1.0	17.7

Values with the same superscript are not significantly different at the 5% level.
Source: Modified from Dominy and Lim (1991).

which chemical changes take place and the invasion and growth of moulds (fungi) and insects. Jones (1987) evaluated the influence of relative humidity on moisture content on feeds (Table 7.13), while Cockerell *et al.* (1971) investigated the influence of temperature and relative humidity on the growth of fungi on stored feeds (Figure 7.9) and Zuercher (1987) studied the relationship between ambient temperature and feed moisture content and the risk of infestation within storage feedstuffs (Figure 7.10). In addition, light and oxygen may also effect deterioration.

Stored feeds, regardless of the original moisture content, will reach an equilibrium with the atmospheric moisture content depending on the relative humidity. New (1987) considered that a safe moisture level is that

Table 7.13 Effect of storage relative humidity on the moisture level (%) in feeds

	Relative humidity (%)		
Feed	*25*	*55*	*80*
Pig starter	8.1	13.4	19.3
Swine grower	10.2	14.6	16.9
Cattle ration	8.1	13.7	17.7
Urea cattle supplement	8.9	13.6	17.6
Dehydrated alfalfa	6.8	12.0	16.3

Source: Jones (1987).

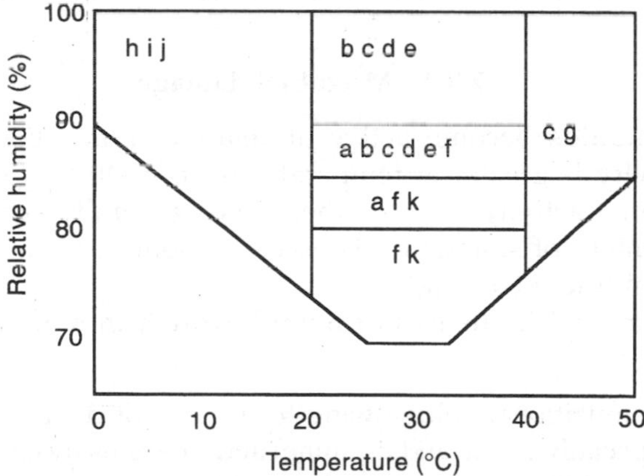

Figure 7.9 Temperatures and relative humidities at which species of fungi commonly found on stored feed materials may be most important. a, *Aspergillus candidus*; b, *A. Flavus*; c, *A. fumigatus*; d, *A. tamarii*; e, *A. niger*; f, *A. glaucus*; g *A. terreus*; h, *Pencillium cyclopium*; i, *P. martensii*; j, *Cladosporium sp.*; k, *Sporendonema sebi*. (From Cockerelil *et al.*, 1971.)

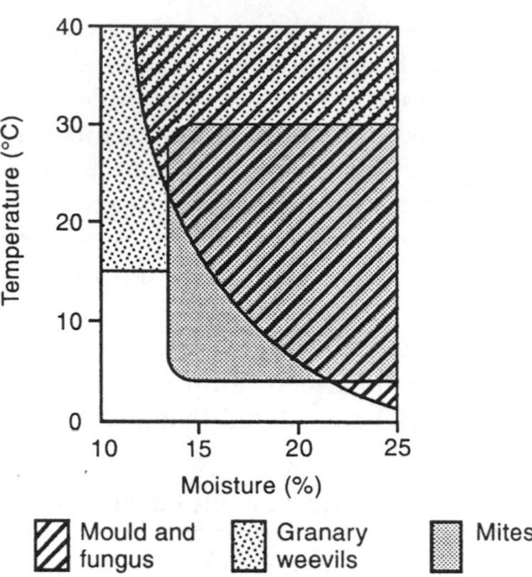

Figure 7.10 Relationship between ambient temperature and feed moisture content and the risk of pest infestation within stored feeds. (From Zuercher, 1987.)

which develops at a relative humidity of about 75%. However, in most tropical areas the relative humidity is much higher, resulting in feeds absorbing moisture during storage. Consequently, the feeds must be stored for a shorter time. High temperatures also affect feeds as temperature alone may oxidize vitamins (especially vitamin C) and will accelerate oxidative processes that have been initiated by microbial activity.

7.8.2 Microbial damage

Generally moulds become active at relative humidities above 70%. Fungal activity is greater at temperatures of 35–40°C (Cockerell *et al.*, 1971). Bacterial activity, on the other hand, generally occurs when the moisture content of storage feeds exceeds about 25%, this occurring at relative humidities of around 90%.

The detrimental influences of mould growth in stored feeds are as follows:

- Reduced nutritional value owing to the loss of dietary lipids, amino acids (especially lysine and arginine) and vitamins by enzymatic digestion (Jones, 1987). Fungi may also assist in the development of lipid ketonic rancidity and non-enzymatic browning (Cockerell *et al.*, 1971).
- Poorer flavour and appearance, making feeds 'lump' and less palatable.

- Certain moulds, in particular *Aspergillus flavus*, produce toxic metabolites or mycotoxins that could cause cancer. The most toxic of these is aflatoxin B. Feedstuffs which are particularly prone to infestation by *A. flavus* are groundnuts, cotton seed and copra (Chow, 1980).

7.8.3 Insect/rodent damage

Insects and rodents can cause considerable damage to feedstuffs. This may be directly through ingestion and contamination (faeces, body parts, *Salmonella*, etc.), or indirectly by producing heat and increasing moisture, thereby making the feedstuffs more susceptible to bacterial and fungal invasion.

7.8.4 Chemical changes during storage

The most commonly occurring chemical change due to storage is the breakdown of fatty acids in the feed resulting in rancidity. Generally, polyunsaturated fatty acids and pure lipids are more prone to oxidation. Rancid fats reduce palatability and can contain toxic compounds which inhibit growth. Carbohydrates can also ferment.

Chemicals produced in the degenerating feeds may reduce amino acid and vitamin availability, vitamin C being particularly susceptible.

7.8.5 Proper storage

Good feed storage should provide protection against high temperature, humidity, moisture and insect and rodent infestations. Feedstuffs should, as far as possible, be stored for a minimum length of time. Materials such as trash fish should be used immediately or kept frozen until used. New (1987) provided guidelines of maximum storage times for some common ingredients used in fish feed manufacture (Table 7.14).

Table 7.14 Maximum permissible storage time for selected feedstuffs

	Tropical zone	*Temperate zone*
Ground ingredients	1–2 months	3 months
Whole grain and oil cakes	3–4 months	5–6 months
Compounded dry feeds	1–2 months	1–2 months
Vitamin mixes (kept cool)	6 months	6 months
Wet ingredients	2–3 months	2–3 months
Frozen materials	2–3 months	2–3 months

Source: New (1987).

For proper feed storage there are a few, but very important 'dos and don'ts'. Generally, feeds should be stored in such a manner that the 'feed sacks' do not touch the floor or the side walls. The store needs to be 100% waterproof, a damp-proof storage facility being ideal. Stacks are best arranged in such a manner that there is ready access to the feeds purchased earliest. Larger stacks tends to reduce or minimize insect damage. However, these have a negative effect in that more heat tends to be generated, resulting in deterioration. Proper ventilation ensuring a continuous draught through the store is most desirable. Remember, storage never enhances feed quality, but proper storage reduces the rapidity at which a feed deteriorates.

CHAPTER 8

Non-nutrient diet components

8.1 INTRODUCTION

Generally, when the composition of a feed is considered, more often than not attention is drawn to those components which provide nutrition to the cultured species. A feed, however, contains some non-nutrients, some of which may even have a detrimental influence on the culture organism, and some of which may be inert and have no effect. Binders used in feeds fall into the latter category, as does the ash component in a feed. In this chapter we will consider non-nutrient diet components in fish feeds and their relevance to present-day aquaculture.

In the early days of feed development, animal products, particularly trash fish and fish meal, constituted the bulk of the feed ingredients used. However, these commodities are becoming increasingly expensive and scarce, particularly in the developing world where the thrust of aquaculture development is greatest (Chapter 10). Nevertheless, because of the relatively high protein requirements of fish and the need to provide the correct amino acid and fatty acid balance, it is generally accepted that a complete practical diet for fish and shrimps cannot be developed without incorporating animal products. However, in order to make aquaculture viable, feed costs need to be reduced. Accordingly, an increased amount of plant products is being incorporated into finfish diets. Some of the most commonly used plant products are soybean meal, cottonseed meal, wheat middlings and rice bran.

The use of plant materials, apart from their nutritional constraints, is not without problems. The most common problem is the presence of antimetabolites of various sorts in specific plant products. This is a subject which has been extensively studied, and the most authoritative work on it is that of Liener (1980). Most of this work is the result of research on plant products as animal feeds.

8.2 ANTIMETABOLITES

It should be evident by now that intensification of aquaculture can take place only with a corresponding development in feed technology.

Nature has endowed many plants with the capacity to synthesize a wide variety of chemical substances that are known to exert a deleterious effect when ingested by animals – antimetabolites. These substances are also referred to as antinutritional factors when they are present in feedstuffs and influence a metabolic pathway when introduced into and/or eaten by an organism.

Most of these substances are heat labile and are relatively easily destroyed by cooking. The effects on the organisms of ingestion of antimetabolites include death due to inhibition of growth, a decrease in food efficiency, pancreatic hypertrophy, hypoglycaemia, liver damage and other pathological conditions. The extent of manifestation of any of these depends on the quantity of antimetabolite(s) ingested, the animal species, size, age and physiological condition.

The antinutritional factors of plant products known to us can be divided into four major groups (after Tacon, 1987b).

Proteins	protease inhibitors, haemagglutinins
Glycosides	goitrogens, cyanogens, saponine, oestrogens
Phenols	gossypols, tannins
Miscellaneous	anti-minerals, anti-vitamins, anti-enzymes, food allergens, microbial/plant carcinogens, toxic amino acids

The specific endogenous antinutritional factors in plant products are summarized in Table 8.1. It is evident from the table that the antinutritional factors found in plants and plant products are diverse, a plant often having more than one endogenous antinutritional factor. Details of some of these antinutritional factors follow.

8.2.1 Individual antinutritional factors

(a) Protease inhibitors

Protease inhibitors are substances that have the ability to inhibit the proteolytic activity of certain enzymes, and are found throughout the plant kingdom, particularly among legumes such as soybean. There may be a number of types of protease inhibitors within a plant, differing in their molecular weight, structure (amino acid sequence) and specific action. In soybean, for example, as many as five protease inhibitors have been reported. Protease inhibitors basically act by binding with chymotrypsin and/or trypsin, rendering them inactive. They are also known to display an effect on metabolism of certain amino acids such as cysteine. Almost all protein inhibitors are heat labile and will be broken down when cooked.

(b) Haemagglutinins (lectins)

Haemagglutinins, or lectins, are proteins that possess a specific affinity for sugar molecules. As the name implies, they cause agglutination of

Table 8.1 Endogenous anti-nutritional factors present in plant foodstuffs

Foodstuff	*Anti-nutritional factor* [a]
Cereals	
Barley (*Hordeum vulgare*)	1,2,5,8,25
Rice (*Oryza sativum*)	1,2,5,8,13,25
Sorghum (*Sorghum bicolor*)	1,4,5,7,18,25
Wheat (*Triticum vulgare*)	1,2,5,8,11,18,22,25
Corn, maize (*Zea mays*)	1,5,8,19,25
Root tubers	
Sweet potato (*Ipomoea batata*)	1,19
Potato (*Solanum tuberosum*)	1,2,4,8,18,19,21
Cassava (*Manihot utilissima*)	1,4,25
Legumes	
Broad, faba bean (*Vicia faba*)	1,2,5,7,22
Chickpea, Bengal gram (*Cicer arietinum*)	1,4,5,8,11,25
Cowpea (*Vigna unguiculata*)	1,2,5,11,25
Rice bean (*V. umbellata*)	2
Grass pea (*Lathyrus sativus*)	1,9
Lima bean (*Phaseolus lunatus*)	1,2,4,5,7
Haricot, navy, kidney bean (*P. vulgaris*)	1,2,4,5,6,11,12,18,25
Mung bean, green gram (*P. aureus*)	1,5,6,11,13,25
Runner bean (*P. coccineus*)	1,2
Black gram (*P. mungo*)	1,5
Horse gram (*Macrotyloma uniflorum*)	1,2
Hyacinth, field bean (*Dolichus lablab*)	1,2,4
Lentil (*Lens culinaris*)	1,2,6,25
Lupin (*Lupinus albus*)	1
Field pea (*Pisum sativum*)	1,2,4,5,6,12
Pigeon pea, red gram (*Cajanus cajan*)	1,2,4,5,25
Sword, jack bean (*Canavalia gladiata*)	1,2,4,6
Velvet bean (*Stizobolium deeringianuum*)	1,22
Winged bean (*Psophocarpus tetragonolobus*)	1,2
Guinea pea (*Abrus precatorius*)	1,2
Carob bean (*Ceratonia siliqua*)	1,7
Guar bean (*Cyamopsis psoraloides*)	1
Alfalfa, lucerne (*Medicago sativa*)	1,6,8,12
Ipil-ipil (*Leucaena leucocephala*)	23
Oil-seeds	
Groundnut, peanut (*Arachis hypogaea*)	1,2,5,6,8,25
Rapeseed (*Brassica campestris napus*)	1,3,5,7,25
Indian mustard (*B. juncea*)	1,3,13,25
Soybean (*Glycine max*)	1,2,3,5,6,8,11,12,14,16,17,25
Sunflower (*Helianthus annuus*)	1,7,20,25
Cottonseed (*Gossypium* spp.)	5,8,10,12,24,25
Linseed (*Linum usitatissimum*)	4,8,13,15
Sesame (*Sesamum indicum*)	5,25
Crambe, Abyssinian cabbage (*Crambe abyssinica*)	3

[a]Compiled from the data of Kay (1979) and Liener (1980).
Anti-nutritional factor: 1, protease inhibitor, 2, phytohaemagglutinin; 3, glucosinolate; 4, cyanogen; 5, phytic acid; 6, saponin; 7, tannin; 8, oestrogenic factor; 9, lathyrogen; 10, gossypol; 11, flatulence factor; 12, anti-vitamin E factor; 13, anti-vitamin B1 (thiamine) factor; 14, anti-vitamin A factor; 15, anti-vitamin B6 (pyridoxine) factor; 16, anti-vitamin D factor; 17, anti-vitamin B_{12} factor; 18, amylase inhibitor; 19, invertase inhibitor; 20, arginase inhibitor; 21, cholinesterase inhibitor; 22, dihydroxyphenylalanine; 23, mimosine; 24, cyclopropenoic fatty acid; 25, possible mycotoxin (aflatoxin) contamination.
Source: Tacon (1987b).

red blood cells. Like protease inhibitors, plant haemagglutinins are also varied. Haemagglutinins are found mostly in seeds of higher plants, but can also be present in tubers and plant saps. Lectins reduce absorption of nutrients in the gut and may cause internal haemorrhage and reduce growth. Lectins are also heat labile.

(c) Cyanogens

The presence of a trace amount of cyanide is widespread in the plant kingdom and occurs mainly in the form of cyanogenitic glucosides. Relatively high concentrations of these molecules are found in certain grains, root crops and fruit kernels. Cyanogenitic glucosides are hydrolysed by specific glucosidase enzymes, when toxic hydrogen cyanide (HCN) is liberated. However, glucosidases are extracellular and gain access to the glucosides only after physical disruption of the cells.

(d) Gossypols

Gossypols are polyphenolic pigments indigenous to the genus *Gossypium* (cotton plant) and certain other members of the family Malvaceae. These pigments are found in discrete bodies known as pigment glands in all parts of the plant. At least 15 gossypol pigments or derivatives are known to exist in cottonseed oils and meals, of which six have been isolated. Gossypol is reactive and exhibits acidic properties. High levels of gossypol cause unfavourable physiological effects such as a reduction in succinic dehydrogenase and cytochrome activity. The reaction between gossypol and protein during processing reduces protein quality, especially by reducing the bioavailability of lysine. General symptoms of gossypol toxicity are depressed appetite and loss of body weight. It is known that gossypol reacts with iron to form the inactive ferrous gossypolate complex. As a result dietary iron, and sometimes other mineral salts, can be used successfully to counter the undesirable effects of gossypol in properly balanced rations for gossypol-sensitive species.

(e) Mimosine

In the 1970s the legume *Leucaena leucocephala* (called 'ipil-ipil' in Asia and 'Koa haole' in Hawaii) was recommended as a forage crop for livestock, and was also considered as suitable for incorporation into fish feeds. However, it is known that its use as a livestock feed is limited by the presence of an unusual amino acid – mimosine. Mimosine is thought to affect the production of thyroxine and hence the growth of the organism. It structurally resembles the amino acid tyrosine and it might function *in vivo* as an antagonist to this amino acid. The mimosine content of ipil-ipil can be reduced by soaking in water, making it suitable for incorporation into fish diets.

(f) Others

Some leguminous seeds (beans) also contain a lipoxygenase enzyme that catalyses the oxidation of polyunsaturated fatty acids to form hydroperoxides, which are responsible for bean flavour. These hydroperoxides and their secondary products can react with amino acids or proteins and certain vitamins to lower the nutritive value of the feed (Gardner, 1985).

In spite of the presence of antinutritional factors, ingredients which have been referred to earlier have been effectively incorporated into experimental fish feeds, and in some instances, particularly in the case of soybean, in commercial diets. The level of potential inclusion is dependent on the stage of development and the species cultured as well as the pretreatment of the ingredient. Wee (1991) reviewed the results of experimental studies on utilization of these ingredients in fish feeds for tropical species, and concluded that with proper pretreatment most can be utilized to replace 20–30% of fish meal protein without compromising growth.

8.3 BINDERS

Often substances are added to feeds to improve the durability, i.e. to preserve the physical nature/form of a feed and/or to improve the stability of a feed in water. Such substances are referred to as binders, and generally do not contribute to the nutrition of the target species. However, some natural ingredients, in particular root crops which are high in carbohydrate (e.g. potato or tapioca starch, sago, sticky rice, etc.), are used in aquaculture feeds for their binding properties. The pelleting processes make some of these carbohydrates more digestible and hence available to fish to some degree.

Many substances are known to increase the water stability. Some of these are specialist chemicals and others are natural products, raw or refined. Some of the specialist chemicals are guar gum, carrageenin, agar, various forms of starches, carboxymethylcellulose, alginates, bentonites and hemicelluloses, etc. Substances such as lucerne meal, salt, molasses, bone, brewer's grain and whey, on the other hand, tend to act antagonistically to pellet stability of dry pellets. These substances are used to increase stability in moist feeds.

8.4 ACCIDENTAL CONTAMINANTS

Accidental contamination of feedstuffs (and hence formulated feeds) can occur for one of three reasons. These are:

1. gradual diffusion from environmental contaminants of persistent chemicals and their subsequent uptake and bioaccumulation in the food chain;

2. short-term higher level contamination originally from industrial accidents and waste disposal; and
3. contamination of the various additives used in manufacture.

The first two sources of contamination contribute to contaminants in natural fish populations, resulting in biological effects, and in rare cases death, in people who eat such fish. One of the most cited cases is the Minamata Bay incident in the 1970s in Japan. A number of deaths occurred as a result of consumption of tuna inhabiting the bay, and were attributed to excessive amounts of mercury in the tuna flesh.

Unfortunately, research on accidental contaminants has centred around study of fish populations in polluted waters and the direct effect of the commonly occurring organic pollutants and heavy metals on fish. A summary of such findings is given by Hendricks and Bailey (1989). Very little is known about contaminants (accidental) in aquaculture feeds, and even less about the effects of such feeds on the cultured organisms and the long-term influences on consumers. Very often sun-dried fish meal tends to be contaminated with sand and salt, resulting in much variability in the quality of fish meal from various sources. The degree of contamination obviously has a direct influence on the growth and production of the target species.

Tacon and De Silva (1983) studied the mineral content of a series of commercial diets used in Europe. These authors found that large differences existed in the mineral composition, particularly the trace elements Fe, Zn, Cu, Mn, Co and Pb, between different feeds. All macroelements were found in excess of the recommended nutrient requirements, as were the trace elements, for which dietary requirements are known.

8.5 NON-NUTRIENT, NON-TOXIC DIETARY COMPONENTS

There are certain constituents of diets which have no apparent nutritive value and are not known to be toxic. The two major components which come under this category are fibre and ash. Where, however, shrimp and other shellfish processing waste is used as a feedstuff, chitin also falls within this category.

8.5.1 Fibre

Finfish have no capacity to digest fibrous material such as cellulose and lignin. This is thought to be the case for all animals. However, animals such as ruminants and other mammalian herbivores have evolved a symbiotic relationship with cellulose-digesting microorganisms. These microorganisms are harboured in a specialized section of their gut, where they digest cellulose, releasing its constituent molecules for absorption of use in the metabolic processes. Rimmer and Wiebe (1987) demonstrated the presence of volatile fatty acids, a phenomenon associated with fermentation, in the gut of two fish species of the genus *Kyphosus*. These species inhabiting coral reefs are known to feed on algae, and the observations of Rimmer and Wiebe (1987) provide indirect evidence of the ability of these species to digest cellulose utilizing a mechanism similar to that documented in ruminants.

All plant ingredients contain a certain amount of fibre. A certain amount of fibre in feeds permits better binding and moderates the passage of food through the alimentary canal. However, it is not desirable to have a fibre content exceeding 8–12% in diets for fish, as the increase in fibre content would consequently result in the decrease of the quantity of a usable nutrient in the diet. When the fibre content is excessive it results in a decrease in the total dry matter and nutrient digestibility of the diet resulting in poor performance. Reduced digestibility will also have indirect influences on the culture system owing to increased levels of faecal output affecting the water quality and hence performance. However, suitable plant material can be included in diets as fillers, and this is often done in experimental studies.

8.5.2 Ash

Ash is a heterogeneous group of materials, including the non-combustible inorganic components of feedstuffs and diets. In feeds, ash may also include contaminants such as silica (sand grains). A small proportion of ash is digested, but like fibre all feedstuffs and practical diets will contain ash. In practical diets, the ash content should be kept low, and it is not desirable for the ash content to exceed 12%. Ash *per se* is not known to be harmful to cultured organisms. In a diet, however, the ash content is increased at the expense of a nutritive component. Like fibre, the ash content affects the digestibility (total and nutrient) of diets, and would indirectly influence the culture.

8.6 STEROIDS AND STEROID-LIKE SUBSTANCES

Steroids are a large group of compounds which have a variety of roles, including (and the best documented of which are) reproductive hormones. Steroids are derived from acetyl coenzyme A via cholesterol. Steroids affect the physiology of fish in numerous ways, and a treatment of this subject is beyond the scope of this book. Readers should refer to Fostier *et al.* (1983) for details. In aquaculture steroids are administered, often with the diet, to induce sex reversal (reviewed by Hunter and Donaldson, 1983). The quantities required to be administered orally to bring about such effects far exceed the quantities found normally in fish diets. Steroids and steroid-like substances occur in all aquafeeds. In the preparation of fish meal, whole fish and/or the by-products of the fish processing industry are used. Consequently, the fish meal and fish diets contain both androgens and oestrogens, the proportions differing depending on the raw material used for fish meal manufacture. Steroid-like substances known as phyto-oestrogens occur in a variety of plants (Stob, 1983); some of them occur naturally and their occurrence is known to rise when the plant is attacked by fungi.

Until now, nine steroids have been assayed from commercial fish diets including two phyto-oestrogens (Pelissero and Sumpter, 1992). However, the main ones assayed are androstenediones and 17α, 20β-dihydroxy-progesterones (Feist and Schreck, 1990). The quantities of steroids in fish diets range between several nanograms (ng) to several hundred ng per 100 g of feed. The phyto-oestrogens or the 'steroid-like' compounds originate from agricultural products used in fish diets.

The effects of steroids and steroid-like substances occurring normally in feeds on target species is not well studied. Pelissero and Sumpter (1992) concluded that, because steroids enhance vitellogenesis, and as vitellogenesis places a very considerable metabolic demand on fish, it is likely that oestrogenic contamination of male or immature fish by xeno-biotic oestrogenic compounds would have an impact on metabolism in these fish, this effect finally resulting in negative repercussions on growth, health and reproduction.

CHAPTER 9

Feeding of cultured finfish

9.1 INTRODUCTION

It has been stressed throughout this book that feed cost is one of the highest, if not the highest, recurring cost in intensive and/or semi-intensive aquaculture. Therefore, in order to make culture operations viable, it is imperative that the diet/feed presented to the target species is not only nutritionally balanced and easily utilized, but also readily ingested whenever it is made available to the organism.

The feeding behaviour of animals is complex. The acts of feeding and ingestion are the final result of a number of interacting factors between the animal and its environment. The senses and the hormonal systems of the animal play an important role in these interactions. In addition, there are social interactions which determine feeding behaviour patterns. Feeding behaviour may be associated with environmental cues such as time of day, tidal amplitude and season, amongst others. In the case of predatory animals, the aggregatory pattern(s) of the prey play an important role in the feeding responses. Needless to say much has been written on the feeding behaviour of different animals with general theories such as the 'optimum foraging theory' being developed.

From a zoological point of view, most fish species are specialized to feed on specific food items, with daily and seasonal rhythms in the feeding cycle. However, aspects of feeding behaviour that interest aquaculturists are quite different from what is commonly presented about feeding behaviour in general zoology and animal behaviour texts. Essentially, what is of importance to an aquaculturist are those elements of behaviour that can be utilized, directly or indirectly, to enhance production. Feeding techniques have two general aims: firstly to encourage rapid and positive consumption, thus increasing feed ingestion, preventing the potential for leaching of water-soluble nutrients and reducing wastage; and secondly to minimize the metabolic energy expenditure associated with feeding, thus giving a greater potential for growth.

The optimal feeding response and performance of cultured species do not depend only on the provision of a suitable diet. The culture environment also influences the feeding response and the performance. It is well documented that when the ambient water temperature is below or above the tolerance range of a species it tends not to feed. Similarly, low oxygen levels in the water influence feed intake. Apart from these, pollutants will have varying influences on food intake. However, one expects the level of pollutants in a cultured environment to be far below the levels which are likely to elicit such unfavourable responses. Considerations on influence of pollutants on feeds and feeding are outside the scope of this book and will not be dealt with.

9.2 APPETITE AND SATIATION

Appetite is the desire for food, and satiation is the fulfilment of this desire. Appetite and satiation are linked. This linkage is shown schematically as:

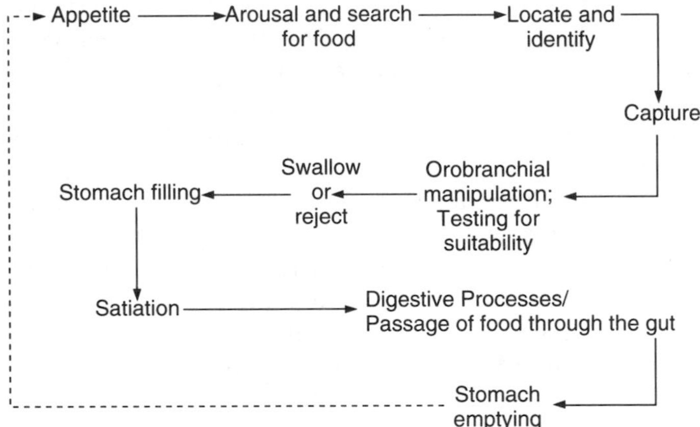

Therefore, appetite and satiation are important to culturists because of the need to ensure that the correct feeding regimes are employed.

Appetite is known to be controlled by metabolic, neurophysiological and hormonal mechanisms (Holmgren *et al.*, 1983). The responses are thought to involve hypothalamic centres of the brain, these being primarily stimulated by gut fullness and/or metabolite levels in the circulatory system. It is also believed that fish adjust food consumption to satisfy the metabolic energy requirements (Jobling and Wandsvik, 1983). However, more recently it has been demonstrated that, once minimum energy requirements are satisfied, the daily consumption, at least in some species, may be related to dietary quality such as the protein content (De Silva and Gunasekera, 1989). Apart from these physiological parameters, learning and social interactions may also influence consumption. In one of the earliest experimental studies on finfish, Brown (1946) was able to detect the influence of social interactions in brown trout. Here, larger animals were thought to exert psychological effects on the smaller individuals, inhibiting their feeding responses.

In fish culture most matching of feeding regimes to appetite and the derivation of feeding tables has been based on studies on growth, FCR and on gut evacuation rates. It is currently generally accepted that stomach distension inhibits appetite while gastric evacuation excites appetite.

In culture systems, the organism may be stressed in numerous ways, primarily because of deterioration in water quality, handling and social interactions. This stress can suppress appetite, the extent of which depends on the magnitude and duration of the stress.

Although it is known that stress factors influence appetite, very few quantified data on this phenomenon are available. Knights (1985) determined the effect of temperature change on appetite. Eel fingerlings were subjected to a low-temperature shock (LTS) by lowering the temperature rapidly to 12°C from 25°C. Food was then offered and the temperature brought up to 25°C. The amount consumed was determined at regular intervals and compared with control values. The results of this experiment are shown in Figure 9.1. The figure shows that return of appetite was markedly delayed following a LTS.

A form of stress stemming from social interactions is the development of social hierarchies. These hierarchies are a common phenomenon in the animal kingdom, and fish are no exception, although they are not always found to a degree noticeable to the observer. Also, hierarchical effects may be expressed to different degrees during the life-cycle. It is generally believed that more aggressive fish in a farmed population affect appetite and/or feeding of subordinates. Dominance and size hierarchies are generally interrelated, resulting in a large variance of body size that may be seen in size frequency distributions at harvesting.

Dominance and size hierarchies under culture conditions prevent subordinates feeding sufficiently, resulting in a large range of sizes at harvesting, and thereby affecting the market price and resulting in feed wastage (under demand feeder conditions).

9.3 FACTORS INFUENCING FEEDING BEHAVIOUR

Having considered briefly what drives a fish to feed and the stresses that operate against the drive to feed, the next step would be to consider the act of feeding *per se*.

A diet becomes acceptable only when it satisfies some or all of the criteria of appearance, smell, feel/texture and taste. The relative importance of these to food intake depends on whether the species is predominantly a visual feeder (e.g. salmonids) or a chemosensory

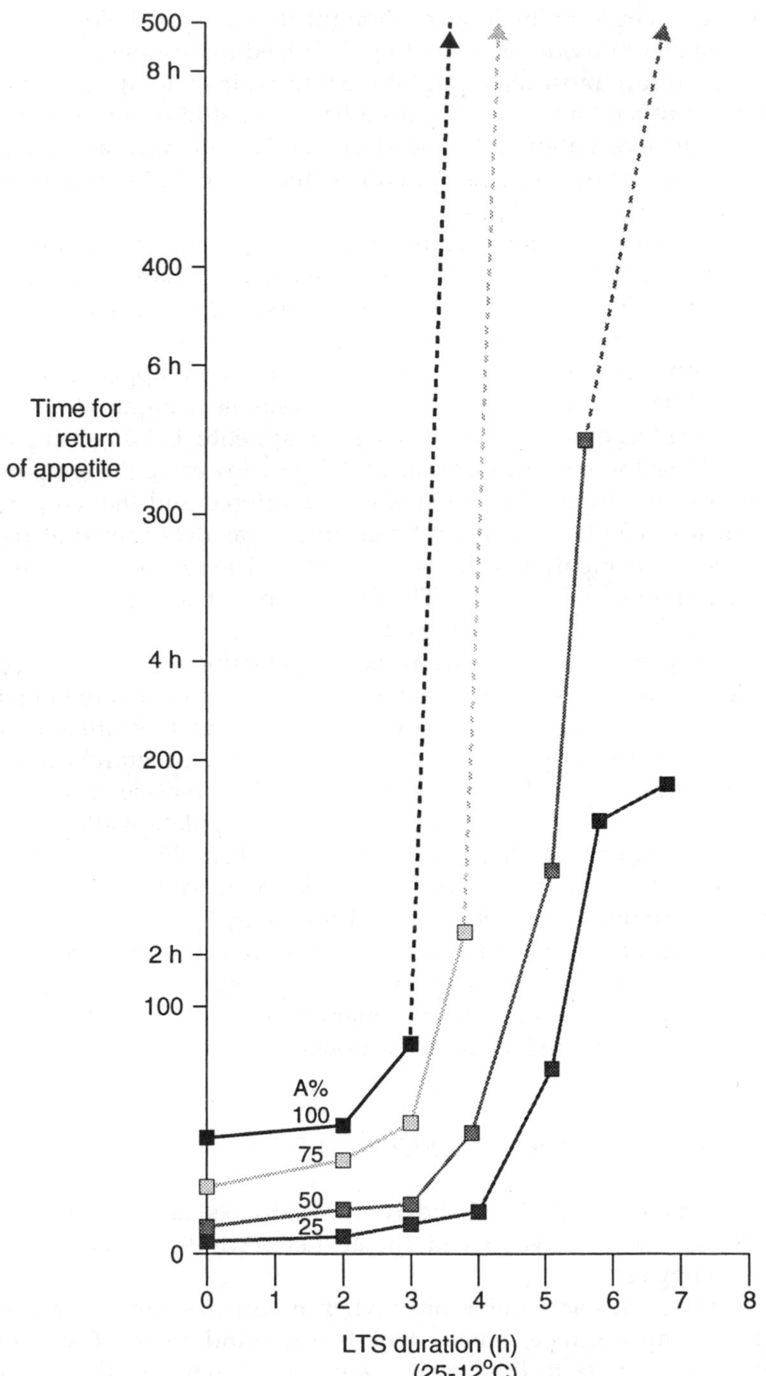

Figure 9.1 Effects of duration of low-temperature shocks (LTS) on appetite of eel, *Anguilla anguilla*. A%, percent appetite regained. (From Knights, 1985.)

feeder (e.g. sole). A generalized feeding response follows the following path.

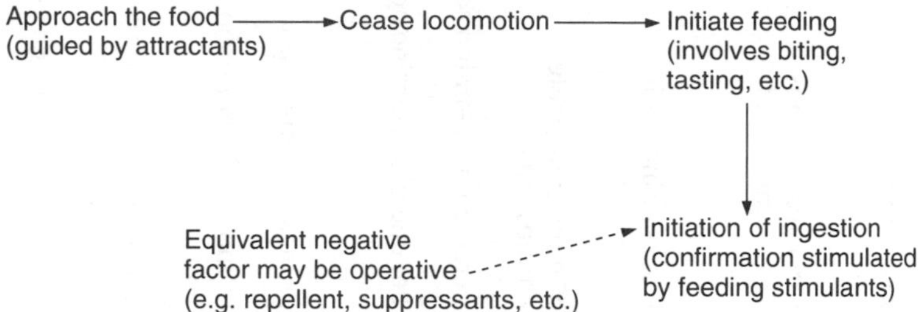

Irrespective of whether the fish is a visual or chemosensory feeder, the taste buds in the buccopharyngeal apparatus determine whether the food is to be ingested or not.

9.3.1 Gustatory stimulants

Of the different types of stimulants, gustatory stimulants are the most extensively studied, and the best known. Gustatory stimulants influence taste and are also thought to be the most important from the point of view of food ingestion. Almost all research on gustatory stimulants has been carried out on cultured carnivorous species. The method used to identify feeding stimulants was based on the **'omission test'** developed by Hashimoto *et al.* (1968). In this test the response to the complete mixture is compared with that of its components, at comparable concentrations. The available evidence indicates that chemical stimulants are effective at or below the family level for any group of fishes. That is a chemical stimulant for one species will not necessarily be a chemical stimulant for all species in that order, but it is likely to be a stimulant for all the species in the same family. Further, there appears to be a relationship between feed stimulants and feeding habits of fish in the wild. For example, inosine and inosine 5′-monophosphate stimulate feeding in the flatfish turbot and brill (Table 9.1). In both species, juveniles feed on molluscs and worms, while the adult turbot feed on fish and adult brill on fish, squid and crustaceans. All these prey organisms contain the earlier-mentioned stimulants.

K. A. Jones (1990) tested a wide variety of chemicals such as amino acids, amides, amines, alcohols, aldehydes and saccharides for their effectiveness as gustatory stimulants in rainbow trout. He found that the chemicals varied in their stimulatory influence, even within the same class/group of chemicals. It should also be pointed out that tests done so far have revealed that L-amino acid mixtures act as stimulants for various species. However, in all cases the corresponding D-amino acids are ineffective. It has been suggested that in general the gustatory system of fishes, as compared with the olfactory system, tends to be more selective in its chemical spectrum and that this accounts for much of the

Table 9.1 Survey of food organisms and feeding stimulants

Order	*Natural food organisms*	*Feeding stimulant*
Clupeiformes		
Rainbow trout (*O. myksis*)	Various invertebrates	Mixtures of L-amino acids
Anguilliformes		
European eel (*Anguilla anguilla*)	Crustacea, molluscs, worms, small fish	Mixtures of L-amino acids
Japanese eel (*A. japonica*)	Crustacea, molluscs, worms, small fish	Mixtures of L-amino acids
Perciformes		
Family Serranidae		
Sea bass (*Dicentrarchus labrax*)	Juvenile: Crustacea, amphipods, small fish Adults: fish	Mixtures of L-amino acids
Family Carangidae		
Yellowtail (*Seriola quinqueradiata*)	Juveniles: cephalopods, fish Adults: fish	Inosine 5′-monophosphate (plus L-amino acids)
Family Pomadasyidae		
Pigfish (*Orthopristis chrysopterus*)	Various invertebrates, small fish	Glycine betaine plus L-amino acids
Family Sparidae		
Red sea bream (*Pagrus major*)	Crustacea, worms	Glycine betaine plus L-amino acids
Pleuronectiformes		
Family Bothidae		
Turbot (*Scophthalmus maximus*)	Juveniles: molluscs and worms Adults: fish	Inosine or inosine 5′-monophosphate
Brill (*S. rhombus*)	Juveniles: molluscs and worms Adults: fish, squid, crustacea	Inosine or inosine 5′-monophosphate
Family Pleuronectidae		
Plaice (*Pleuronectes platessa*)	Worms, molluscs, crustacea	Complex mixture of chemicals
Dab (*Limanda limanda*)	Crustacea mainly	
Family Soleidae		
Dover sole (*Solea solea*)	Worms, molluscs, crustacea	Glycine betaine plus L-amino acids
Tetradontiformes		
Puffer (*Fugu pardalis*)	Worms, molluscs, crustacea	Glycine betaine plus L-amino acids

Source: Mackie and Mitchell (1985). Please note that rainbow trout has been erroneously included as belonging to Clupeiformes in the original table, whereas it belongs to Salmoniformes.

species specificity seen in feeding behaviour in fish (Goh and Tamura, 1980). Some of the most effective stimulants and/or chemoattractants are found in shellfish extracts.

The direct application of gustatory/chemical stimulants in aquaculture has been most effective for sluggish, fastidious feeders such as the sole (*Solea solea*) and weaning species such as the eel in early stages of the life-cycle. Dover sole culture became commercially and technically viable with the discovery that the addition of Nephrops or scallop waste to the feed induced this species to ingest artificial feeds. It has since been discovered that these wastes contain inosine and inosine 5′-monophosphates which act as feeding stimulants.

9.3.2 Feeding deterrents

There is an increasing tendency to use non-conventional feedstuffs in fish diets. There have been reports that, for example, substitution of a proportion of fish meal with leaf protein concentrate in diets for rainbow trout results in reduced palatability (Gwiazda *et al.*, 1983). This may be the results of a feeding deterrent, chlorogenic acid (a phenolic germination inhibitor), present in the leaf material. Plants produce a wide array of such feeding deterrents. Therefore, nutritionists should always be alert to this when new diets using plant materials are formulated.

It is not only plant materials that contain feeding depressants. Terrestrial and aquatic oligochaetes are now used extensively in waste management, and are considered as a suitable protein source for incorporation in fish feeds. However, work by Tacon *et al.* (1983b) has demonstrated that the earthworm, *Eisemia foetida*, used for feed in the poultry and pig industries, is unpalatable to brown trout.

9.3.3 Weaning

The change-over from an indigenous to an exogenous food supply in fish is a crucial stage, particularly for marine fish. In general, marine larvae are smaller at hatching and the yolk sac phase is shorter. Marine fish larvae are therefore generally difficult to rear on artificial diets. As a result, live food remains the most practical solution. The prolonged use of live food is expensive, and the quality unreliable. This has led to the development of a wide variety of methods for weaning fish from live foods to artificial feeds during the larval or early juvenile phases. The use of feeding stimulants in the process of weaning to artificial foods is still in its infancy, and is a difficult area of experimentation.

Kamstra and Heinsbroek (1991) studied the influence of feed attractants incorporated into a trout fry crumble in start-feeding of glass eels. They observed that the attractants, in this case extracts of natural foods or a mixture of synthetic amino acids, were most effective at suboptimal levels of feeding. At optimal levels of feeding, about 5% of the body weight per day, the attractants did not influence feed intake. In general, our knowledge on feeding attractants at weaning remains scanty, and it will be some time before attractants are utilized on a large scale at weaning in aquaculture.

9.4 FEEDING PRACTICES IN AQUACULTURE

So far, it has been repeatedly stressed that in aquaculture, as in any form of husbandry, feeding is crucial to viability and success. Therefore, the act of feeding, in a way, can be pinpointed as the single most important element in the culture practice.

The practice of feeding in an aquaculture system involves the following three stages:

- determining how much should be fed (ration size);
- determining how many times the organism should be fed in the day (feeding frequency) and the optimum time(s) of feeding; and
- efficiently broadcasting the predetermined ration to the system (feeding). Each of these stages is discussed in detail below.

9.4.1 Ration size

Ration size, or feeding rate, defines the amount of feed/diet made available to the cultured organisms. Determining the optimum ration size is one of the most difficult tasks in an aquaculture operation. The theoretical aspects of determination of ration size were dealt with earlier in Chapter 3. In this chapter the practical aspects are considered.

An optimum ration is one which gives the best growth and FCR. Such a ration, if properly dispensed, will result in minimum wastage and minimal deterioration of the water quality. Underfeeding, or a lower ration, results in poor growth and production. Overfeeding results in wastage and water quality deterioration. The latter may, in turn, influence growth, however good the feed.

Ration size is variable. A juvenile fish requires more energy for metabolism per unit weight (Chapter 2) and has the potential to grow faster than an adult fish. Therefore, juvenile fish need a higher ration. Ration size therefore, needs to be modified according to the size and age of the cultured organism. Water quality, particularly temperature, also affects feeding rate and therefore ration size. For this reason allowances must be made for water quality when determining optimal feeding rate. For instance, when the ambient temperature is higher or lower than the normal range of optimal activity it will be wasteful, and in fact may even be detrimental, to feed at the normal rate.

The ration size is normally calculated as a percentage of the biomass present. This percentage to be fed is not fixed, with the percentage decreasing as the organisms grow, but the absolute amount fed increases by virtue of the fact that the stock has grown and the total biomass has increased.

Applying a feeding rate accurately depends on the accurate estimation of the biomass in the system (average weight × number). The average weight can be obtained by regular sampling. Accurate records will not only help the culturist to maintain the correct feeding schedules but will also give an excellent guide as to the performance of the system.

Generally, manufacturers of compounded feeds give a feeding guide for their products. The reliability of such guides is variable. Guides for most salmonids, which are generally cultured intensively, are based on years of culture experience and research findings. They are therefore, generally more reliable. Tables 9.2 to 9.5 reproduced from various authors provide a few examples of feeding tables for selected species. It needs to be reiterated that these tables are only guidelines and, in practice, daily observations on the behaviour of the fish need to be made, and the feeding level adjusted accordingly. However, even after making allowances for variables such as temperature (Figure 9.2), oxygen level, etc. and using the correct feeding tables, it must be realized that fish do not feed at the same intensity, day in day out. The daily feed intake varies under laboratory conditions (De Silva and Gunasekera, 1989), and this variability is greater under culture conditions. This variability was demonstrated by Russell (1992) using a commercial feeder, Aqua Smart 1, to monitor the feed intake and the corresponding wastage for Atlantic salmon cultured in cages in Tasmania, Australia (Figure 9.3). Figure 9.3 gives a clear example of the daily variability in the feed intake and the

Table 9.2 Feeding table for rainbow trout fed dry diets

		Amount of feed (per cent body weight/biomass) per day				
Fish size (g)	*Crumble and pellet size*	*7°C*	*9°C*	*11°C*	*13°C*	*15°C*
0.38	No. 1	3.4	3.9	4.8	5.8	6.4
0.77	No. 1	3.3	3.8	4.7	5.6	6.1
1.43	No. 2	3.0	3.6	4.5	5.1	5.8
2.5	No. 2	2.8	3.2	4.0	4.9	5.1
5.0	No. 3	2.6	3.0	3.8	4.5	4.7
7.7	Nos. 3 and 4	2.3	2.8	3.6	3.9	4.1
11.1	No. 4	2.0	2.4	2.9	3.2	3.8
25.0	2.4 mm	1.7	1.9	2.1	2.6	3.2
33.3	2.4 mm	1.6	1.8	1.9	2.2	2.9
50.0	3.4 mm	1.4	1.6	1.8	2.1	2.5
66.7	3.4 mm	1.3	1.5	1.7	2.0	2.4
100.0	4.8 mm	1.2	1.4	1.6	1.8	2.0
200.0	4.8 mm	1.1	1.3	1.5	1.7	1.9
500.0	6.4 mm	0.9	1.0	1.1	1.3	1.6

Source: As adapted by New (1987).

Table 9.3 Feeding rates for channel catfish fed floating feed

Fish size	*Per cent biomass to be fed daily at various temperatures (°C)*					
(g)	*15*	*18*	*21*	*24*	*27*	*30*
4.4	2.0	2.5	3.1	3.5	4.0	4.4
10.5	1.7	2.2	2.7	3.1	3.5	3.9
20.5	1.5	2.0	2.4	2.7	3.1	3.4
35.4	1.4	1.8	2.1	2.5	2.8	3.1
56.2	1.2	1.6	1.9	2.2	2.5	2.8
83.9	1.1	1.4	1.7	2.0	2.3	2.5
163.9	0.9	1.2	1.4	1.7	1.9	2.1
283.2	0.8	1.0	1.2	1.4	1.5	1.7
449.7	0.6	0.8	1.0	1.1	1.3	1.4
553.1	0.6	0.7	0.9	1.0	1.1	1.3

Source: Foltz (1982).

Table 9.4 Feeding rate for tilapia with commercial pellets

Fish size (g)	*Feeding rate (per cent biomass/day)*
<10	9–7
10–40	8–6
40–100	7–5
>100	5–3

Source: New (1987) (based on sales literature, President Enterprises Corporation, Taiwan).

Table 9.5 Experimental feeding guide for carp

	Feeding rates (per cent biomass)					
Temperature (°C)	*Pellet size 1.5 mm*	*1.5 mm*	*2.7 mm*	*4 mm*	*5 mm*	*5 mm*
	Animal weight >5	*5–20*	*20–50*	*50–100*	*100–300*	*300–1000*
<17	6	51	4	3	2	1.5
17–20	7	6	5	4	3	2
20–23	9	7	6	5	4	3
23–26	12	10	8	6	5	4
>26	19	12	11	8	6	5

Source: New (1987).

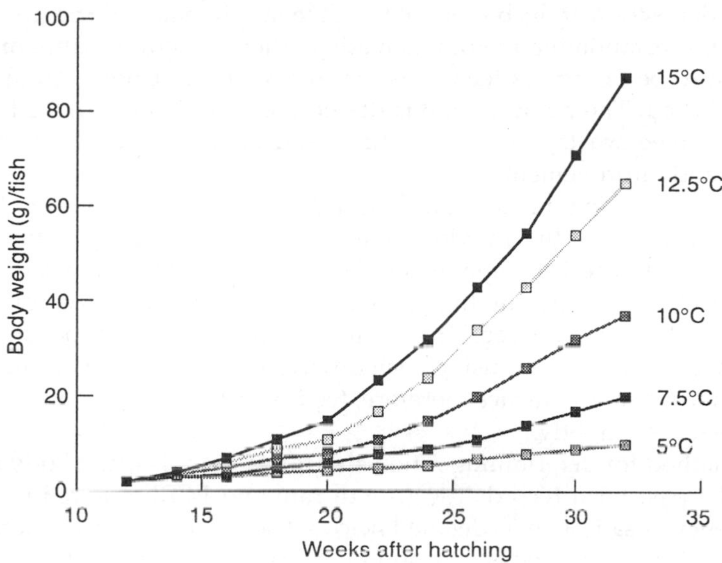

Figure 9.2 Growth of rainbow trout at different water temperatures. (From Cho *et al.*, 1985.)

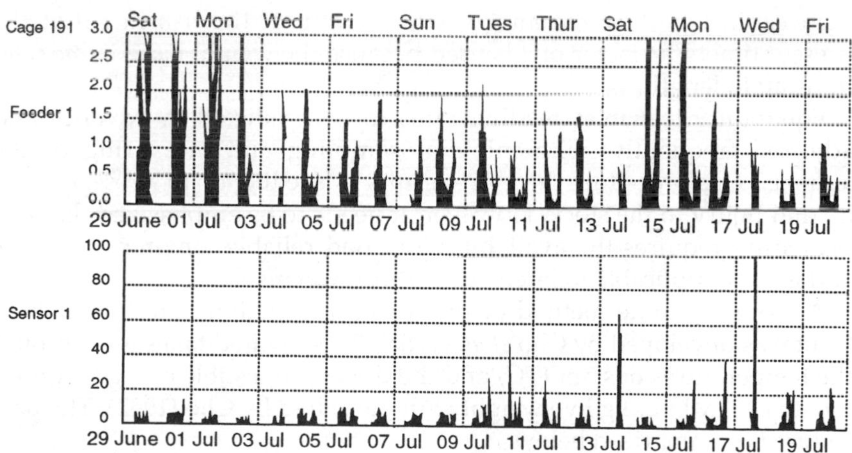

Figure 9.3 Daily variation in feed volume (feeder 1) and waste (sensor 1) as determined for cage farmed Atlantic salmon using an Aqua Smart I automated feeder. (From Russell, 1992.)

waste deposition at the bottom of the cage in salmonid culture (the lower graph), substantiating the points made earlier. Generally, salmonids are known to be voracious feeders compared with most other groups that are cultured. The point to note is the potential variability in feed intake and the feed wastage in aquaculture, and hence the scope for further study and improvement.

Ideally, feeding rates should be adjusted daily. However, this is impractical using the previous methods as the feeding rates can be adjusted with precision only when the biomass is known accurately. The use of electronically intelligent demand feeders may overcome this problem to some degree. The systems, however, are expensive and cost–benefit analysis for using such devices on a large scale has not been yet determined. In practice therefore, feeding rates are generally adjusted weekly or fortnightly.

A method for determining ration size for salmonids using body length and the past farm records for growth rate and FCR, thus reducing the frequency of sampling, is detailed below. The rate of increase in length of salmonids when reared at constant temperature is somewhat constant in the first 1.5 years of life. Also, the average weight at a given length for particular temperature regimes is well documented. As such, the daily feed required can be calculated as follows:

$$\text{Body weight to be fed daily } (\%) = \frac{\text{FCR} \times 3 \times A \times 100}{B} \tag{9.1}$$

where FCR is the food conversion ratio (Chapter 7), A is the daily increase in length in cm and B is the length of the fish in cm at the present time. The factor of 3 is used because the weight has a cubic relationship to length.

Equation 9.1 enables feeding rates to be predicted in advance and adjusted frequently, with minimal sampling and measuring of the animals. This procedure therefore results in a reduction in stress, mortality and injury to the stock. However, it needs to be emphasized that this procedure requires the availability of good reliable long-term records, and is only applicable to intensive culture systems.

A more elaborate method of predicting the feeding rate for rainbow trout was developed by Cho *et al.* (1985). This method takes into account the temperature constant (TC) and the dietary digestible energy requirement of 15 MJ per kg live weight gain measured by Cho (1982). The procedure is summarized below.

A temperature growth constant (TC) is calculated using the previous year's growth data, and is influenced by the genetic make-up, water quality and other management practices. It needs to be checked and recalculated for different lots of fish.

$$\text{TC} = \frac{\sum (T \times \text{days})}{W_t^{1/3} - W_o^{1/3}} \tag{9.2}$$

where T is the temperature in °C, W_o is the initial body weight and W_t is the final body weight in grams.

The TC calculated for the previous year's data is then used to calculate the expected live body weight (W_{te}) of the fish.

$$W_{te} = \left[W_o^{1/3} + \frac{\sum (T \times \text{days})}{\text{TC}} \right]^3 \quad (9.3)$$

Following this the expected live weight gain (TGN) may be determined.

$$\text{TGN (g)} = (W_t - W_o) \times n \quad (9.4)$$

where n is the number of fish.

Finally, the expected total feed intake (TFI) is estimated using the following equation:

$$\text{TFI (kg)} = \frac{0.015 \times \text{TGN}}{\text{DE per kg feed}} \quad (9.5)$$

where DE is MJ of digestible energy per kg of feed.

Changes in biomass due to mortality also affect the feeding rate. Therefore, unless some estimate of population density and therefore mortality in the culture vessel can be made, there is a chance of providing an incorrect ration. This is particularly difficult in ponds with turbid water. Again, accurate records of the previous year's production will allow estimates of mortality to be made.

Excessive feeding does not necessarily result in higher growth. Beyond a certain level, excessive feeding has no influence on the growth. However, it would result in a poor FCR. There is a feeding level at which an organism grows best and the FCR is minimal (Figure 9.4). Feeding practices should strive to attain such a level/range, which can be considered as the optimal feeding range.

Generally, very small fish and shrimp are fed in excess. In catfish culture fry are fed several times per day, up to 50% of the biomass, while young shrimp and post-larvae are fed up to 100% of the biomass per day. This high rate is needed to compensate for the rather low ingestion rate. Obviously, such high feeding rates result in deterioration of water quality. It is because of these reasons that hatcheries are generally required to have almost an unlimited supply of good-quality water for optimal functioning.

Finally, none of the methods described can substitute for the alertness and sharpness of the farmer or the feeder. Some culturists, for example the freshwater prawn farmers in Thailand, feed 'to demand' rather than in accordance with a feeding rate table, with the daily feeding rate being adjusted according to the previous day's feeding. This is reasonably effective where the farmer is experienced in culture activities. Simple

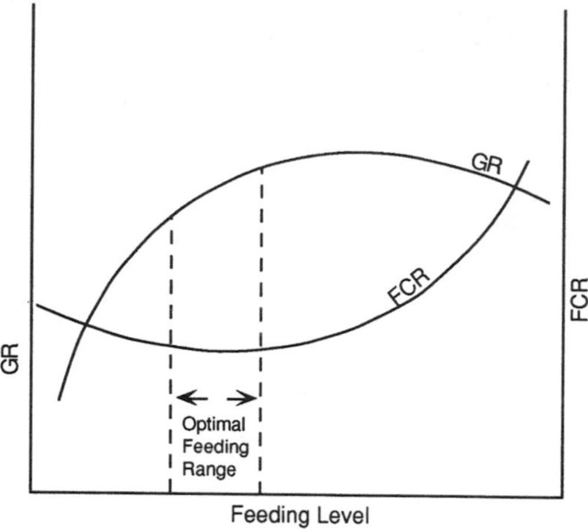

Figure 9.4 A theoretical representation of the changes in growth ratio (GR) and food conversion ratio (FCR) in relation to the feeding level.

comparative observations such as the following need to be practised by the farmer:

- Do the fish move towards the feed as quickly as yesterday?
- Are the fish not interested in the food at all?
- Is the water greener than yesterday?
- Is only a proportion of the fish coming to the feed?

These questions are crucial in determining whether fish are doing well, the ration is sufficient or overfeeding is occurring, and whether the fish are in good health. Such alertness, combined with maintenance of reliable and accurate past records and simple water quality measurements such as temperature, will go a long way to make the operation successful and cost-effective.

9.4.2 Feeding frequency

The feeding frequency is important to ensure maximal FCR and dress weight of the cultured organism. Therefore an important step in the feeding strategy is to determine the optimal frequency of feeding. Of course, this will not be applicable to culture practices using demand feeders.

Determining the optimal feeding frequencies is a grey area in the nutrition of cultured species. Chiu (1989) summarized some of the experimental data available on feeding frequency at which optimal growth was observed in different finfish species (Table 9.6). The vari-

Table 9.6 Feeding frequency required to promote optimum growth and feed efficiency in different species of fish

Species of fish	*Optimum feeding frequency (feeds/day)*	*Initial fish size (g)*
Cyprinus carpio	3	0.17
Chanos chanos	8	0.60
Oncorhynchus mykiss	2	7.6–16.0
Epinephelus tauvina	1	70
Heteropneustes fossilis	1–2	4.5–9
Ictalurus punctatus	2	–
Channa striatus	1	0.66
Clarias lazera	Continuous	0.5
Oreochromis niloticus	1	6.8

Source: Chiu (1989).

ability in the feeding frequencies shown in the table is probably more indicative of the uncertainty of the results rather than of the biological variability *per se*.

Piper *et al.* (1982) were among the first to address the problem of feeding frequencies in any great detail. They put forward five basic rules for feeding:

- For optimum growth and feed conversion, each feed should ideally be 1% of the body weight. Therefore if the ration for the day is 5% of the body weight per day, fish need to be fed five times, 1% of the body weight each time.
- Survival is not significantly influenced by feeding frequency once the transition from an endogenous to an exogenous food supply is complete.
- Higher feeding frequencies reduce starvation and stunting, thereby resulting in uniformity in size.
- Dry feeds need to be distributed more frequently than moist feeds.
- At least 90% of the presented feed should be consumed within the first 15 minutes of the feeding time.

(a) Salmonids

Salmonids, in general, are good feeders. They feed to satiation and do not tend to eat again until the stomach is almost completely evacuated. Therefore, a feeding frequency of once or twice a day is often more than sufficient.

The feeding frequencies recommended by Piper *et al.* (1982) for salmonids are given in Table 9.7. This table shows that fry are fed more often, with frequency of feeding decreasing as the fish grow. Sometimes

Table 9.7 Suggested feeding frequencies for salmonids

	Fish size (g)								
Species	*0.3*	*0.45*	*0.61*	*0.91*	*1.82*	*3.6*	*6.1*	*15.1*	*>45.1*
Coho salmon	9	8	7	6	5	3	3	–	–
Chinook salmon	8	8	8	6	5	4	3	–	–
Rainbow trout	8	8	6	6	5	4	4	3	2

Source: Piper *et al.* (1982).

a 24-h lighting regime is provided for the first few days to encourage the fry to take dry feed, and frequency can be as high as 20 times per day. Fish over 12 cm in length (about 23 g) are fed twice a day.

(b) Catfish

The bulk of the information available on catfish is for channel catfish, perhaps the most important catfish cultured intensively in the USA. Channel catfish fry commence feeding 5–6 days after hatching. Feeding frequency is 8–10 times at the first feeding to three times per day when they are about 7 cm in length. Juveniles grow best when fed twice a day – mid-morning and late afternoon. It has been found that a reduced feeding frequency is advantageous when the water temperature is very high (above 29°C) and/or low (below 15°C).

(c) Common carp

Common carp do best on frequent feedings. However, the optimal feed frequency needs to be assessed for individual culture situations. Omar and Günther (1987) reported that in mirror carp (a strain of *C. carpio*) increasing the feeding frequency from four to six times daily significantly improved body weight gain and feed utilization.

(d) Tilapia

Tilapia tend to feed throughout the day, with minor peaks of activity in the early hours of the morning and at dusk in the wild. New (1987) suggested that manual feeding several times per day is the most appropriate for intensively grown tilapia.

It should be evident that there are no hard and fast rules in regard to feeding frequency under culture conditions. This point is further stressed by the work of Chua and Teng (1978). When in doubt, therefore, feed as frequently as economics permit (Table 9.8).

Table 9.8 Feeding rate for tilapia (*T. nilotica*) in tanks and cages at 27–31°C fed a 46% protein commercial fish feed

Fish size (g)	*Per cent biomass fed/day*
Up to 5	30 reducing to 20
5–20	14 reducing to 12
20–40	7 reducing to 6.5
40–100	6 reducing to 4.5
100–200	4 reducing to 2
200–300	1.8 reducing to 1.5

Source: Pullin and Lowe-McConnell (1982).

9.4.3 Feeding

It has been stressed that in any aquaculture operation feeding is the most important daily activity, utilizing a considerable proportion of staff time irrespective of the technology used. Even when automated feeders are used feed has to be delivered to these, the equipment needs to be regularly serviced and their performance checked as frequently as possible.

Cultured organisms may be fed by one of two ways: by broadcasting (manually) or by feeders.

(a) Broadcasting

Broadcasting or hand feeding is the most common form of feeding in the semi-intensive culture practices of the developing world. It is also used in intensive culture practices to varying extents. It is labour intensive. Hand feeding essentially involves the dispersing of a known quantity of food into the system.

Finfish are easily conditioned to feeding and react to the first sign or appearance of food, making them suitable for broadcast feeding. Shrimp are sluggish feeders. Therefore, in shrimp culture it is imperative that feed is broadcast to cover maximal pond area. More often in shrimp farms, feed trays are provided, which are periodically checked. In some semi-intensive cultures for major carp species, feed is kept in a gunny bag with a few small holes, and hung on a pole in the centre of the pond. Fish soon learn to nip at the feed at their will. Both these methods result in substantial loss of feed and nutrients. For certain species (e.g. snakehead, *Ophicephalus striatus*) feed is broadcast in a pre-determined area in a pond equivalent to a feeding platform (Figure 9.5). Snakehead do not feed on floating pellets, and often feed in the form of a dough is kept on a platform, just below the surface, the fish soon learning to feed effectively. Feed-dispensing practices for semi-intensive finfish culture are dealt with in detail by New (1987).

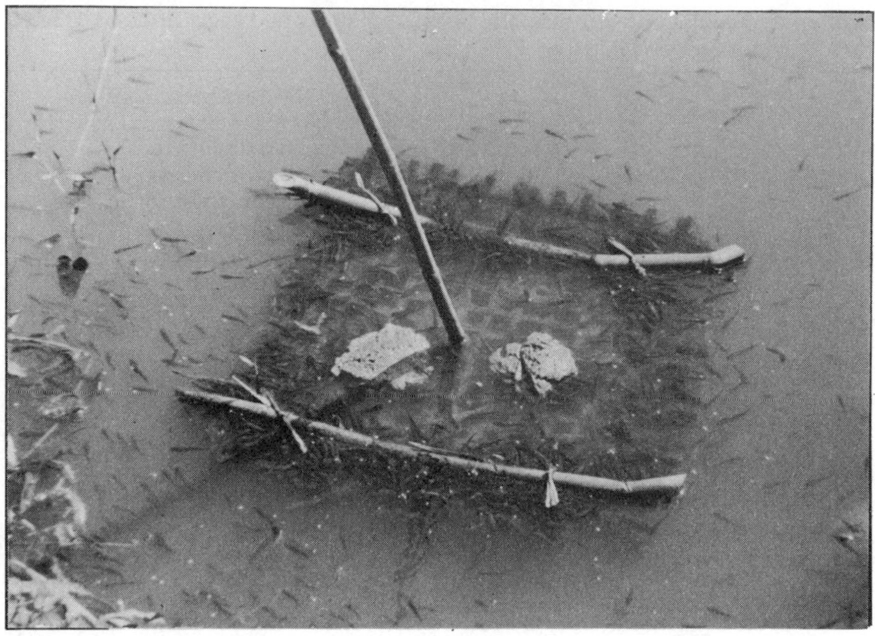

Figure 9.5 A 'feeding platform' in snakehead culture in Thailand. Feed is dispensed in the area demarcated by the planks in the pond.

(b) Feeders

As in most industrial and agricultural practices, intensification demands cost-cutting measures. Labour costs in most developed countries are high and feeding is labour intensive. Therefore, the aquaculture industry over the last two decades has developed automated, mechanical feeders to dispense feeds. Use of automated feeders, in addition to reducing the labour cost in the farm, is claimed to result in an improved FCR and minimizes feed handling. Automated feeders also permit feeding at any time of the day and every day of the year. This is particularly advantageous where there are labour constraints, and in off-shore cage culture operations.

Generally, all mechanical feeders have a hopper, a regulator, a dispenser and a controller. Each of these components may vary in structure and design, resulting in a large array of mechanical feeders which are of suitable design, for most aquaculture applications. These automated mechanical feeders are of two basic types: non-demand and demand type.

Non-demand feeders

In the non-demand type of feeder a predetermined quantity of feed, at predetermined time intervals, is dispensed. The quantity of feed dispensed is determined by the power of the motor or the pressure of the air stream as the case may be, and by the size of the pellets. The designs vary a great deal. White (1991) classified automated, non-demand mechanical feeders into four basic types: drop feeders, auger feeders, disc feeders and pneumatic feeders.

Drop feeders

Drop feeders use belt dispensers or rotating discs for dispensing feed. Rotating discs permit more controlled delivery and can be programmed to dispense discrete quantities of feed at defined rates at predetermined time intervals. Drop feeders, however, do not have a regulator. As a consequence these do not dispense precise quantities of feed. These feeders are therefore used mainly to feed crumbles or starter feeds to juvenile stock where feeding rates are not critical.

Auger feeders

As the name suggests these feeders dispense feed through the use of an auger. The quantity of feed dispensed is regulated by the period of operation or the number of turns of the auger. Auger feeders can be designed to broadcast feed up to distances of 4–5 m using spring-loaded paddles connected to the auger motor.

Disc feeders

Disc feeders are principally used for dispensing pelleted feeds (>3 mm) over a wide area. The design of disc feeders varies, but generally the disc is mounted at a set distance below the hopper. When activated, feed falls on the spinning disc and is broadcast over the desired area. Early disc feeders did not dispense an accurate quantity of feed, the quantity dispensed being determined by the gap between the hopper and the disc. However, more recent designs have overcome this problem by incorporating regulating devices.

Pneumatic feeders

Pneumatic feeders utilize either a high- or low-pressure air stream to dispense feed. The quantity of feed delivered into the path of the air stream is determined using a regulator. High-pressure feeders are driven by a source of compressed air, while low-pressure pneumatic feeders utilize a fan. The former are often stationary, mounted next to ponds, and the latter are often mobile. Blow feeders are commonly used in shrimp culture to distribute feed over a maximum area of the pond bottom.

Demand feeders

Demand feeders differ from other feeding mechanisms in that they are activated by the stock, and not the controller. As such, demand feeders permit *ad libitum* feeding, and do not require accurate knowledge of the biomass of the stock. The basic design of a demand feeder (Figure 9.6) is

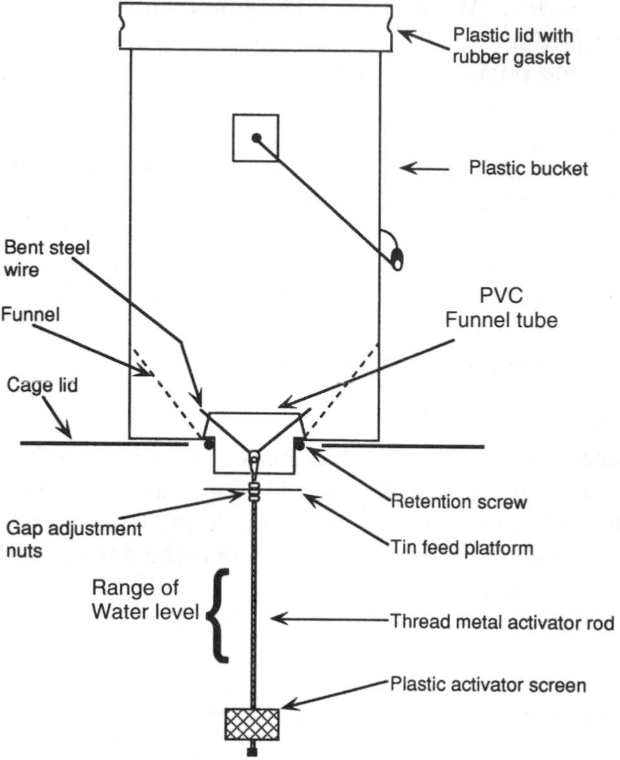

Figure 9.6 A demand feeder (not to scale).

essentially identical to that of a disc feeder except that the feed-releasing trigger system is activated by the stock rather than by a mechanical or electronic controller. The main disadvantage of demand feeders, however, is that relatively few individuals (usually the dominant individuals in the group) learn to activate the trigger, and they continue to show their dominance even when they are satiated. Dominant individuals can prevent subordinates activating the trigger, preventing some individuals obtaining sufficient amounts of food. It is also considered that demand feeding prevents overfeeding, and hence wastage of food. However, there is no conclusive proof of this.

With increasing intensification of some culture practices such as, for example, utilization of off-shore cages for culture, there has been a resurgence of research on demand as well as automated feeders. Recently Alanärä (1992) investigated the influence of time restricted demand feeding on rainbow trout. He concluded that two feeding periods per day each of 2 h duration are sufficient for optimal growth. Obviously, variations in demand feeding to suit particular species and a culture practice are in the offing.

Both non-demand and demand feeders are very useful tools for replacing hand feeding, and are therefore increasingly used in aquaculture operations. The selection of different types of mechanical feeders depends primarily on the design of the system, e.g. for large ponds a feeder which broadcasts feed over a larger area, such as a pneumatic feeder. An ever-increasing number of models of feeders are available, most of which tend to spread the feed to a wider area, preventing or minimizing the influences of dominant individuals. The electronics incorporated into feeders is also becoming more and more sophisticated, leading towards a better control of the quantity of feed dispensed. Integration of information regarding the feeding responses and water quality criteria into the regulating mechanism may also be utilized. White (1991) attempted to analyse the suitability of different types of automated, mechanical feeders for different culture practices (Table 9.9). He concluded that disc feeders have the widest applicability, being suitable for a range of operations, such as shrimp and finfish culture.

Blyth and Purser (1992) showed that information on the level of feeding and feeding time can be incorporated into the operation of automatic feeders, resulting in significant savings in feed usage. This resulted in advanced automatic feeders, referred to as intelligent feeding systems, being developed. Such feeders are expected to predict feed intake more accurately and thereby reduce wastage.

Recently, Juell *et al.* (1993) described a method in which automatic feeders are controlled by a hydroacoustic 'food detector' where the latter is used as an indication of reduced appetite. In a controlled study in salmon cages they found that salmon grown in cages with a hydroacoustic food detector performed better than a carefully hand-fed group. Juell *et al.* (1993) argued that, unlike in demand feeding, hydroacoustic food detection constitutes a negative feedback system based on indirect measurement of an unconditioned group response (in demand feeders, fish are conditioned to bite the trigger) to reduced feed motivation, or loss of appetite. These authors asserted that this is similar to termination of feeding in hand feeding when the farmer visually evaluates that the fish are satiated.

9.5 COMPARISON OF FEEDING PRACTICES

Most of the research concerning the effectiveness of hand feeding and non-demand and demand feeders has been performed on salmonids. Each of these feeding practices offers advantages and disadvantages, which are summarized in Table 9.10. Generally, little experimental work has been based on the use of all three basic feeding methods concurrently. Recently, however, Tidwell *et al.* (1991) studied the production of rainbow trout, using different feeding practices. It was found that fish

Table 9.9 Suitability of various types of feeders under a range of culture applications. A score from 0 to 3 is assigned to each feeder type to indicate its suitability given consideration of the variation in feed sizes, typical surface areas over which feed must be broadcast and current industry management techniques for each application

Application		*Feeder type*				
		Drop	*Demand*	*Auger*	*Disc*	*Pneumatic*
Crustacean Culture						
Juvenile Stock						
	Tank culture	1	0	3	2	0
	Small pond culture	0	0	3	2	1
Grow-out Stock						
	Tank culture	1	0	3	2	0
	Small pond culture	0	0	2	3	1
	Large pond culture	0	0	0	0	3
Finfish Culture						
Juvenile Stock						
	Tank culture	2	2	3	1	0
	Raceway culture	2	2	3	2	1
	Small pond culture	1	1	2	3	1
Grow-out Stock						
	Tank culture	2	2	2	2	1
	Raceway culture	1	2	2	2	1
	Small pond culture	1	2	2	3	1
	Large pond culture	0	1	0	3	3
	Cage culture	0	1	0	3	3

Key: 0, unsuitable; 1, limited application; 2, effective; 3, highly suitable.
Source: White (1991).

Table 9.10 Summary of advantages and disadvantages of common feeding practices in aquaculture

Type of practice	*Advantages*	*Disadvantages*
Hand feeding to satiation	Best assurance of maximum feeding effectiveness May decrease size variation Higher food consumption Higher growth rates	Labour intensive May result in higher FCR Carcass fat levels may increase
Use of feeding charts (in conjunction with automatic feeders)	Less labour intensive	Large pond sizes preclude accurate estimates of fish biomass May not accurately predict the amount of feed that can be consumed in relation to temperature Capital intensive Size variation higher
Demand feeders	Labour saving Permit the fish to feed *ad libitum*	All individuals may not be able to feed to satiation Hierarchical influences are manifested to a greater degree

hand fed to satiation consumed 66% more feed than when fed according to feed chart and 163% more than fish fed by demand feeders. Fish fed to satiation also had a higher growth rate, harvest weight, protein gain and no significant increase in body lipids.

9.6 RECORD KEEPING IN FEEDING

Records of feed usage should be part of the general records kept in an efficiently run operation. Some of the details on management of an aquaculture facility such as temperature, salinity, water exchange rates, etc., may not be specific to feed use. These factors, however, are essential for the interpretation of a particular feeding programme.

Records relating to feeding that should be kept may be divided into two categories. Firstly, it is important to record the type of feed(s) used. Details that should be noted for purchased feeds are date delivered, manufacturer's name, product name, batch number, amount purchased and delivered, cost per unit weight and special notes (manufacturer's specifications, etc.). In the case of on-farm produced feed the following should be kept: feed formulation number, date and time made, and source and cost of each ingredient.

The other details to be recorded fall into the category of the feeding procedure, and the effectiveness of the feeding. These include: date and times of feeding, amount fed per feeding, growth rate (from regular sampling), reaction of the organism(s) to the feed, any mortality [if possible number and size(s)], quantity harvested, size distribution at harvest and any special comments.

9.7 FERTILIZATION AND SUPPLEMENTARY FEEDING

The great bulk of aquaculture production in the world takes place in semi-intensive and/or extensive production systems which employ a fertilization and a supplementary feeding strategy. This trend is likely to continue in the future, with semi-intensive culture practices proving to be the most profitable even for shrimp culture (Csavas, 1990). These systems would also be sufficiently resilient to counteract changes in markets and temporary setbacks due to diseases.

In the earlier sections in this chapter, we dealt with aspects on feeds and feeding in essentially intensive operations. Unlike intensive culture, in semi-intensive culture the feed supplements the natural food produced in the culture system. A proper balance therefore needs to be achieved between the supplemental feed and the natural food availability. However, in semi-intensive finfish culture, fertilization is utilized to enhance the production of natural food organisms in the system, particularly in the early phases of stocking, instead of providing an external feed.

9.7.1 Fertilization

Fertilization is a common practice that is utilized in semi-intensive and extensive aquaculture in the tropics. The role of fertilization (organic manure or inorganic fertilizers) in finfish culture has been adequately reviewed by Wohlfarth and Schroeder (1979) and Wohlfarth and Hulata (1987). Although pond fertilization is an ancient and much researched practice, questions still remain as to the dynamics of pond systems in relation to fertilization. This makes it difficult, if not impossible, to standardize and provide guidelines in respect of the amount and type of fertilizer to be added to a system, when to add fertilizer and the frequency of application, and the expected return from such additions.

The reasons for this uncertainty are manifold. Firstly, the release of nutrients from the different types of fertilizers occurs at different rates and is dependent on temperature, soil conditions and other water quality parameters. In the case of organic manuring the microbial flora also plays an important role. Utilization of nutrients in the production cycle(s) will also deviate between systems. This is affected by physical conditions such as temperature and the indigenous composition of the biological community, as well as by the interactions between species and finally on the species cultured. In spite of whatever uncertainty there is with this practice, fertilization is a useful, cheap and a common practice adopted in semi-intensive finfish and shrimp culture in the tropics (Figure 9.7).

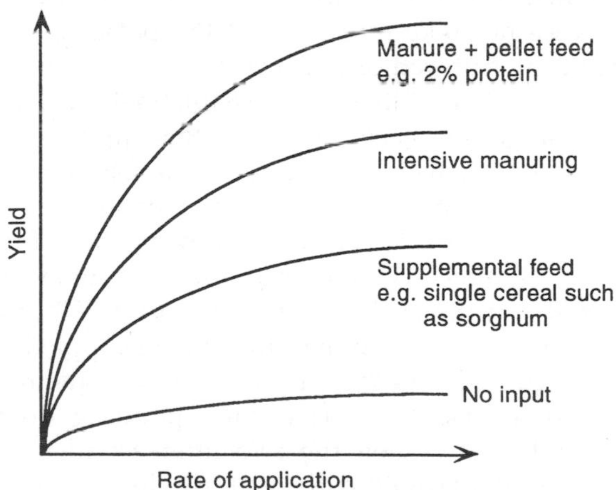

Figure 9.7 A schematic representation of the influence of manure application in comparison to other inputs on yields in semi-intensive aquaculture.

9.7.2 Natural food – supplemental feed interactions

Under normal conditions semi-intensive aquaculture systems are stocked when a luxuriant growth of natural food is present. This is achieved through correct fertilization. This natural food, if available in sufficient quantities, is nutritionally adequate for growth of the stock. However, fertilization does not provide enough natural food for most target species to reach their full growth potential. Therefore, there exists a need for feed supplementation to achieve profitability in semi-intensive aquaculture.

The nutritional value of natural food organisms in a pond is sufficient to support excellent fish growth. They are a rich source of protein, often containing 40–60% protein on a dry matter basis. According to Albrecht and Breitsprecher (1969), the mean protein, carbohydrate and lipid content of fish food organisms in ponds are 52.1, 27.3 and 7.7% respectively. The calorific value ranges from 6.7 to 23.8 kJ/g (1.6–5.7 kcal/g). Prus (1970) found that the nutrition value ranges from 28% to 35% of protein with a 96 mg/kcal protein to energy ratio. There is no information available particularly on the digestibility of pond organisms by target species, although a study by De Silva *et al.* (1984b), has shown that the food of tilapia (*O. mossambicus*) from quasinatural impoundments, even when it contains a high proportion of detritus, is well digested.

In semi-intensive culture there are complex interactions between the natural food organisms and the supplementary feeding practice(s). The possible qualitative interactions in a well-prepared pond which is to be subjected to supplementary feeding is schematically depicted in Figure 9.8. From this figure it can be seen that as the standing crop of the pond increases it becomes necessary to supply either high-energy or high-protein supplemental feeds to maintain optimal growth. The above simplification does not take into account the possible changes in the feeding habits of the cultured organisms (Weatherly, 1963; Werner and Hall, 1963; Odum, 1970; De Silva *et al.*, 1984b). Spataru *et al.* (1980) have also shown in their studies on common carp that supplementary feeding can affect feeding habits and food selection, the fish tending to select a narrower range of food items when on supplementary feeding regimes.

There is some experimental evidence to show the need for supplemental feeds. As the difference between standing crop biomass of the cultured organism(s) and the critical standing crop (CSC) of natural food organisms increases, the deficit in natural protein supply increases. This is illustrated in Figure 9.9. As there is an increase in the rate of yield of the cultured organism(s), there is a corresponding decrease in the amount of natural food. When the two intersect then three options become available to the farmer: cull some fish; leave the system as it is and accept a reduction in the growth rate; or provide a supplemental feed. This was shown for common carp, under monoculture: a pelleted

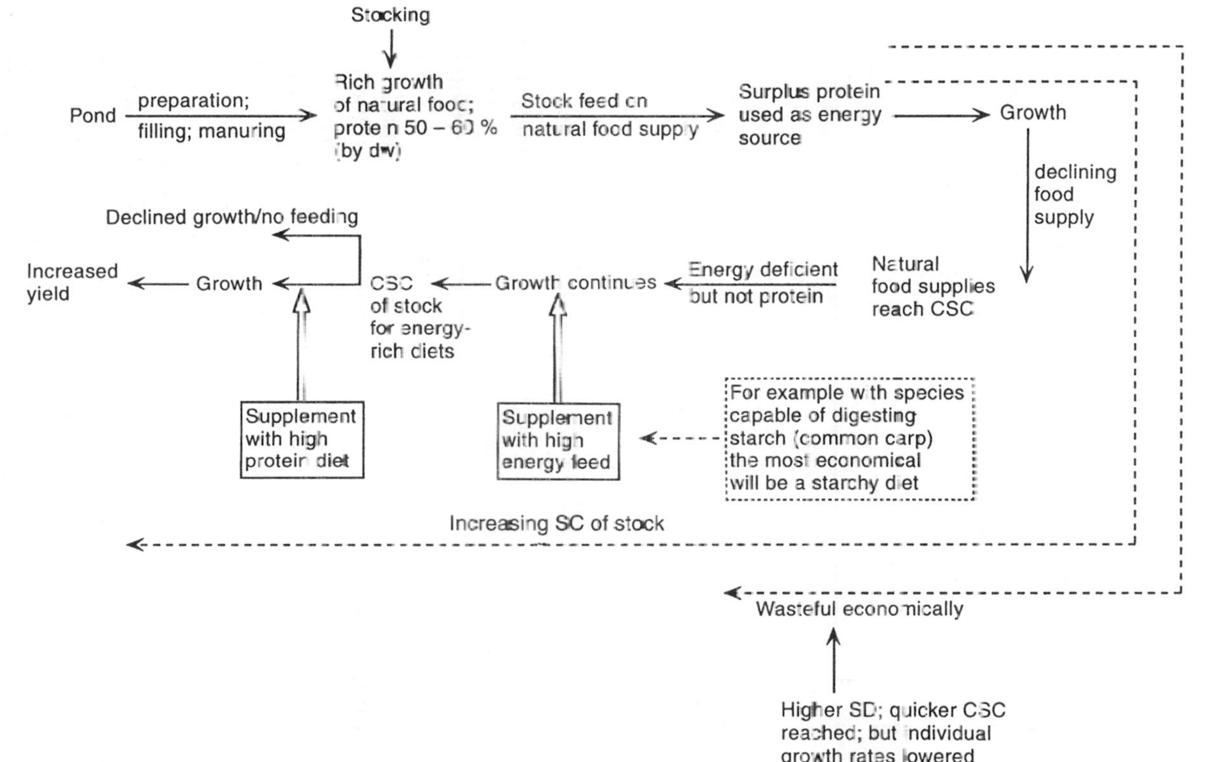

Figure 9.8 The qualitative changes that occur in a semi-intensive culture pond. Also indicated are the potential utility of supplementary feed strategies. SC, standing crop; CSC, critical standing crop; SD, stocking density.

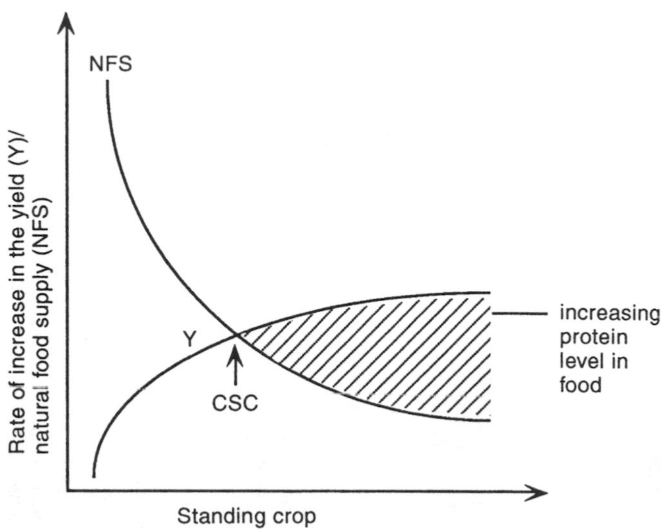

Figure 9.9 Changes in the biomass of natural food organisms and fish yield in relation to the standing crop of the stocked species, and the ensuing protein needs of the supplemental feed(s).

feed of sorghum (9% protein) or diets of 22.5% and 27.5% protein did not influence growth until a SC of 800 kg/ha was reached (Hepher *et al.*, 1971; Hepher, 1975). However, when the SC was increased further the 9% protein diet was not effective. The 22.5% diet became ineffective when the SC reached 1400 kg/ha. The provision of a protein-rich diet is also supported by work on Nile tilapia, *Oreochromis niloticus* (Hanley, 1991). In this case, dietary lipid level did not influence growth rate or FCR, and it was recommended that diets for tilapia should be compounded on the basis of protein rather than energy consideration.

Sumagaysay *et al.* (1990), working on milkfish, *Chanos chanos*, demonstrated that supplementary feeding had a positive influence on yields at higher stocking rates. In addition, it was shown that the yields and the profits in semi-intensive milkfish culture were related to the quality of the supplementary feed, and the existence of a relationship between standing crop, growth and the quality of the supplementary feed, as for common carp, was confirmed (Sumagaysay *et al.*, 1991). Furthermore, it has been demonstrated that the gastric evacuation rate of milkfish feeding on a natural diet increases when compared with fish fed a supplemental diet, and that when a quality supplemental diet is provided (34.3% protein) daily ration of fingerling milkfish (35 g) to marketable size (116 g) ranged from 0.60 to 19.68 kcal per fish per day (Sumagaysay, 1993). The author reckoned that any deviation in the daily ration from the above estimates may indicate that insufficient dietary energy is obtained by fish from natural food alone.

Now that a need for feed supplementation in semi-intensive culture systems has been established, it is important to discuss the utilization efficiency of the feeds used. Work on the semi-intensive culture of common carp found that, at a SC of about 500 kg/ha, sorghum as a supplementary diet was inadequate (Hepher, 1988a). Consequently, the FCR of the fish fed the diet of sorghum increased sharply, as opposed to those maintained on a pelleted feed containing 25% protein (Figure 9.10). Such a sharp increase in the FCR is critical from an economic point of view.

It was suggested by Hepher (1988b) that in such circumstances one of three options is open to the farmer: improve the diet resulting in a cost increase which may not necessarily be paid off by a corresponding increase in yield; reduce the SC, thus decreasing the yield; or increase the natural productivity by fertilizing.

It can be seen from these suggestions that the FCR of supplemental feeds is dependent upon the natural food supply as well as the level of SC and other factors that are normally associated with intensive culture systems.

It is apparent, then, that feed supplementation in semi-intensive aquaculture cannot be done in a random fashion if the maximum return is to be gained. In fact, because of the interactions between the natural food

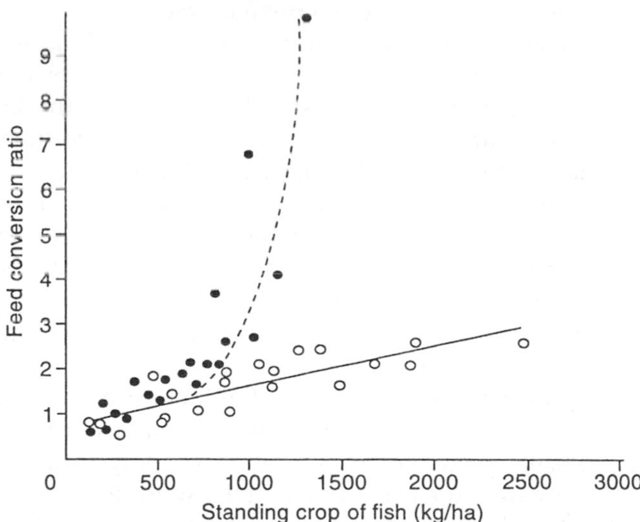

Figure 9.10 Supplement feed conversion ratio at different standing crops of common carp (SD 2000/ha). Solid circles, fish fed on sorghum; open circles, fish fed on protein-rich pellets. (From Hepher, 1978.)

supply, fertilization and the growth of the stock in semi-intensive culture, feed supplementation is more complex than providing nutritionally wholesome diets under intensive culture conditions. Therefore, to be economically successful, the semi-intensive farmer must be alert to the many facets of the system, supplying the correct type and quantity of supplementary feed as required.

9.7.3 Supplementary feeds

In addition to the use of fertilizers for the production of natural food organisms within the water body, an external diet can also be given as a supplementary source of nutrients to semi-intensively cultured organisms. The use of supplementary diet feeding, combined with fertilization, allows higher stocking densities, facilitates growth and results in higher production.

As with fertilizer application alone, the success of a particular supplementary feeding regime in one location need not necessarily apply to another. The benefits of supplementary feeding will also be dependent on composition of the supplementary feed and the natural productivity of the water body. Unlike 'complete' diet feeding, a supplementary feed may consist of a single ingredient or a mixture of ingredients. It is generally not required to add most of the micronutrients, such as vitamins and minerals, to the formulation of supplementary feeds.

In Asia, where aquaculture is mostly a rural occupation, the selection and utilization of supplementary feeds is linked to other agricultural activities in the particular region. As such, the bulk of the ingredients used as supplementary feeds are agricultural by-products, or by-products of the animal husbandry industry. A list of commonly used feed types and the FCR values reported in respect of each is given in Table 9.11. It can be seen from this table that fish are not capable of efficiently utilizing most of these ingredients, in particular the grasses. This relatively inefficient utilization of grasses and other cellulose-rich materials is to be expected because fish do not possess the enzyme cellulase to digest these feed types.

There is little information available in the literature on the types and extent of usage of the different supplementary feeds on farms. However, the almost explosive growth of a semi-intensive, polyculture practice of Indian major carps over the past decade in the south-eastern Indian state of Andhra Pradesh has supplied some data. In a survey covering 189 farms in four districts of the state, it was found that nine major ingredients and seven feed types are used by the farmers. On average 27 000 kg/ha/year of feed was utilized as supplementary feed (Tables 9.12 and 9.13, Figure 9.11). This provides an indication of the diversity of supplemental feeds that are utilized in semi-intensive finfish culture,

Table 9.11 The FCR reported for some of the commonly used supplementary feeds

Feed of animal origin	*FCR*	*Feed of plant origin*	*FCR*
Daphnids	4–6.4	Soybean	3–5
Mysis	2–3.9	Wheat flour	7.2
Gammarus	3.9–6.6	Barley flour	2.6
Prawn and shrimp	4–6	Corn	4–6.0
Earthworm (fresh)	8–10	Wheat bran	6.13–7.32
Clams (fresh)	1.3	Barley bran	7.0
Snail, fresh	22.0	Irish potato	20–30
Snail, fresh-dried	10.2	Cereals	4–6
Chironomids	2.3–4.4	Groundnut cake	2–4
Housefly maggots	7.1	Ground maize	3.5
Locust, fresh	10.7	Ground rice	4.5
Locust, dried	5	Oilpalm cake	6–12
Silkworm pupae, fresh	3–5	Manioc leaves	10–20
Silkworm pupae, dried	1.25–2.1	Manioc rind	50.7
		Manioc flakes	17.6
		Manioc flour	49.4
		Banana leaves	25.0
Freshwater fish	4–8	Napier grass	48.0
Fresh sea fish (trash)	6–9	Rye grass[a]	17–23
Fish flour	1.5–3.0	Sudan grass[a]	19–28
Fresh meat	5–8	Elephant grass[a]	30–40
Meat flour	1.99–2.02	Hybrid grass[a]	25–30
Dried blood powder	1.51–1.68	Lucerne[a]	25–30
Liver, spleen	5.5–8.0	Clover[a]	25–30
		Chinese cabbage[a]	35–40
		Water hyacinth[a]	50
		Water lettuce[a]	50
Mixed diets			
Fresh sardine, mackerel, shad, dried silkworm pupae	5.5		
Liver, sardine, silkworm pupae	4.5		
Silkworm pupae, silkworm faeces, grass, soybean cake, pig manure, night soil	4–8		
Raw silkworm pupae, pressed barley, *Lemna* and *Gammarus*	2.55–4		
Two-thirds groundnut cake, one-third manioc leaves	3.5		
One-half manioc leaves, one-half ground rice	11.0		
Manioc leaves and fresh manioc root	26.8		
Fish flour, rice flour	2.5–3.0		
Meat flour, potato	3.5–4		
Fresh silkworm pupae, wheat flour	10.4		
Fish flour, soybean cake, yeast	1.7–2.8		
Fish flour, cottonseed meal, yeast	1.56–3.4		

[a] Food quotients for herbivorous fish species in China (Yang, 1985).
Source: Ling (1967).

Table 9.12 Usage of supplemental ingredients for Indian carp in semi-intensive culture in Andhra Pradesh, India

Feed ingredients	*Usage by farmers*		*Total input (kg/ha/year)*	
	No.	*%*	*Range*	*Mean*
Rice bran (RB)	16	8	2000–7000	4180
Deoiled bran (DOB)	188	99	4000–43 000	18 430
Groundnut cake (GNC)	156	83	600–12 000	5310
Cottonseed meal (CSM)	120	63	1000–10 000	3730
Sunflower meal (SFM)	42	22	1000–15 000	3730
Soybean meal (SBM)	8	4	2000–7000	4200
Millets (MIL)	6	3	1000–4000	2220
Sorghum (SOR)	6	3	1000–4000	2470
Deoiled cake (DOC)	13	7	900–4670	4620
Eggs	3	2	100–200	160
Salt (%)	83	44	1–5	2.1
Minerals (%)	21	11	0.3–3	1.7

even within a single region, and also the variation of farmer preference for particular ingredients and/or ingredient mixes. Our knowledge on use of supplementary feeds at farmer level *per se* is very limited. Until the recent survey by Verrina *et al.* (a, b, in preparation) there had been no documentation of the use of common salt by Andhra Pradesh farmers as an additive in supplementary feed(s). However, this is just one example, and perhaps there are many more such ingredients of which the scientific community is unaware and whose scientific basis of usage is unexplained. This further exemplifies the need for a deviation from the traditional nutritional research based on dose–response approach as was advocated by De Silva and Davy (1992).

Over the last decade the terms supplementary feeds and supplementary ingredients have been almost completely replaced by the term non-conventional feedstuffs. No longer is there an emphasis on these ingredients *per se* but on utilization of such ingredients in compounded feeds. Comprehensive accounts on such feed resources are given by Devendra (1985), Tacon and Jackson (1985), Pantastico (1988) and Wee (1991). The proximate composition of feed ingredients, their suitability for incorporation into practical compounded feeds and other relevant information are dealt with in these reviews. Wee (1991) summarized the results of some of the experimental work based on the utilization of feed ingredients in fish feeds for tropical species. He showed that most oil-seed meals and grains when incorporated into compounded feeds at levels up to about 35% of dietary protein resulted in growth equal to more than 80% of that reported for control diets in which fish meal was used as the primary source of protein.

Table 9.13 Combination of supplemental feed ingredients commonly used in semi-intensive culture of Indian major carps in Andhra Pradesh

Item	*Farmers*		*Total input (kg/ha/year)*		*Other ingredients added*
	No.	*%*	*Range*	*Mean*	
DOB	15	8	5000–33 000	20 340	NIL
DOB+CSM	7	4	19 000–40 000	25 280	GNC
DOB+DOC	6	3	10 000–33 000	24 830	CSM/GNC/SFM
DOB+GNC	142	75	5000–50 000	27 650	CSM/SFM/SBM/SOR/MIL
DOB+RB	1	1	22 000	22 260	GNC/CSM
DOB+SF	2	1	34 000–59 000	46 800	CSM/GNC
RB+DOB	15	8	20 000–39 000	26 790	GNC/DOC/CSM/SFM/SOR/MIL
NIL	1	1	0	0	NIL
TOTAL	189	100	5000–59 000	27 000	

Abbreviations for the feed ingredients are given in Table 9.12.
Source: Data from Veerina *et al.* (a and b in preparation).

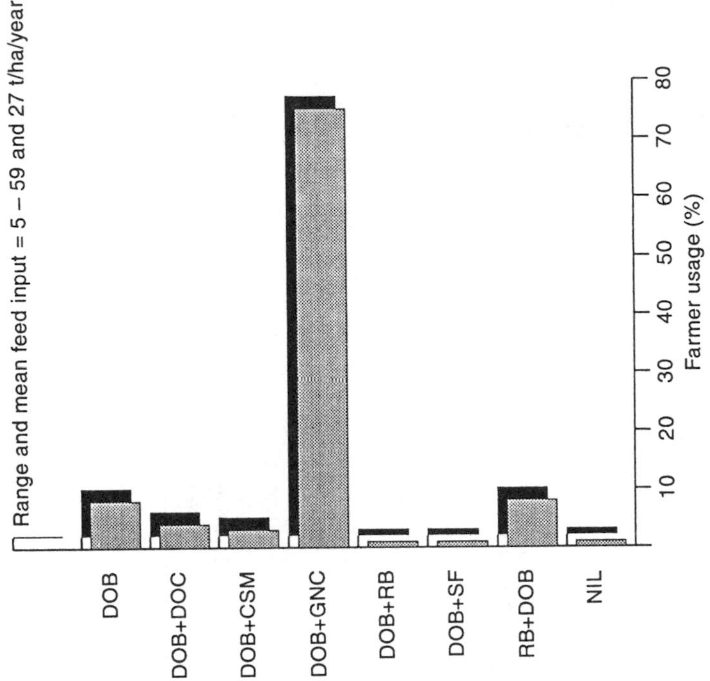

Figure 9.11 Extent of use of different supplementary feed combinations in semi-intensive carp culture practices in four districts in Andhra Pradesh, India. The abbreviations are the same as those in Table 9.12. (From Verrina *et al.* in preparation.)

These findings have given some indirect indications of the effectiveness of use of agricultural by-products in the simple mix forms commonly used by the rural farmer. These producers at present account for nearly more than 50% of the cultured finfish production in the world.

9.8 WEANING

From the point of view of aquaculture, weaning implies the transformation of the young to an external food source, mostly artificial diets. This is a crucial stage in the life-cycle of all animals, especially fish, which in most cases receive no parental care. It is at this stage that high mortalities are encountered especially with marine and/or catadromous species such as milkfish, eel and flatfish species.

Modern aquaculture generally relies on producing its own young, and thus ensuring appreciable survival rates of young fry at weaning is essential. There are a number of factors that operate at this time. At hatching most finfish and shrimp species are small in size. In the case of finfish, yolk reserves on which they exist for a short period of time are present. The mouth is not well developed, nor is the digestive system. The digestive enzymes are not all operative at the time of hatching, and it has been suggested that enzymes associated with the ingested food may aid in the digestion. Thus food material ingested is limited to a narrow size range, and must be easily digestible. These aspects were dealt with in detail in Chapters 4 and 5. Most finfish larvae in the wild feed on microscopic plants and animals. In aquaculture these food organisms, known as live foods, are provided in the initial stages of the culture operation.

A typical feed pattern of a marine larva as it grows may be considered as:

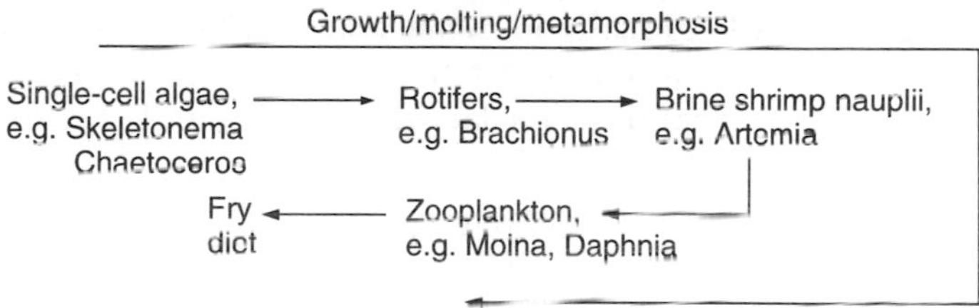

However, preparation of live food on a large scale is expensive and time consuming, and the quality of the food cannot be guaranteed. This is particularly the case of marine fish culture. Moreover, no one live food will provide adequate nutrition for the cultured organism to achieve maximum growth. Providing a wide array of live foods to maintain a nutritional balance becomes difficult and costly. It is therefore advantageous, especially for intensively cultured marine species, to wean larvae off live food and on to artificial diets.

The most crucial factor in the weaning process appears, however, to be the provision of prepared or live food of the preferred particle size. Research has shown that fish prefer food particles to be smaller than their mouth size, and larvae often feed on particles smaller than they are physically capable of ingesting. This is exemplified by the preferred food particle size (dry feed) for juvenile Atlantic salmon, European eel and common carp, being approximately 0.3–0.4, 0.4–0.6 and 0.3–0.5 of the mouth size respectively (from Thorpe and Wankowski, 1979; Knights, 1983; Hasan and Macintosh, 1992, in order). Hasan and MacIntosh (1992) also reported that in common carp the size of food particles most quickly ingested increases with body size. These authors suggested that the faster ingestion time recorded for particles of the preferred size reflects a combination of food size preference and effect of particle size on the feeding efficiency of the fish.

Recent research and developments in larval weaning, particularly in respect of marine finfish and shrimp, suggest that weaning larvae on to an artificial feed should be done directly. To wean larvae on to artificial diets effectively a number of strategies need to be followed. Feeding should be as frequent as possible, by hand or automatically. Diets should be changed gradually by mixing increasing proportions of new with old. Changing directly from one diet to the next in a single step should be avoided. Water currents to disperse the food as extensively as possible (Figure 9.12) should be generated. The particle size of the food should be equivalent to that of live food which has proved satisfactory and, finally, the foods provided should be attractive in taste.

Research on weaning, particularly on factors that would enhance weaning, is being carried out on a number of cultured species. For example, it has been shown that continuous lighting enhances weaning in the larval rabbit fish, *Siganus guttatus* (Duray and Kohno, 1988), and in first feeding fry of Atlantic salmon, *Salmo salar* (Steffansson *et al.*, 1990).

A considerable amount of research has been performed on enhancing the suitability of artificial diets for larvae of finfish and shellfish by the

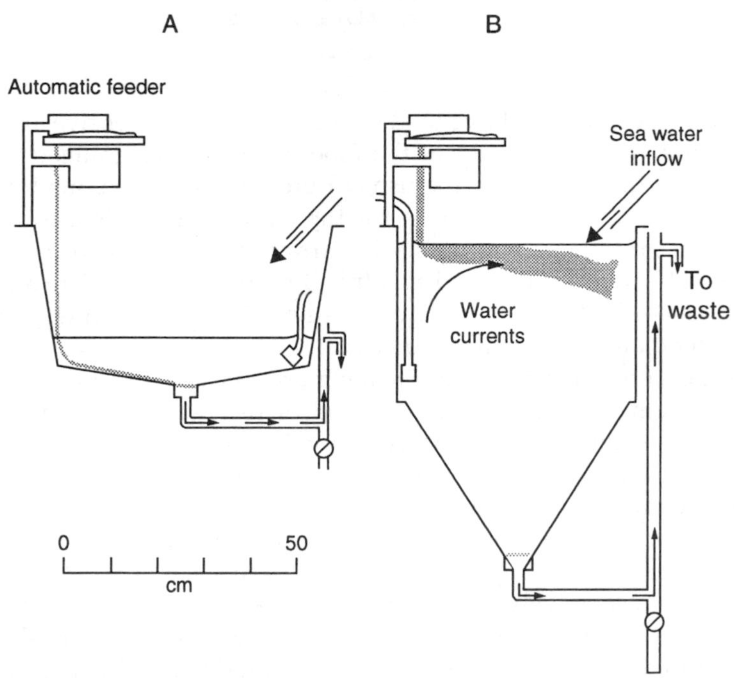

Figure 9.12 The use of water current to disperse feeds in weaning tanks, as suggested by Knights (1985). Note that in A the shape of the vessel, the placement of the aeration and the water inflow combine to prevent dispersal of food throughout the tank, as is shown in B.

process of microencapsulation. However, these diets are costly and have not been universally accepted, with culturists still tending to use live foods in spite of the difficulties associated with their culture.

Information on the suitability of artificial feed(s) during weaning is controversial. Rottman *et al.* (1991) compared the influence of live foods and dry diets for intensive culture of two carp species and found that those larvae fed on the live foods performed better. Larvae maintained on the freshwater rotifer *Brachionus* performed better than those fed on *Artemia* nauplii. Dabrowski and Kaushik (1985) reported that coregonid larvae (*Coregonus schinzi palea*) performed better on a dry diet than on a diet of *Artemia* nauplii. and small cladocerans. Jana and Chakrabarti (1990) demonstrated that Indian major carp larvae performed better on an exogenous supply of live plankton as compared with inducing a plankton bloom in the system. Rutledge and Rimmer (1991) reported the success of rearing, and thus weaning, newly hatched sea bass (*Lates calcarifer*) in fertilized earthen ponds. The growth rates reported by these authors were superior to larvae reared intensively on live foods and therefore enabled early transfer to freshwater and weaning on to a pelleted feed. The list of varying observations goes on and on, and it is difficult to be conclusive on this aspect of nutrition. The differences may be brought about by ontogenic and phylogenic differences in species under consideration. Generalization, therefore, on aspects of weaning is not possible.

CHAPTER 10

The feed industry

10.1 INTRODUCTION

The fish feed industry, also popularly referred to as the 'aquafeed' industry, is one of rapid development and growth. This is a new industry, and as yet its size is not directly correlated to the extent of the aquaculture industry throughout the world. However, growth of the aquafeed industry in Asia over the last decade is beginning to extend commercial fish feed production towards realizing its true potential.

A feed industry develops where there is a sustained demand for its product. The culture of carnivorous fish species is almost totally dependent on the use of artificial diets. As such, the aquafeed industry has become established in countries where carnivores are cultured. The production of carnivorous fish in the world (1985–89) is summarized in Table 10.1. It can be seen from this table that the early development of the aquafeed industry based around carnivorous fish has neglected the majority of aquaculture production.

However, this does not mean that non-carnivorous species are not given artificial diets. In Asia, which is the biggest finfish producer, carnivorous fish only accounted for 7% of production in 1989, as compared with the rest of the world, where 52% of finfish produced are carnivores (Table 10.1). It is therefore apparent that non-carnivorous species dominate aquaculture in Asia, and in fact 39.5% of the total aquafeed market in the Asia–Pacific area was directed to the requirements of this section of the industry in 1990 (New and Csavas, 1993). New and Csavas (1993) estimated that non-carnivorous finfish culture in Asia would require 34% of a total aquafeed market of 2.6 million tonnes by the year 2000. It is also worth noting that rural Asian aquaculture is largely yet to utilize pelleted feeds to any great extent, generally relying on farm-made feeds comprising simple mixtures of agricultural by-products and/or slaughterhouse waste made into crude dry or moist pellets. There is therefore a huge potential for increase of the aquafeed industry if it can cater to farmers from these areas.

Recent development in the aquafeed industry has occurred in places where shrimp culture has become predominant. Aquaculture of shrimp is expanding in the Asian countries of Thailand, Indonesia and India and

Table 10.1 Trends in finfish aquaculture production 1985–89 (all quantities are in millions of tonnes)

	1985	*1986*	*1987*	*1988*	*1989*
Total finfish production	5.047	5.743	6.550	7.150	7.327
Non-carnivorous fish					
World	4.471	4.918	5.600	6.097	6.166
Asia	3.694	4.312	4.964	5.414	5.487
World (%)	88.6	85.6	85.5	85.3	84.1
Asia[a]	82.6	87.7	88.6	88.8	88.9
Asia (%)[b]	73.2	75.1	75.8	75.7	74.9
Carnivorous fish					
World	0.533	0.823	0.948	1.050	1.158
Asia	0.347	0.352	0.394	0.439	0.427
World[c] (%)	6.9	14.3	14.5	14.7	15.8
Asia[a]	65.0	42.8	41.6	41.8	36.9
Asia (%)[b]	6.9	6.1	6.1	6.1	5.8

[a] Percentage Asian contribution to the world production of the commodity.
[b] Percentage Asian contribution of commodity to world finfish production.
[c] Percentage contribution to the world total finfish production.
Source: Based on data from FAO (1990, 1991).

the South American country of Ecuador. A concurrent expansion of the feed industry has already taken place in these regions.

In this chapter we have also included sections which are not a direct facet of the industry, but which in the long term may have a bearing on the industry. In addition, we have included aspects on 'farm-made' feeds, a facet which is rarely mentioned in texts, but which is nevertheless a very important facet because the great bulk of finfish culture relies on 'farm-made' feeds.

10.2 STATUS OF THE AQUAFEED INDUSTRY

New (1991) made an attempt to estimate the feed markets for commercially fed species. A summary of the feed market estimates of New (1991) is given in Table 10.2. It can be seen from this table that Europe, N. America and Asia dominate the market for feed for carnivorous fish, while Asia alone dominates the non-carnivorous fish and shrimp feed markets. Of the carnivorous species trout, catfish, salmon and eels provide the bulk of the market, requiring 29%, 25%, 19%, and 14% respectively. Common carp (43%), tilapia (27%) and milkfish (30%) account for the non-carnivorous feeds.

New (1991) estimated the 1988 shrimp feed market to be about 513 000 tonnes, of which 86% was used in Asia (Table 10.1). The total shrimp feed production was estimated to reach 653 000 t in 1990, with Asia

Table 10.2 Estimates of feed markets in 1988 in 1000 t for different groups of organisms

Group	*N. America*	*S. America*	*Asia*	*Europe*	*Other*	*Total*
Carnivorous fish	407	12	397	448	8	1272
Non-carnivorous fish	8		339	50	25	422
Shrimp	18	54	440		1	513
Total	433	66	1176	498	34	2207

Source: Based on data from New (1991).

Table 10.3 Estimates of farmed shrimp production and shrimp feed production

Country	*Heads-on production (1000 Mt)*		*Proportion of global production (%)*		*Estimated commercial shrimp feed production in 1990 (1000 Mt)*
	1989	*1990*	*1989*	*1990*	
Eastern Hemisphere					
China	165	150	29.2	23.7	60
Indonesia	90	120	15.9	19.0	192
Thailand	90	110	15.9	17.4	176
India	25	32	4.4	5.1	13
Philippines	50	30	6.9	4.7	48
Vietnam	30	30	5.3	4.7	12
Taiwan	20	30	3.6	4.7	60
Bangladesh	–	25	–	4.0	10
Japan	30.1	3.5	5.3	0.5	7
Others	–	5.0	–	0.8	2
Total Eastern	500.1	535.5	88.5	84.6	580
Western Hemisphere					
Ecuador	45	73	8.0	11.5	44
Other South American	7	10	1.2	1.6	6
Honduras	3.4	4.5	0.6	0.7	7
Other Central American	6.1	5	1.1	0.5	8
Mexico	2.5	4	0.5	0.6	6
USA	–	0.9	–	0.2	2
Others	0.7	–	0.1	–	–
Total Western	64.7	97.4	11.5	15.4	73
Global Total	564.8	632.9	100.0	100.0	653

Source: New (1991).

retaining its dominance (Table 10.3). New and Wijkstrom (1990) estimated the shrimp feed market in Asia, to be around 1.132×10^6 t in the year 2000, implying a growth of 73% from the 1990 level. A concurrent growth of the fish feed industry can also be expected to cater to the increasing intensification in the Asian finfish culture practices (Csavas, 1990), although not to the same extent as for shrimp feed.

Table 10.4 Aquafeed market in the Asia–Pacific region in 1990 and the year 2000 (all values are in tonnes)

Species group	*1990*	*2000*
Carnivorous finfish	471 640	817 485
Non-carnivorous finfish	554 206	885 575
Crustacean	532 875	919 794
Total aquafeed market	1 558 721	2 622 854

Source: Based on New and Csavas (1993).

New and Csavas (1993) revised the estimates of the total aquafeed market in the Asia–Pacific region. These predictions, summarized in Table 10.4, indicate that the aquafeed market in the Asia–Pacific region is expected to grow nearly 68% in the next decade, a relatively high rate of growth for any industry.

In the following sections patterns of growth of the aquafeed industry in Europe, Canada, Japan and SE Asia will be detailed.

10.2.1 Europe

Salmonids represent the major group of freshwater finfish cultivated in most countries in Europe, primarily represented by rainbow trout and Atlantic salmon. In most western European countries, particularly those of the European Economic Commission, fish feeds in the form of dry pellets are produced locally. Extruded pellets are manufactured by a few big companies mainly for seawater culture of salmonids and other marine fish. Specific water-stable diets are manufactured for shrimp culture.

According to Kaushik (1990), there has been a relative decrease in the cost of feed in overall production cost of rainbow trout. Owing to competition between feed manufacturers and increasing costs of other charges involved in trout farming, feed cost is currently estimated at 30% of the overall cost for trout production (particularly applicable to France).

10.2.2 Canada

Prior to the mid-1980s, fish feed was imported into Canada from the US. However, in the second half of this decade a viable feed industry developed in Canada. The development of this industry was associated with salmon farming in net-cages, which boomed on both coasts as a result of Norwegian initiatives. Cho *et al.* (1990) estimated the total requirement for 1988 to be 48 000 t, and predicted this to increase to 90 000 t by 1990, based on a FCR of 2.0 (Table 10.5).

Table 10.5 Fish feed production and future fish feed (dry) requirements in Canada. All values are in tonnes

Year	*British Columbia*	*Prairie provinces, Ontario, Quebec*	*Maritime provinces*	*Canada total*
1980	1500	3800	800	6100
1985	2500	4900	2400	9800
1990[a]	66 000	6500	17 500	90 000
1995[a]	110 000	8700	27 000	145 700

[a] Estimated.
Sources: Cho *et al.* (1990).

10.2.3 Japan

The fish feed industry in Japan developed particularly for eel, carp, trout and ayu. The freshwater production of these species has plateaued, as has the fish feed production for those species (Figure 10.1). It can be seen from this figure that the total annual production is relatively steady at approximately 140 000 t.

In the coastal areas, the predominant feeding is low-value fish, commonly known as 'trash fish'. It is estimated that 1.7 million tonnes of trash fish are used annually. However, because of adverse effects of this feeding method on water quality there is an increasing trend to utilize dry diets. The development of dry diets for marine species has neverthe-

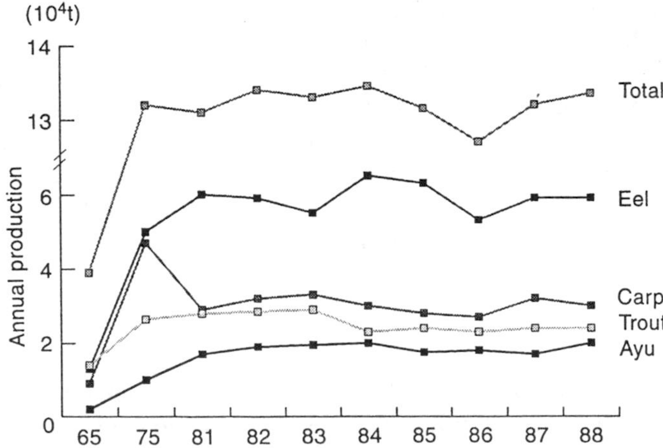

Figure 10.1 Trends in the production of formulated feeds for finfish culture in Japan. (From T. Watanabe *et al.*, 1990.)

less been a slow and a tedious process, and the adoption of research findings by farmers even slower! Japan offers specific examples in regard to direct influence of long-term nutritional research findings on culture practices, environmental degradation minimization and profitability (T. Watanabe, 1991).

10.2.4 Asia (South-East)

The intensification of certain culture practices for milkfish, carp and tilapia and combined with the 'boom' in the shrimp industry has had a major impact on the development and growth of the aquafeed industry in SE Asia (Table 10.6). It can be seen here that demand for fish feed has increased in Indonesia, while remaining relatively static in the other countries. The demand for shrimp feed, on the other hand, has escalated in the Philippines, Thailand and Indonesia being around 460 000 t in 1988. The number of feedmill plants increased to keep pace with the demand. In spite of this expansion some countries still had to import shrimp feed, with Indonesia importing nearly 80% of its shrimp feed requirements in 1988.

It should be borne in mind that there are none or very few fish feed manufacturers *per se* in SE Asia. Most fish feed manufacturers, are animal feed manufacturers, and fish feeds are often only of secondary importance to them. This does not mean that multinational companies are not involved in fish feed manufacture in this region.

Table 10.6 Market demand for fish and shrimp feed (in 1000 Mt) in ASEAN Countries from 1985 to 1988. The number of shrimp feedmills is given in parentheses

	1985		*1986*		*1987*		*1988*	
Country	*FF*	*SF*	*FF*	*SF*	*FF*	*SF*	*FF*	*SF*
Indonesia	6.0	0.6 (2)	14.4	3.6 (8)	18.0	9.6 (11)	27.0	60.0 (19)
Malaysia	0.1	1.2 (1)	0.3	1.8 (4)	0.6	2.4 (4)	1.0	3.0 (5)
Philippines	0	0.6 (1)	0.1	4.8 (4)	0.2	18.0 (8)	0.7	54.0 (12)
Thailand	18.0	0.6 (1)	24.0	6.0 (4)	24.0	49.0 (6)	18.0	100.0 (15)
Total	24.1	3.0 (5)	38.8	16.2 (20)	42.8	79.0 (31)	46.7	217.0 (53)

FF, fish feed; SF, shrimp feed.
Source: Adapted from Boonyaratapalin and Akiyama (1990).

10.3 PROBLEMS OF THE AQUAFEED INDUSTRY

Aquaculture has been considered as the answer to the shortfall from the global capture fisheries production that is likely to occur by year 2000. This certainly is one of the main arguments that lured international organizations and national governments into aquaculture.

The trend in aquaculture production (globally) is that the main sectors of growth are seen in finfish and crustacean culture. The intensification of finfish and crustacean culture is bound to make them more dependent on artificial, compounded feeds. Therefore, aquaculturists using artificial diets are in a sense competing with people and terrestrial livestock farmers for basic feed ingredients. As pointed out earlier, intensification of culture practices is gradually making even the culture of non-carnivorous finfish species dependent on commercial feeds, thereby enhancing the competition for primary feed ingredients.

Of the ingredients commonly used in aquaculture diets, fish meal is known to be the best source of essential amino acids required by finfish and shrimps. It is also believed to act as a feed attractant, and is claimed to have unidentified growth promoters. In all compounded feeds used in aquaculture, fish meal remains the main protein source, although partial replacement of fish meal in fish diets has been possible (Chapter 9). However, the fact that commercial feeds still contain significant quantities of fish meal itself provides indirect evidence that partial, let alone complete, replacement of fish meal is not yet effective in intensive culture.

The use of fish meal in aquaculture diets essentially involves the conversion of wild caught fish into cultured animals. This conversion, however, is far from efficient. Based on an average FCR of 1.5, it can be calculated that, to produce 1 kg of eel, 1.5 kg of feed is required. At an inclusion level of 50%, 0.75 kg of fish meal will be used. Since about 5 kg of raw fish (trash fish) is required to make 1 kg of fish meal, approximately 3.75 kg of raw fish is required to culture 1 kg of eel. Wijkstrom and New (1989) estimated 'fish meal equivalent' used for aquaculture production during years 1984, 1985 and 1986 (Table 10.7). They concluded that, at the current levels of culture, carnivorous finfish and shrimp culture industries consume nearly 8% of the total fish meal production of the world. According to these authors, by year 2000, 1.3×10^6 t of fish meal will be required to produce 2.4×10^6 t of carnivorous fish and shrimp, accounting for 6.5×10^6 t of wild-caught fish. This seems to contradict the original presumption in this section that aquaculture is the answer to the shortfall from the global capture fisheries. It appears that the decrease in wild-caught fish may in fact limit or curtail the growth of the aquaculture industry.

The International Fish Meal Manufacturer's Association, on the other hand, has predicted a 5% decline in fish meal production over the next

Table 10.7 Estimated usage of fish meal in carnivorous fish and shrimp production (1000 t)

	1984	*1985*	*1986*
Global fish meal (GFM) production	5809	6032	6516
Aquaculture production			
Carnivorous fish	657	725	785
Shrimp	165	184	206
Total	822	909	991
Fish meal equivalent (FME) required for feed	444	490	523
Of which fish meal (FM)	302	337	363
FM as a percentage of GFM production	5.3	5.6	5.6
FME as a percentage of GFM production	7.6	8.1	8.0

Source: Wijkstrom and New (1989).

decade, to approximately 6–6.5 × 10^6 t in the year 2000. This would mean aquaculturists would require 15–17% of the world fish meal supply – a doubling of the 1986 level. Obviously, in a scenario in which the supply is limited, prices are bound to rise and the competition for fish meal will be greater. According to Wijkstrom and New (1989), the current technology with regard to aquafeeds is such that shrimp and carnivorous fish producers will run into a cost-price freeze. It is therefore unlikely that the share of these commodities will increase much beyond 20% by the year 2000. These authors suggested that to escape the 'fish meal trap' the marine protein component in aquatic feeds should be replaced by alternative ingredients, and sea ranching expanded. They were also careful to warn that the 'fish meal trap' may only be the first of several 'ingredient traps' for aquaculture in the future.

Unfortunately, in contrast to fish meal, no reliable statistics are available on other 'ingredient traps' that the industry is likely to confront in the future. In most semi-intensively managed operations in Asia, particularly in India and China, fish are fed with agricultural by-products mixed in different proportions, sometimes with fish meal added (section 9.7). These ingredients, including rice and wheat bran and oilcake residues, were considered waste products and their disposal was problematic. The growth of the livestock industry the world over and increasing other alternative uses have made these ingredients less and less available to aquaculturists. This leads us to a situation where incorporating such ingredients into compounded feed might no longer be economical or cost-effective. Therefore, even in respect of semi-intensive systems, which utilize diet supplementation, we might not be completely independent of a potential 'ingredient trap'.

10.4 ALTERNATIVE FEED INGREDIENTS

Dietary ingredients were dealt with in detail in Chapter 7 in reference to diet formulation. However, in this section we thought it is appropriate to reconsider some aspects in relation to the feed industry at large.

Often when one refers to alternative and/or non-conventional feed ingredients in aquafeeds, it is with reference to an ingredient that is capable of acting as a partial or complete substitute for fish meal. Fish meal, as pointed out earlier, has a well-balanced spectrum of essential amino acids and some essential fatty acids, good organoleptic properties and is reputed to contain unidentified growth-promoting factors. It is useful at this stage to compare fish meal as a dietary ingredient with some others.

The proportion of essential amino acid in the protein of selected feed ingredients, expressed as a percentage of crude protein, is given in Table 10.8. It is evident from this that all feedstuffs have high levels of crude protein, but that the amino acid balances differ. Corn gluten and blood meal, in contrast to herring meal (fish meal), are imbalanced with respect to leucine. Corn gluten is also imbalanced with respect of lysine and tryptophan, while blood meal is very low in isoleucine.

When the chemical scores of protein of these feedstuffs (Table 10.9) are compared, herring meal would fully meet the essential amino acid requirements of trout. The apparent chemical score data show that there is no single feedstuff that provides a possible alternative to fish meal.

Table 10.8 Proportions of essential amino acids as a percentage of the protein of certain feedstuff proteins

	Met	*Cys*	*Lys*	*Trp*	*Thr*	*Ile*	*Hist*	*Val*	*Leu*	*Arg*	*Phe+ Tyr*
Herring meal (68% CP)	3.05	1.00	7.90	1.11	4.00	4.16	2.65	7.92	7.08	7.80	6.60
Corn gluten meal (60% CP)	3.16	1.83	1.67	0.50	3.33	3.83	2.00	4.50	15.67	3.17	11.66
Blood meal (80% CP)	1.25	1.75	6.63	1.25	4.75	1.00	3.81	6.50	12.87	2.94	9.02
Meat and bone meal (50% CP)	1.34	0.66	5.20	0.52	3.26	3.40	1.92	4.50	6.40	6.70	4.98
Poultry by-product meal (58% CP)	1.79	1.72	4.43	0.95	3.50	4.02	2.78	4.57	7.59	6.62	4.66
Soybean meal (48.5% CP)	1.44	1.53	6.59	1.57	3.71	5.15	2.68	5.57	7.62	7.84	8.38

Source: Cowey (1990).

Table 10.9 Apparent chemical score of protein from certain feedstuff proteins in terms of essential amino acid requirement of salmonids

	Met	*Lys*	*Trp*	*Thr*	*Ile*	*Hist*	*Val*	*Leu*	*Arg*	*Phe+ Tyr*
Requirement as per cent protein[a]	2.9	5.0	0.5	2.2	2.2	1.8	3.2	3.9	4.0	5.1
Herring meal	105[1]	158	222	182	189	147	248	182	195	129[1]
Corn gluten meal	109	33[1]	100	151	174	111	141	402	80[2]	229
Blood meal	43[1]	133	250	216	45[2]	212	203	330	74	177
Meat and bone meal	46[1]	104[2]	104[2]	148	155	107	141	164	168	98[2]
Poultry by-product meal	62[1]	89[2]	190	159	183	154	143	195	166	91[2]
Soybean meal	50[1]	132[2]	314	169	234	149	174	195	196	164

[a] Values from NRC (1981) for Chinook salmon except Arg and Met which are values for rainbow trout from Kim *et al.* (1983, 1984). Superscripts 1 and 2 refer to first and second most limiting amino acids respectively.
Source: Cowey (1989).

Non-conventional feed ingredients are generally referred to as those ingredients with a potential of either partially or wholly replacing fish meal in aquafeeds. New (1987) documented the major categories of ingredients used in aquaculture feeds. Research on such ingredients has been on-going for nearly 15 years. This research has been done on both cold- and warmwater species, particularly finfish species. Wee (1991) reviewed the information available on possible use of non-conventional feed ingredients in compounded diets for finfish.

Feedstuffs referred to by New (1987) and Wee (1991), amongst others, have been used either singly or in combination in rural aquaculture practices in Asia. In spite of the research findings, the degree to which fish meal is substituted in aquafeed industry at present is minimal. Cowey (1990), when discussing trout diets, attributed this to the fact that some of the materials used as substitutes for fish meal are no more plentiful than fish meal and at least as expensive.

10.5 REDUCING FEED COSTS

Like the earlier section this one deals with an aspect which is not directly related to the feed industry. However, the section includes information which in the long term may have a bearing on the feed industry, hence the decision to include it here rather than in Chapter 9.

Most of the nutritional research and other efforts over the last two decades have been directed towards determining the nutrient requirements of culturable species and developing low-cost feeds. Presently, at least in developed countries, research effort is expended on developing high nutrient density diets, with a view to minimizing environmental degradation. In livestock rearing, husbandry practices have been developed which have enabled substantial savings of feed costs. Foremost amongst these are

- reducing or minimizing physical wastage; and
- adopting proper feeding practices such as feeding at the most suitable time of the day at the most optimal frequency.

In contrast to agriculture, in aquaculture inherent difficulties are imposed by the culture medium on optimal 'feed management'. For example, even under the best of conditions a certain degree of feed wastage is unavoidable in aquaculture. Wastage becomes higher with shrimp culture. However, overall the research effort expended on feed cost-saving through use of different feed types in conjunction with proper husbandry practices in aquaculture has been rather meagre.

10.5.1 Mixed feeding schedules

More recently, De Silva (1985b) has shown that, at least in semi-intensive culture practices, feed cost-savings could be made by alternating feeds of high and low protein contents. This preliminary laboratory work has been tested and proven to hold true under pond conditions. The concept of a mixed feeding strategy, as a means of saving feed cost in semi-intensive culture, was based on the observation that the digestibility of a feed varies from day to day, following an apparent cyclic pattern (two to three days of high digestibility alternating with a day or two of low digestibility). *O. niloticus* were fed a high-protein diet alternated, for differing numbers of days, with a low-protein diet. It was observed that certain alternate feeding schedules resulted in an almost equal performance of the fry reared continuously on a high-protein diet. The feed cost saving that resulted from adoption of the appropriate mixed feeding schedule approximated 30%.

Subsequently, the utilization of mixed feeding schedules on Indian major carps (*Catla catla* and *Labeo rohita*) was tested in the laboratory (Nandeesha *et al.*, 1994a). The results of these experiments also suggested that significant feed cost savings could be made by using mixed feeding schedules for Indian major carps. These laboratory experiments have now been extended to field trials on Indian major carp polyculture practices (Nandeesha *et al.*, 1994b), and the results have shown that up to 30% of feed costs can be saved without prolonging the grow-out period and/or performance of the target species (Figure 10.2). Apart from direct feed cost-savings, the other beneficial effect of use of mixed feeding schedule is the reduced nitrogen input into the culture system. This in turn reduces possible eutrophication in the ponds.

Undoubtedly, there are practical problems of adoption of mixed feedings schedules 'on farms'. Farms will have to store two different types of feeds and each feed dispensed according to a strict predetermined schedule. However, these problems are not insurmountable, particularly when the extra effort is likely to be significantly cost-effective. On the other hand, the mixed feeding schedule husbandry practice(s) can be easily adopted by rural fish culturists by minor adjustments to the proportion of ingredients in the feeds that are currently being used.

10.5.2 Economically optimal dietary protein level

In the previous chapters it has been pointed out that the gross dietary requirements in the species group are not significantly different between individual species. De Silva *et al.* (1989) used published data on the protein requirements of tilapia species to show that when the minimum energy requirements are met the growth performance to dietary protein level depicts a second-order curve (Figure 10.3). Based on this relationship they deduced the economically optimal dietary protein level for tilapia fry to be between 30% and 32% by dry weight, which is lower than the dietary protein level which gives maximum growth. These authors, using the appropriate FCR and PER values, went on to demonstrate that, in spite of the lower rate of growth obtained, it would be more economical to rear tilapia fry on the lower protein diet, which they termed the 'economically optimal dietary protein level'. An economically optimal dietary protein level, when used, results in substantial savings in feed costs. Based on available data the concept has also been extended to Indian and Chinese carps (De Silva and Gunasekera, 1991). These findings also indirectly illustrate the fact that, in most studies, the optimal protein requirement has been grossly overestimated, and rarely have two independent studies on the same species given comparable results. Some of the differences observed between experimental studies can be accounted for by differences in the experimental conditions as well as differences in the experimental diets. Santiago and Reyes (1991) working on bighead carp (*Aristichthys nobilis*) fry found the optimal protein requirements to be around 30%, a level much lower than that reported by earlier workers but close to the economically optimal dietary protein level deduced by De Silva and Gunasekera (1991) for Chinese major carps.

The concept of an economically optimal dietary protein level has yet to be tested in true culture conditions. Increasingly, evidence is becoming available that the dietary protein requirements for most species have been overestimated, thereby indirectly reinforcing the concept. It is of interest to note that Luquet (1989), in his evaluation of results of Lovell (1980), Wannigama *et al.* (1985) and Newman *et al.* (1979), pointed out the validity of using diets of lower protein content in practical culture than that suggested by nutrient requirement studies. Luquet (1989)

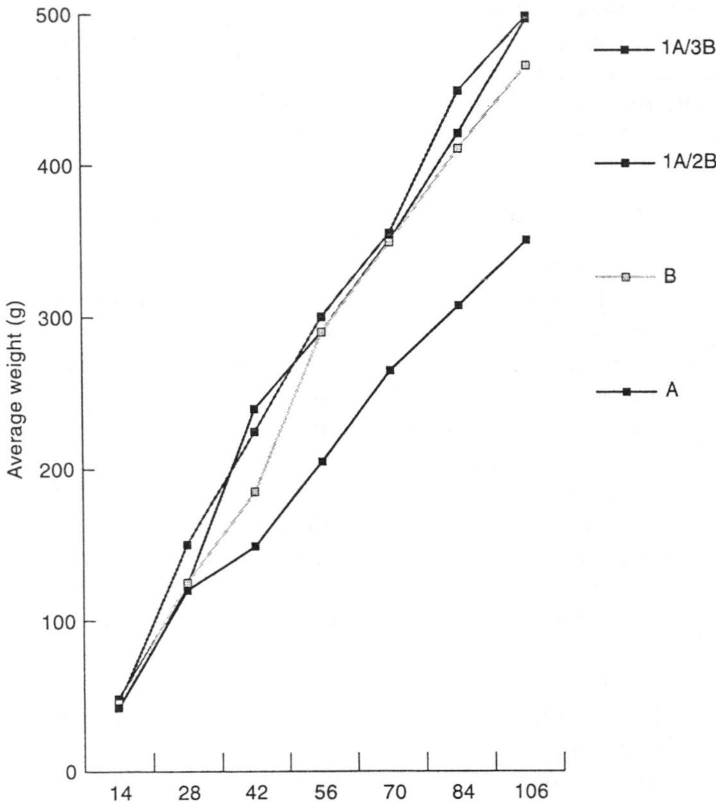

Figure 10.2 The mean weight of common carp and rohu cultured (106 days) on low-protein diet (A), high-protein diet (B) and two mixed feeding schedules of 1A/2B and 1A/3B. In the mixed feeding schedules the numerical value refers to the number of days each of the diets were presented. For example, in the 1A/2B schedule diet A (low protein) is alternated with 2 days of diet B (high protein). (From Nandeesha *et al.*, 1994a.)

went on to recommend that under mass culture the concept of economically optimal protein dietary level needs to be kept in mind. In this instance, however, the concept was not defined and the suggestion was based on the fact that the target species is capable of obtaining a part of its nutrition from natural production. In this respect it differs significantly from the concept developed by De Silva *et al.* (1989), which was based on results of laboratory studies; if proven to be true under field

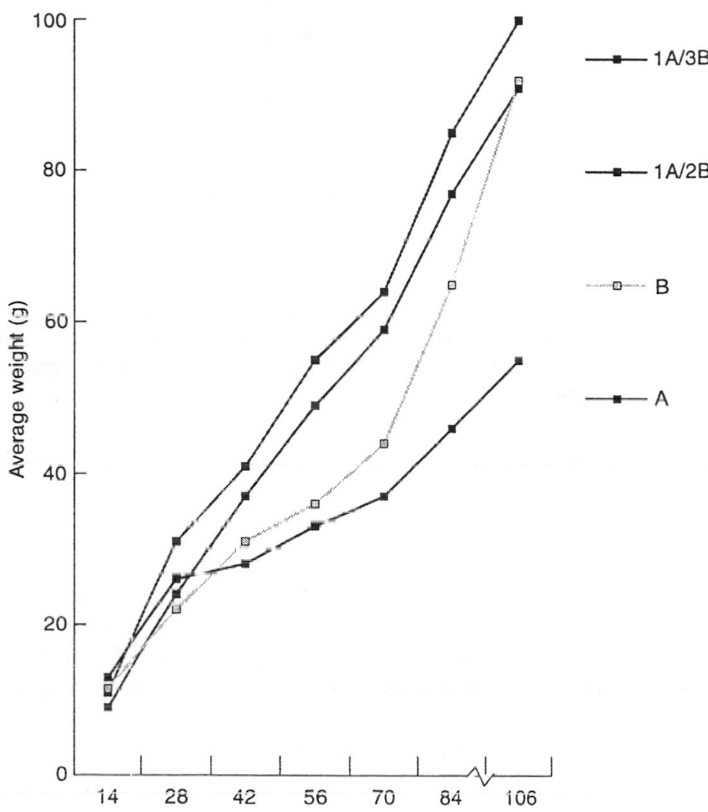

Figure 10.2 *(continued)*

conditions, the economically optimal dietary protein level will be lower than the level which results in maximum growth even if the target species is not obtaining any nutrition from the natural pond food production. Urban and Pruder (1991) have also demonstrated that a diet producing poorer growth can be the diet producing the maximum profit. This is very similar to the concept of the economically optimal dietary protein level.

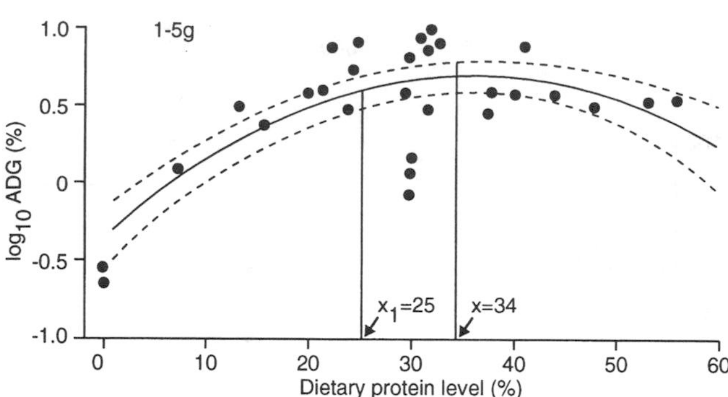

Figure 10.3 The fitted second-order polynomial curve of growth rate (per cent daily growth) of different species of tilapias of 1–5 g body weight in relation to dietary protein level. X, the dietary protein level at which growth is maximum; X_1, economically optimal dietary protein level; ADG, average daily growth. (Modified from De Silva *et al.*, 1989.)

10.5.3 Protein-sparing feeds

In fish feeds the most expensive component is the protein source, which generally is mainly fish meal. In addition to being used for growth and repair, protein is also used extensively for providing energy in routine metabolism by fish, hence the notion that fish require a higher level of protein in their diet than farm animals such as poultry. This notion is being increasingly refuted (Bowen, 1987; Lovell, 1989b), and was discussed in detail in Chapter 3.

Increasingly attempts are being made to incorporate other energy sources into fish diets, with the expectation that these sources will spare the bulk of dietary protein for growth and repair. Dietary energy sources that are known to spare proteins are carbohydrates in eels (Degani and Viola, 1987) and lipids in rainbow trout (Medland and Beamish, 1985; Beamish and Medland, 1986) and in hybrid tilapia (De Silva *et al.*, 1991). All studies have indicated that efficient utilization of dietary protein is achieved by increasing the other source of diet energy (carbohydrates or lipid) provided that the protein intake is above a critical threshold. It has been shown that the lipid levels in the diets of rainbow trout and hybrid tilapia can be increased by up to 24% and 18% respectively without sacrificing performance and/or carcass quality. By increasing the inclusion level of these alternative energy sources in fish diets, the cost of the feed can be reduced as protein sources are more expensive in general. However, increased lipid level reduces storage life of feeds and imposes difficulties in pelleting. In some respects, the development of high-energy diets for intensive culture practices, which are known to have less harmful influences on environmental quality, is also an extension of the concept of protein sparing.

Viola *et al.* (1992) reported the possibility of sparing protein by lysine supplementation in intensive common carp culture in Israel. Apart from the protein sparing these authors reported a 20% reduction in nitrogen discharge into the environment as a result of lysine supplementation. The results reported by Viola *et al.* (1992) are different to those of studies on amino acid supplementation to improve the quality of a feed of known amino acid deficiencies. The above concepts, if they are in due course adopted by aquaculturists, will result in changes in the feed industry *per se*. There are likely to be changes in basic formulations and in the costing.

10.6 TRENDS IN FEED DEVELOPMENT

Apart from replacing fish meal with alternative feed sources, there are other developments that could be expected in the aquaculture feed industry. These advances are particularly aimed at aquafeeds for high-value, intensively cultured species. The envisaged developments are likely to occur in three areas. Diets will be formulated with an emphasis on minimizing environmental degradation, enhancing the quality of the produce and enhancing larval survival and growth, particularly marine species.

The major developments with respect to finfish diets are likely to be directed towards minimizing the environmental degradation that is brought about by uneaten food (pellets) and faecal material. Already in Scandinavian countries, legal restrictions have been imposed on the quality of feeds, and such restrictions are likely to become more and more stringent with time. For example, in Finland, fish feeds used in farms situated on lakes must not contain more than 0.75% total phosphorus. In Denmark, from January 1992, the gross energy content of the dry matter of the feed must be at a minimum of 25 MJ/g (6.0 Mcal/kg) and at least 78% of this energy must be metabolizable (Kiaerskou, 1991). Obviously, such legislation is bound to bring about changes in feed quality and manufacturing standards. These aspects are dealt with in more detail in Chapter 11.

Diet development will also proceed to produce feeds that result in a product with specific flesh characteristics desired by the consumer. Redder flesh is desirable in shrimp and salmonids and is achieved by introducing carotenoids in the feeds. Equally, diets will be developed to make the flesh firmer and less fatty, thereby not only increasing its market acceptability but also its shelf-life.

Nutritional enrichment of larval feeds, particularly with respect to shrimp and marine finfish, also seems imminent. It is conceded that the major breakthroughs in the larviculture of many fish species is the (n-3) PUFA enrichment of *Artemia* and the rotifer *Brachionus* and the bioencapsulation technology into feed development. Such feed advancements have resulted in increased survival, higher growth and greater resistance to stress. It is likely that this technology will become more prevalent in the foreseeable future, and provides an impetus for intensification of culture practices as well as commencement of commercial culture of new marine finfish species such as the halibut.

Increasing research effort is also being expended on feed attractants, as well as growth-promoting substances. In the next decade one would perhaps envisage, as a routine, incorporation of such additives into the feeds. These additives, unlike anabolic steroids, are likely to be devoid of long-term effects on the stock or the consumers.

It is unlikely that major changes will occur in manufacturing technology. As indicated earlier, fish or aquafeed manufacturing will generally remain a subsidiary to livestock feed manufacture. Therefore, unless major changes occur in the manufacture of livestock feeds, aquafeed manufacturing is unlikely to change greatly. However, there could be changes in the market, such as expansion in the market for extruded feeds in preference to steam-pelleted feeds, this trend perhaps being more prevalent in countries which have stringent environmental protection laws.

10.7 APPLICATION OF RESEARCH FINDINGS

A scan of the scientific literature relevant to aquaculture over the last decade would reveal that nearly 25% of the articles deal with aspects of nutrition. A fair proportion of these are on the development and/or suitability of new diets; many concerned with replacement of fish meal. An obvious question that needs to be addressed is: 'To what extent are these research findings translated into the commercial aquaculture industry, or in fact, ever extend beyond the laboratory?' There are no direct answers to this question, but it is appropriate that it be addressed here.

The question is best approached from two perspectives. The bulk of aquaculture production in the world occurs in the developing world, and was intended as a source of cheap animal protein for human consumption. However, this use is declining in developing countries. Most developing countries now consider aquaculture to be a means of earning foreign exchange and providing much-needed employment opportunities. Governments in developing countries tend to encourage culture of high-valued species, at times even at the expense of valued ecosystems such as mangroves. In spite of such trends, the great bulk of aquaculture production is bound by traditional practices and has been governed and directed by trial and error. Fish farmers in the developing world are, by and large, subsistence farmers. They are reluctant to move away from traditional practices and adopt new technologies. However, in the light of competition for common resources needed for fish farming, this attitude is changing, albeit slowly. Even in one of the most technologically advanced countries, Japan, where farming activities are still bound by tradition, it is estimated that there is a 10-year lag before new diets are adopted by traditional fish farmers.

In most of Asia, it has been suggested that the lack of understanding and rapport between the researcher and the farmer has been partially responsible for the reluctance of the farmer to adopt new findings. De Silva and Davy (1992) have suggested that the most appropriate approach would be for the researchers to evaluate the current feeding practices on farms and attempt to investigate improvements in existing feeding practices and feeds rather than develop diets which are completely alien to the farmer. This would involve evaluating the type and quality of ingredients used and improving and/or developing diets based on these, with minimal new inputs. This strategy is being followed increasingly now, and hopefully the 'farmer involvement' from the outset will encourage the translation of research findings into practice.

In the developed world, where aquafeed manufacture is a component of the larger animal feed industry, the problems are different. Here the translation of research findings is primarily determined by the economic return to the feed manufacturers. Apart from the direct economic returns, the feed manufacturers will have to evaluate the cost-effectiveness of introducing new formulas into their already established production lines. They would have to investigate the overall benefit(s) of changing production schedules because a cheaper alternative ingredient becomes available for a certain time during the year, and so on. From this it can be seen that the answer is not at all straightforward. The introduction of a particular dietary formula into the aquafeed industry depends not only on its biological suitability but also on a host of intermingling factors.

10.8 FARM-MADE FEEDS

The earlier topics dealt with aspects relevant to the commercial feed industry. In view of the dominant rural nature of the aquaculture industry and because a great proportion of these practices are semi-intensive in nature, there is a 'farm-made' feed industry which often goes unnoticed. Recently, New *et al.* (1993) defined farm-made feeds as 'compounded feeds in pellet or other forms, consisting of one or more artificial and/or natural feedstuffs, produced for the exclusive use of a particular farming activity, not for commercial sale or profit'. Although, farm-made feeds do not constitute a particular industry *per se* they constitute a substantial component of the aquaculture industry. According to New *et al.* (1993) the farm-made feed component is substantial, as well as very variable in quality. For example in China, where aquaculture production in 1991 was nearly 6.93×10^6 t, the aquafeed industry accounted for only 15% of the total aquaculture feed usage (Ping, 1993).

In Chapter 9, supplemental feeds based on simple ingredient mixes which are used in certain culture practices were discussed. The degree of sophistication utilized in preparation of farm-made feeds is also very variable, often depending on the size and nature of the activity, the degree of intensity and the species cultured. Examples of the machinery used for on-farm preparation of moist and dry pellets are shown in Figures 10.4 and 10.5 respectively. Examples of farm-made feeds are given in Table 10.10. The most obvious difference is the variability in the type of ingredient used, this being determined mostly by local availability and price.

In the preparation of farm-made feeds simple precautionary measures could lead to an improvement in feed quality. In farm-made feed preparation moist pellets are often sun dried. Direct sun-drying destroys some of the vitamins, particularly vitamin C, which is often added in the form of a commercial premix. Drying under a suitable, transparent cover will prevent or at least minimize the destruction of some of the vitamins,

Figure 10.4 On-farm pellet-making machinery for snakehead culture in Chockchai Farm, Supanburi Province, Thailand. Note the raw material used on the left-hand side.

Figure 10.5 An improvised siever used for making dry pellets in a small (4 ha) farm in Thailand.

which generally account for 10–15% of the cost of the ingredients. Needless to say, in the case of farm-made feeds, quality assurance is completely lacking. This together with lack of proper record keeping does not permit the farmer to evaluate feed-related performance of the stock.

It has to be conceded that, however simple the methods used in farm-made feed preparation may be, farmers continue to use a particular feed because it is profitable. Farmers decide to use a particular feed, generally after trial and error, which has an economy of scale that suits them. Thus improvements to farm-made feeds need to be achieved by improving current practices, not by attempting to impose an entirely new feed formulation developed in the laboratory.

Table 10.10 Ingredients used and other relevant information of selected farm-made feeds in Thailand and India

Ingredient	%[a]	%[a]	%[b]	%[b]
Trash fish	70	–	–	–
Chicken offal	20	–	–	–
Fish meal (50–60% CP)	–	15	25	–
Clam meat	–	–	–	50
Soybean meal/cake	–	30	25	–
Cassava chips	–	10	–	–
Wheat flour	–	–	15	–
Corn	–	22	–	–
Rice bran	10	15	–	15
Groundnut cake	–	–	25	30
Leucaena	–	5	–	–
Dicalcium phosphate	–	1	–	–
Animal fat	–	2	–	3
Vitamin premix	–	+	2	2
Feed form	Soft pellet, prepared daily	Moist pellet, prepared daily	Soft balls	Soft balls
Species cultured	Snakehead	Cat fish	Carp	Carp

Sources: [a] Jantrarotai (1993).
[b] Nandeesha (1993).

CHAPTER 11

Feeds, feeding and the environment

11.1 INTRODUCTION

During the early stages of development, aquaculture was seen as non-polluting, an environmentally non-degrading activity. However, in the 1980s, the decade of the environment, the consequences of environmental damage and manipulation became a global concern. Agricultural and industrial activities began to be carefully evaluated and the environmental protection standards were tightened. At the time when aquaculture was rapidly expanding through rather indiscriminate use and opening of wetlands or intensification of on-going culture practices, the traditional conception of the non-polluting nature of aquaculture began to change.

Aquaculture requires rather stringent water quality, except perhaps when culturing activity is done in sewage or human wastes (Piedrahita and Tchobanoglous, 1987). This enhanced the view that aquaculture is a non-polluter and would in fact be an environmental cleaner. However, no longer does every aquaculturist hold this view. It is accepted that aquaculture, like any other man-induced activity, influences the environment, the extent to which it does so being primarily determined by the intensity of the culture practice coupled with the nature of the systems.

11.2 MACROINFLUENCES ON THE ENVIRONMENT

The general effects of aquaculture on the environment were aptly dealt with by Pillay (1992). We will focus our attention in this section on the gross biological changes brought about by aquacultural practices as such changes are generally diet and/or nutrition related.

In inland culture activities, the effluent from the farms is carried to water bodies, streams, lakes, the coast, etc., which will bring about a nutrient loading in these water bodies. The environmental loading resulting from culture activities varies widely (see Table 11.1), depending primarily on the type of feed and the husbandry practices, as well as the extent of mortalities.

Table 11.1 Selected estimates of carbon (C), nitrogen (N) and phosphorus (P) output from various *Oncorhynchus mykiss* aquaculture operations

Total C	
	50.88 kg/kg food fed/year
	0.734 kg/kg fish /year (average five farms)
	0.28–0.162 $kg/m^2/year$
	0.027 $kg/m^2/year$
Phosphorus	
PO_4P	
	0.012 kg/kg fish/year
Total	
	0.036 kg/kg fish/year
	4.161 kg/kg food fed/year
	0.151 kg/kg/fish production/year
	1035×10^{-4} to 0.080 $kg/m^2/year$
Nitrogen	
NH_3-N	
	0.012 kg/kg fish/year
Organic N	
	0.015 kg/kg fish/year
Total N	
	0.045 kg/kg fish/year
	0.077 kg/kg fish/year
	5.329 kg/kg food fed/year

Source: Iwama (1991).

Self-polluting problems in aquaculture attracted early attention, particularly in cage culture, in which uneaten food and faeces accumulated beneath the cages. Over the years this accumulation resulted in the build-up of a sludge, which caused local water quality problems. The main concerns, however, were triggered when it became apparent that such problems were not always localized and could have far-reaching consequences. The culture of yellowtail, *Seriola quinqueradiata*, and red sea bream, *Pagrus major*, in cages in enclosed bays in Japan and the associated feeding practices have been correlated with the occurrence of red tides (T. Watanabe *et al.*, 1990). These red tides resulted in detrimental effects on both the general environment and the culture activities themselves, causing high mortality. T. Watanabe (1991) summarized the damage to culture activities from red tides in the Seto Sea area.

Similarly, culture activities in inland waterways may cause eutrophication of the associated water bodies. Stirling and Dey (1990) investigated the impact of intensive rainbow trout cage farming on the phytoplankton and periphyton in a shallow 71-ha, unstratified lake in Scotland. They found that *Microcystis* dominated the phytoplankton and that algal composition and water quality criteria indicated that the lake was highly eutrophic. The tendency to eutrophicate is a common problem that is associated with cage culture in enclosed water bodies and is diet related. These authors concluded that the overall impact of cage fish culture on lake phytoplankton did not differ from cultural eutrophication due to sewage and agricultural run-off. Stirling and Dey (1990) suggested that the greatest environmental threat is prolonged settled weather, when lakes and reservoirs are prone to stratification. Turbulence enhances oxygenation through the water column, increases optical depth and therefore limits algal development. Therefore, a 'well-mixed' lake can continue to provide an adequate environment for cage culture even under a condition of advanced eutrophication. Stratification, on the other hand, reduces optical depth and promotes massive blooms of cyanobacteria. The breakdown of buoyancy control of the cyanobacteria within such blooms results in serious surface accumulations and catastrophic deoxygenation within such blooms. Other problems, such as potential toxicity to fish and earthy odour of the flesh due to the chemical geosmin, could arise from the decomposition of the algae after a bloom.

Bergheim *et al.* (1991) considered the ecological changes in benthic communities associated with net cage farming of salmonids in Norwegian waters. According to these authors, the first signs of ecological change are the appearance of opportunistic species such as the polychaete worms *Capitella* spp., *Scolelepis* spp. and *Polydora* spp. They also found that a series of faunal zones became identifiable, each zone having a distinct macrofaunal community and reflecting the reduced input of organic waste with increased distance from the culture cages. Gowen and Bradbury (1987) reviewed the impact of salmon farming in coastal waters, and some of these aspects are considered in detail by them.

Salmonid culture is one of the most, if not the most, intensive form of culture that is seen, and not surprisingly it is also the most intensively researched. Apart from this, salmonid culture is essentially confined to developed countries in the temperate zone where environmental regulations are becoming increasingly stringent. Further details regarding the environmental loading of wastes from specific culture activities are provided by Foy and Rosell (1991), Merican and Phillips (1985) and Pridmore and Rutherford (1992) amongst others.

11.3 FEEDS, FEEDING AND THE ENVIRONMENT

According to Cho *et al.* (1991), fish do not pollute, but rather feeds and feeding pollute. Therefore, there is only one source of fish culture waste, namely feed. This view appears to be extreme and too simplistic.

However, there is general consensus that nutrition-related factors are the main causes of pollution of the environment due to aquaculture.

Of all culture activities cage culture influences the immediate macroenvironment to the greatest extent. This influence arises from deposition of faecal matter and uneaten food, under the cages and in the immediate vicinity.

Minimizing feed loss is the most obvious and the most effective way of reducing pollution due to aquacultural activities. Bergheim *et al.* (1991) quoting the work of Seymour (1990) argued that improvements in FCR could be made by feeding when it can be used most efficiently. This is an area which unfortunately has received little attention from researchers. However, work by Seymour (1985) has shown that fish exhibit cycles of appetite development associated with gut evacuation, whilst De Silva and Perera (1983) demonstrated the existence of a rhythmicity in digestibility. It needs to be pointed out that feeding practices are also dictated by weather, day length and staff working hours. These prerequisites tend to impose limitations on the utilization of ideal feeding practices for a species throughout its culture cycle.

Feed loss also occurs from small particles and dust in the feed, which are generally not consumed by the fish. Estimates of such feed losses are not readily available. Seymour and Johnson (1990) (quoted by Bergheim *et al.*, 1991) estimated that for the Norwegian salmonid industry, such losses amounted to 2% and 9% for extruded and pelleted feeds respectively. These amounts are roughly equivalent to 10 t of dry waste annually for a middle-size cage pen farm in Norway.

11.4 FEED QUALITY AND ENVIRONMENTAL POLLUTION

The principal causes of pollution in aquaculture practices are uneaten feed and partially digested feed or faecal matter. These components become polluters by loading the environment with an excess of phosphates and/or nitrates and other additives (such as antibiotics) which have been incorporated into the feed(s). In Figure 11.1 the typical flow of the bioelements carbon, phosphorus and nitrogen through a net pen farm culturing salmonids is shown. This computation is based on a FCR of 1.5. It is evident that only a small proportion of the bioelements are incorporated into fish flesh, the rest being released into the environment in one form or another.

In freshwater culture, eutrophication results mainly from the loading of phosphorus into the environment. Feed ingredients of animal origin have a high phosphorus content, with fish meal (1.5–3%) and bone meal (3.5–5.5%) being two examples. In contrast, ingredients of plant origin have phosphorus contents generally ranging from 0.3–0.4% (grain) to

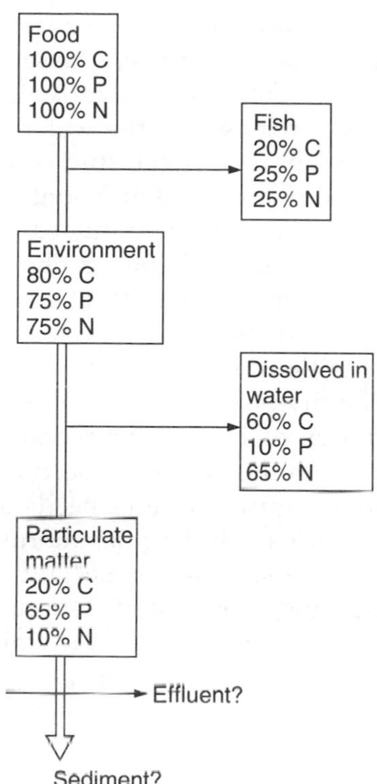

Figure 11.1 The flow of bioelements through salmonid net pens. (Based on Hakanson, 1986; Kryvi, 1989.)

0.5–1.4% (vegetable origin). The phosphorus content of the feed ingredients is not the only consideration, as the form in which it is present also affects its potential as a polluter. Phosphorus in the feed ingredients occurs in a number of forms. In ingredients of animal origin it occurs in the inorganic form as well as phosphate complexes of protein, lipid and carbohydrate. These forms are more available to the fish. In seed plants it is generally present as a phytate. Phytates are generally unavailable to finfish and monogastric animals because they lack the enzyme phytase, which digests the phytates to phytic acid, releasing the phosphorus.

Fish excrete phosphorus in a variety of forms, generally divided into soluble and particulate forms. The soluble forms, organic phosphorus and phosphate (PO_4^{3-}), affect water quality directly. The particulate forms, on the other hand, accumulate in the sludge and release phosphorus slowly. Dissolved reactive phosphorus is normally regarded as the most important fraction affecting water quality, because it is most available for phytoplankton growth. The form of phosphorus in the feed will affect the amount and form excreted as well as the amount of phosphorus that is subsequently biodegraded from the sediment. Persson (1991) estimated that the average amount of organically bound phosphorus fraction in feed and faeces released to the water is approximately 80% and 60% respectively. Phillips *et al.* (1993) studied the pattern of leaching and the amounts of three phosphorus fractions (total, total dissolved, and dissolved reactive) leached in standard and low-phosphorus salmonid diets (22.7 mg/g and 11.2 mg/g phosphorus respectively). These authors reported variable leaching of phosphorus from faeces, while the leaching from the diets was similar for the different diets. The leaching tended to be initially linear (up to about 10 min), and then almost levelled off. The dissolved reactive phosphorus accounted for 50–60% of total phosphorus losses from feeds and 30–40% from faeces.

It should be obvious from the preceding section that the phosphorus content of effluents from culture activities needs to be minimized. Any strategies to be adopted to reduce the phosphorus loading will require cooperation between farmers, nutritionists and feed manufacturers. Ketola (1985) discussed some of the critical aspects of phosphorus nutrition in relation to pollution. It appears that the main key lies in selecting feed ingredients with high phosphorus bioavailability and low water solubility. Phosphorus used for diet supplementation should possess similar qualities.

Several basic improvements in the manufacture of diets have taken place over the last decade to minimize common pollution effects. Firstly, emphasis has been placed on the use of easily digestible ingredients in diets. Diets have been formulated to be well balanced in respect of all the nutrients, thus waste due to overdosing of any one nutrient has been minimized. Finally, emphasis has been placed on the mechanical production of the pellets to reduce possible loss of nutrients.

Within the above guidelines a number of manipulations are possible. Most feed formulations have tended to reduce the protein content in the diet and use relatively non-polluting energy sources such as lipids. Accordingly, the protein-sparing property of fish has been utilized to increase the fat level, resulting in a decrease in the nitrogen discharge per kg of fish produced (Alsted, 1991). Cho *et al.* (1991) refers to such feeds as high nutrient density (HND) feeds. HND feeds result in better food conversion, and hence less production of faeces and metabolic NH_3.

The best-studied example of the influence of intensive aquaculture on the environment and steps that have been taken to reduce this influence through modification of feeds comes from the salmon farming industry in Norway. The salmonid farming industry in Norway has grown from 8000 t in 1980 to about 150 000 t in 1989, the vast majority of this increased production coming from net pen farms located in the fjords along the Norwegian Coast (Seymour and Bergheim, 1991).

Table 11.2 Total production of salmonids in Norway and associated sources of organic waste, 1979–89

Year	*Production*			*Sources of waste* ($\times 10^3$ t)				
	Harvest ($\times 10^3$ t)	*Total* ($\times 10^3$ t)	*Feed use* ($\times 10^3$ t)	*FCR*[a] (*dry wt*)	*Feed*[b] (*dry wt*)	*Faeces*[c] (*dry wt*)	*Slaughtering*[d] (*wet wt*)	*Mortalities*[e] (*wet wt*)
1979	6.8	8.6	22.1	2.6	6.6	3.1	0.8	0.7
1981	12.9	15.6	35.2	2.3	10.6	4.9	1.6	1.2
1983	22.1	26.4	48.3	1.8	14.5	6.8	2.8	2.1
1985	33.8	45.1	88.2	2.0	26.5	12.3	4.2	3.6
1987	56.2	80.1	145.7	1.8	43.7	20.4	7.0	6.4
1989	115.0	150.0	302.0	?	90.6	42.3	14.4	?

[a] FCR = Food conversion rate in kg (dry) feed/kg growth (wet wt).
[b] Estimated as one-third of used feed.
[c] Estimated as 20% of consumed feed (used feed less wastage × 20%).
[d] Estimated as 12.5% of weight of harvested fish.
[e] Estimated as 8% of total fish production.
Source: Seymour and Bergheim (1991).

The total nutrient loading from net pen farms has been estimated as 500 kg BOD_7; 52 kg total N and 9 kg total P per tonne of fish produced (based on a FCR of 1.5), and has been equated to the pollution load from a human population of 2.7×10^6. It is believed that a typical Norwegian net pen farm with an annual production of 200 t and with well-controlled feeding techniques provides an annual loading of 2 t of P, 18 t N and 100 t oxygen consumption as BOD_7. However, over the years, through better feeds (hence better FCR) and husbandry techniques, pollution loading has been decreased relative to the feed inputs (Table 11.2). Maroni (in Seymour and Bergheim, 1991) estimated, for example, that when the FCR increases from 1.0 to 1.5 the pollution loading increases 86% for aquatic matter (as chemical oxygen demand), 70% for total N and 86% for total P.

In an attempt to reduce pollution the Norwegian salmonid industry has made a significant shifts towards use of extruded pellets (Figure 11.2). Extruded pellets have a smaller surface area exposed to the water and hence are generally more stable in water. Seymour and Bergheim (1991) developed a new commercial extruded pellet for Atlantic salmon which remained up to 84% intact after 24 h in water. Commercial pressed pellets (non-extruded types) of the same size disintegrate by up to 50% after 17–53 min. The higher stability in water combined with the increased digestibility of extruded pellets is estimated to be able to reduce the solid waste (dry) per tonne of fish produced from 320–450 kg to 250 kg. However, the increased cost of making extruded pellets and their lower keeping qualities (shelf-life) owing to higher fat levels appear

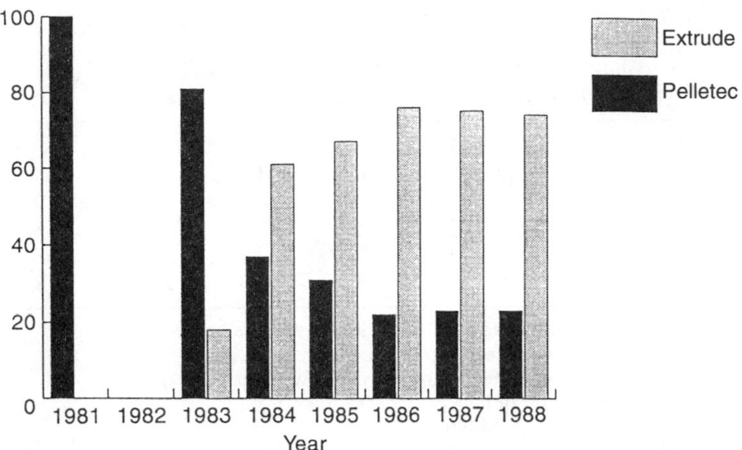

Figure 11.2 Trends in the use of extruded and pelleted feed in Norwegian salmon farming industry during the 1980s. Data from a feed manufacturer. (From Seymour and Bergheim, 1991.)

to have restricted the wide use of extruded pellets in the salmonid industry in other countries.

11.5 FEEDS AND LEGAL ASPECTS

As discussed earlier, aquaculture, like most other agriculture-related activities, influences the environment either directly or indirectly. Environment is common property and needs to be protected by law. Therefore, introduction of legislation to protect degradation of this common property is to be expected. As pointed out earlier, in fish culture the primary source of pollution is related to food and feeding. Accordingly, a good proportion of the legal framework pertaining to the industry in the developed world has been introduced in relation to feeds and feeding. This legislation aims at cohabitation of sustainable fish culture with cleaner waters.

The first legal requirements related to feeds were introduced for yellowtail culture and common carp culture in Japan. Here the use of trash fish was prohibited in yellowtail culture, while use of a low-protein diet was imposed in carp culture (T. Watanabe *et al.*, 1990). In the latter case, the use of diets containing more than 35% protein was banned. Ironically, the introduction of a new low-protein diet did not compro-

Table 11.3 Summary feed specifications in Denmark as laid down in recent legislation

	To 1/1/1990	*From 1/1/1990*	*From 1/1/1992*
Feed conversion ratio (maximum)	1.2	1.1	1.0
Gross energy (MJ/kg) in dry matter (minimum)	23.4	23.9	25.1
Energy digestibility (%) (minimum)	70	74	78
Nitrogen dry matter (%) (maximum)	9	9	8
Protein (per cent of feed)	50.6	50.6	45.0
Phosphorus (per cent of dry matter) (maximum)	1.1	1.1	1.0
Phosphorus feed (%) (maximum)	1.0	1.0	0.9
Dust (%) (maximum)	1	1	1

Source: Kiaerskou (1991).

mise growth or production, but it did decrease the nitrogen loading by about 40%.

A wide array of feed specifications are being introduced by different countries to counter the environmental degradation problems. The regulations in force in Denmark are given in Table 11.3. It is evident from this table that the regulations tend to emulate what has been discussed previously, leading to an overall reduction in the nitrogen and phosphorus loading of the environment.

What should concern all aquaculturists is the lack of and/or the desire to introduce such legislation in developing countries in which it is a major industry. In most of Asia, there is a general trend towards intensification and hence a greater dependence on artificial feeds. As pointed out on a number of instances earlier in this book, there has been a large increase in shrimp culture in Asia, taking it to greater production rates than the world salmonid industry. The FCR in shrimp culture in the best of conditions is higher than 1.5. Therefore, the waste loading from such feeding practices leaves much concern from an environmental point of view.

In ensuing years fish nutritionists will have to move away from their traditional approaches. They can no longer be formulators of nutritionally wholesome diets, but need to consider fresh strategies in diet development and feed cost reduction. The 'pure' nutritional approach is no longer applicable. In Figure 11.3 an attempt is made to trace the interrelationship amongst the various factors and strategies. The answers are not straightforward, and will never be so in the future. Aquaculture, like other food-producing industries, has now reached a stage where the various specialists are no longer able to work in water-tight compartments

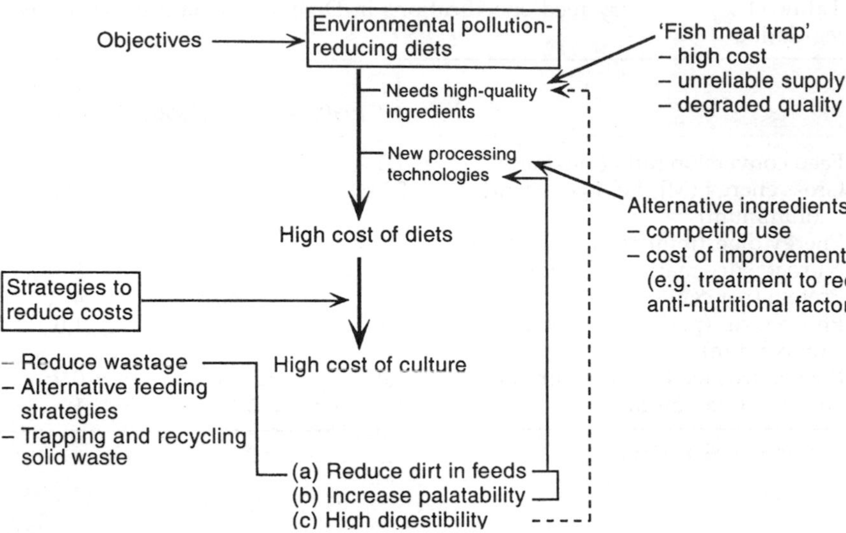

Figure 11.3 Representation of the present and future interrelated factors that are of concern to fish nutritionists at present as well as in the future.

CHAPTER 12

Methods used in studies on nutrition

12.1 INTRODUCTION

As in other fields of expertise, methods used in studies of fish nutrition differ with the objective of the study. However, a number of common measurements do exist. These common denominators need to be easily determinable and readily influenced and/or affected by the nutritional status of the organism. The most obvious of these widely used and applicable parameters is growth.

Nutritional studies can be very broadly categorized into four general areas. These are studies pertaining to the nutritional requirements of cultured species, the nutritional quality of feedstuffs, the nutritional quality and suitability of compounded feeds and nutritional deficiency studies. All these categories use some common methods of evaluation and analysis. As such, the rest of the chapter is presented in a manner so as to avoid repetition, without losing clarity or detail of these methods.

With increasing market competition, particularly in respect of luxury or high-value species, the organoleptic qualities of the final produce are becoming increasingly important. Until recently, the chemical quality of the carcass was considered as sufficiently indicative of the final suitability of a diet. This trend is rapidly changing. Increasingly, the final product is being tested for its organoleptic properties by trained testing panels before acceptance of a diet.

12.2 NUTRITIONAL REQUIREMENTS OF CULTURED SPECIES

It has been pointed out in previous chapters that finfish require 40 or so nutrients for optimal growth and well-being. As you are aware, some of these nutrients are classified as macronutrients (proteins, lipids, carbohydrates and energy) and the others are classified as micronutrients (the 10 essential amino acids, vitamins, mineral and essential fatty acids).

Generally, the optimal requirement for these nutrients is determined by feeding small groups of the target species a series of test diets containing graded levels of the nutrient being investigated. Such trials are conducted in a number of replicates, usually over a period of 10–12 weeks, and the growth response measured. The optimum requirement is considered to be the minimum level providing optimum growth. The growth response in such experiments can vary depending on the nutrient and the species in question. There are two main types of responses, a peak growth and a breakpoint (Figure 12.1). In amino acid requirement studies the 'breakpoint' is considered to reflect the requirement, whereas in protein requirement studies very often the point at which the growth response peaks is considered as the optimal requirement. When using the 'peak growth' method, a dose–response curve does not necessarily reflect the almost insignificant differences in weight gain below and beyond the maximum point. Equally, it does not consider the ability of the animal to adapt to a range of dietary nutrient levels. In studies of the protein requirements of rainbow trout, Zeitoun *et al.* (1976) pointed out that these levels reflected the ability of the animal to adapt to a range of protein levels between a deficiency on the left side of the curve and toxicity on the right. They adopted a statistical approach, determining a range of nutrient levels that can yield a growth response within a certain confidence range of the maximum level. This approach even permitted economic considerations to be introduced into diet formulations.

The dietary requirements of essential amino acids (EAAs) have also been determined by monitoring the specific amino acid levels in tissues (Kaushik, 1979) and the oxidation of radioactively labelled amino acids which were administered orally or by injection (Walton *et al.*, 1982). In all cases amino acids in the test diets are supplied almost entirely in the form of crystalline amino acids or in combination with purified protein sources such as casein and gelatin. Ogino (1980) determined the quantitative EAA requirement of fish on the basis of the daily deposition of individual amino acids in the carcass. In this method the test diet con-

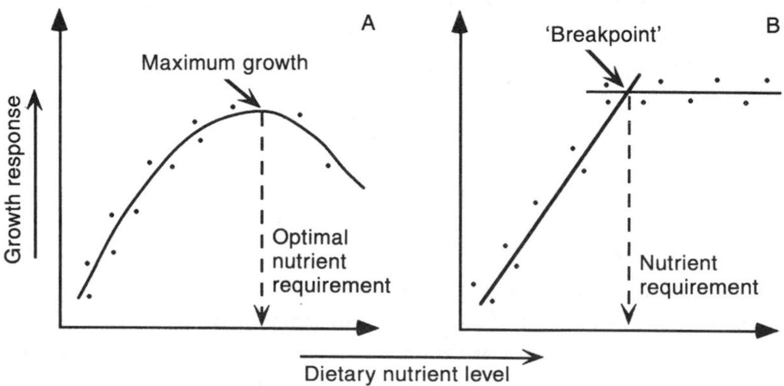

Figure 12.1 The two types of growth responses that are generally observed in nutritional requirement experiments.

tained a 'purified' protein source of high biological value. This method is not recommended, since it does not account for limiting amino acids in the protein source.

In general, only the nutrient to be tested is varied in the test diets for requirement studies. Ideally this is achieved by using pure ingredients. Such ingredients include casein and gelatin as protein sources, and pure oils as lipid sources. However, several practical problems exist in regard to pelletability and acceptability of diets based on pure ingredients may also arise, especially in the case of crystalline amino acid diets. For this reason, requirement studies on major nutrients for most species are often conducted using practical diets, usually with fish meal as the major protein source. The use of practical diets also allows nutritionists to combine requirements studies with experiments determining the utilization of selected nutrient sources.

In nutritional requirement studies, the formulation of the test diets containing graded levels of the test ingredient is the most crucial step. Obviously this is done by varying the amount of the ingredient which is the primary source of the nutrient to be tested. This is exemplified in Table 12.1 in which the diets formulated to test the protein requirement of a species are given. In these diets the protein source, fish meal, is

Table 12.1 Composition of diets (g/100 g diet) containing varying levels of dietary protein fed to the bighead carp fry

	Diet number						
	1	2	3	4	5	6	7
Ingredient							
Fish meal	32.6	40.7	48.9	57.0	65.2	73.3	81.5
Dextrin	53.4	45.7	38.1	30.4	22.9	15.3	6.6
Cod liver oil	2.9	2.3	1.8	1.3	0.7	0.2	0
Calcium phosphate (monobasic)	1.7	1.4	1.2	1.0	0.9	0.7	0.5
Filler (Celufil)	0	0.3	0.5	0.7	0.9	1.1	1.9
Others[a]	9.5	9.5	9.5	9.5	9.5	9.5	9.5
Estimated crude protein (%)	20	25	30	35	40	45	50
Estimated digestible energy (kcal/100 g)[b]	290	290	290	290	290	290	290
Proximate analysis (per cent dry matter basis)							
Crude protein	23.1	28.0	33.2	38.0	43.3	47.2	54.8
Crude fat	6.3	8.3	8.2	9.5	8.4	8.3	6.7
Crude fibre	1.2	1.4	1.6	1.5	1.7	2.2	2.7
Ash	6.5	7.8	8.8	9.9	11.4	13.3	12.4
NFE (by difference)	62.9	54.4	48.2	41.1	35.2	29.0	23.4

[a] Corn oil, 5.00%; vitamin premix, 1.50%; mineral premix, 1.21 %; and binder (carboxymethyl cellulose, 1.79%)
[b] Adopted from values for channel catfish: 2.5 kcal/g protein, 8.1 kcal/g fat and 2.5 kcal/g carbohydrates (NRC, 1977).
Source: Modified after Santiago and Reyes (1991).

varied to provide the required protein level. As the fish meal level is altered the digestible and metabolizable energy content of test diets needs to be kept constant. This is achieved in the present case by balancing the dextrin, cod liver oil and filler levels in the diets.

It is not the intention of the authors to go into further details on test diet formulations in this chapter. It is sufficient to point out that the number of variables to be tested should be kept to a minimum, and the formulations should be based on the same ingredients, keeping the variability of inclusion in each diet to a minimum.

12.3 NUTRITIONAL QUALITY OF FEEDSTUFFS

The nutritional quality of feedstuffs is assessed in two steps using both chemical and digestibility studies.

12.3.1 Chemical characteristics

The chemical characteristics looked for in feedstuffs were dealt with in detail in Chapter 7, these characteristics being the protein, lipid, ash, fibre and energy content. Ideally, the amino acid and/or fatty acid profile of the ingredient should also be known. Most laboratories are not equipped to perform these analyses. However, the amino acid content of most feedstuffs can be found in the literature. The amino acid profiles of most materials do not vary to a great extent from region to region, and may therefore be used with some confidence. The methods employed in the analysis of protein, lipids, carbohydrates, fibre and the energy content can be obtained from books on standard methodologies, the most commonly used one being that of the American Association of Official Analytical Chemists.

12.3.2 Digestibility studies

Ideally, before deciding to include a particular ingredient in a compounded feed, its digestibility should be determined. The methodological aspect pertaining to digestibility estimations of diets and ingredients were dealt with in detail in Chapter 4. For this section there are several important points to note when determining digestibilities. Firstly, the digestibility of an ingredient should not be estimated by feeding the ingredient alone to the animal. The ingredient digestibility is best determined by preparing a test diet, including 15–30% of the ingredient, with a reference diet of known digestibility. Finally, whenever possible, attempts should be made to use endogenous markers.

12.4 NUTRITIONAL QUALITY OF COMPOUNDED FEEDS

A compounded feed is formulated in accordance with the intended nutritional level required for the culture operation. This formulation is performed by taking into account the nutritional quality of the ingredients. However, even with best attempts to balance the protein, lipid and energy contents during the formulation, the levels of these constituents in the final compounded diet deviate somewhat from the expected values. Such deviations are most likely due to improper and/or inadequate mixing, minor changes in quality of ingredients as a result of processing and analytical errors.

The most important nutritional aspect associated with manufacturing compound feeds is to minimize the difference between the original formulation and the final product. The unavoidable variations in the final product make it imperative that the feed be analysed for its major constituents, and any nutritional interactions in the case of optimal requirements studies should be considered.

12.4.1 Digestibility

As indicated earlier, it is expected that digestibility evaluations of the potential major ingredients were performed prior to the formulation, and that the individual dry matter and nutrient digestibilities of the ingredients were found to be suitable. The next logical step would be to evaluate the digestibility of the formulated compounded diet in order to determine whether it falls within the expected range. Details on digestibility estimations were dealt with earlier (Chapter 4).

12.4.2 Water stability and sinking rates

Before commencing feeding trials, it will be useful to determine the characteristics of the compounded diet in water. This is performed by conducting simple studies to determine the time taken for the pellets to disintegrate in water and their sinking rates. Water stability tests may be conducted by determining the dry matter or nutrient(s) recovered in the pellet after exposure to either running water for a short time (10–30 min) or still water for a longer time (greater than 2 h).

12.4.3 Performance of the cultured species

The most crucial evaluation of a compounded feed is measured by its direct impact on the species to which the feed is given. This is performed using a number of criteria.

Foremost among these criteria are growth, food conversion ratio (FCR) or food conversion efficiency (FCE), protein efficiency ratio (PER), protein retention rate and carcass quality. Each of these criteria involves the measurement of various parameters. As FCR, FCE, PER and NPU have been discussed in detail in Chapter 2, only growth and carcass quality will be detailed here.

(a) Growth

Growth can be either somatic or reproductive. Somatic growth entails an increase in the size of the body, while reproductive growth entails an increase in the size of the reproductive organs or the gonads. In both cases, growth refers to an increase in size, often measured as change in weight or in length. In this section we will deal only with somatic growth.

Expression of growth

In order to set up a growth trial to determine the quality of a compounded feed, it is imperative that the animals used for a particular experiment are from the same stock and are of similar size and age. It is also important to provide optimal rearing conditions to all animals. However, even if these criteria are met, individuals within a group or population grow at different rates, primarily as a result of genetic variability.

In studies on nutrition, the mean rate of growth of the group or population is of most interest and value to the investigator. However, in order to make comparisons between groups or populations on different treatments the measured values for growth rate should be expressed in a uniform manner allowing for individual variation. Equally, expression of growth has no real meaning if it is not related to time. Accordingly, growth of an organism is expressed either as percentage average daily growth rate (%ADG) or as percentage specific growth rate (%SGR). Of these two expressions the latter is preferred. These values are estimated using the following formulas:

$$\%\text{ADG} = \frac{W_{t2} - W_{t1} \times 100}{W_{t1} \times (t_2 - t_1)} \tag{12.1}$$

$$\%\text{SGR} = \frac{\text{In}\,W_{t2} - \text{In}\,W_{t1} \times 100}{(t_2 - t_1)} \tag{12.2}$$

where W_{t1} is mean weight at time t_1 or initial mean weight, W_{t2} is mean weight at time t_2 or final mean weight and ln is natural logarithm.

In a nutritional trial, whether in the laboratory or in an experimental pond or culture system, growth of the cultured organism(s) needs to be monitored at regular intervals. In laboratory experiments this is done by weighing the organisms(s) once a fortnight. It must be remembered though that weighing stresses the organism, and all precautions should be taken to minimize this stress. Weighing in air or in water causes stress of different types. Depending on the objectives of the experiment it may be practical to weigh all the individuals in a treatment together, therefore minimizing handling. This is known as group weighing. If it is necessary to obtain individual weights, it is generally desirable to anaesthetize the fish to reduce stress and to improve the accuracy of the weighing. Some nutritionists actually prefer not to weigh during the course of the experiment, only weighing the animals at the start and at the end of the trial.

Field evaluation

It needs to be emphasized here that not all nutrition experiments are conducted within the laboratory. Often the performance of diets needs to be evaluated in real culture situations. Obviously, this would require the sampling strategies to be modified. In such studies there are no hard and fast rules on measurement strategies. However, useful information on growth-related aspects on pond experimentation is given by Wee (1989).

A diet developed in the laboratory and found to be suitable for a species is of little value and/or use unless its performance is tested under field conditions. In pond trials the biggest difficulty faced by researchers is obtaining pond facilities to provide sufficient replicates. Even when the pond facilities are available, variation in soil characteristics and other parameters may introduce an unknown, and often unquantifiable, variable into the experiment. As a result, the observed effects may not necessarily be a true manifestation of the quality of the diet, being clouded by factors other than those being tested. Therefore, any interpretation of the data should be done with extreme caution.

(b) Carcass quality

Carcass quality is generally governed by the relative proportion of moisture, protein, lipid and ash. For this reason it is routine practice to test the carcass quality at the end of feeding trials by assessing the carcass composition. There are numerous studies to show that dietary protein, carbohydrate and fibre and lipid levels do influence the final carcass composition. Therefore the quality of the feed may influence the final carcass composition and hence flesh quality of the final produce. Thus, it is imperative that a feed is not only nutritionally adequate but also gives an acceptable final product. It should be noted that carcass composition is also affected and/or influenced by other factors such as feeding frequency, ration size and water quality.

12.5 NUTRITIONAL DEFICIENCY STUDIES

One aspect of the formulation and preparation of complete diets is to evaluate, specifically, the requirement of micronutrients, such as vitamins and minerals. Often lack or deficiency of a micronutrient results in reduced growth, poor performance and pathological conditions.

In deficiency studies, it is essential that semipurified diets are utilized so as to enable the experimenter to simulate deficiencies of the ingredient under investigation. An example of a set of experimental diets utilized to determine the nutritional requirement of vitamin B_6 in sea bass is given in Table 12.2.

It is also necessary to conduct micronutrient requirement and/or deficiency studies for a minimum period of 10 weeks or until deficiency symptoms begin to manifest themselves. Some symptoms may still not become apparent during this period, for example vitamin B_{12} deficiency symptoms in channel catfish are manifest only around the 20th week. However, this is more an exception than the rule.

Table 12.2 Composition of the experimental semipurified moist pelleted diets (%)

	Diet number			
Ingredient	1	2	3	4
Casein (vitamin-free)	50	50	50	50
Gelatin	10	10	10	10
Cod liver oil	6	6	6	6
Soybean oil	3	3	3	3
Alpha-starch	5	5	5	5
Cellulose	8	8	8	8
Sodium CMC	5	5	5	5
Vitamin mixture[a]	2	2	2	2
McCollum's salt mixture[b]	4	4	4	4
Amino acid mixture[c]	7	7	7	7
Vitamin B_6 HCl (mg/kg dry diet)	0	5	10	20
Water (ml)	80	80	80	80

[a] Vitamin mixture (mg/100 g dry diet): thiamin HCl, 5; riboflavin, 20; choline chloride, 500; nicotinic acid, 75; Ca pantothenate, 50; inositol, 200; biotin, 0.5; folic acid, 1.5; vitamin B_{12}, 0.1; menadione, 4.0; α-tocopherol acetate, 40; vitamin A (IU), 1000; vitamin D_3 (IU), 200; butylated hydroxytoluene, 1; ascorbic acid, 100; cellulose, 999.85. Total 2 000.00.
[b] McCollum's salt mixture no. 185 plus trace elements (unit/100 g mineral mixture): calcium lactate, 32.70 g; K_2HPO_4, 23.98 g; $CaHPO_4 \cdot 2H_2O$, 13.58 g; $MgSO4_4 \cdot 7H_2O$, 13.20 g; $Na_2HPO_4 \cdot 2H_2O$, 8.72 g; NaCl, 4.35 g; ferric citrate, 2.97 g; $ZnSO_4 \cdot 7H_2O$, 0.3 g; $CoCl_2 \cdot 6H_2O$, 100 mg; $MnSO_4 \cdot H_2O$, 80 mg; Kl, 15 mg; $AlCl_3 \cdot 6H_2O$, 15 mg; $CuCl_2$, 10 mg. Total 100 g.
[c] Amino acid mixture (g/100 dry diet): L-phenylalanine, 0.6; L-arginine HCl, 1.3; L-cystine, 0.7; L-tryptophan, 0.2; L-histidine HCl: H_2O, 0.2; Dl-alanine, 1.3; L-aspartic acid Na, 1.0; L-valine, 0.7; L-lysine HCl, 0.6; glycine, 0.4 (Yone, 1976).
Source: Boonyaratapalin (1991).

Table 12.3 The basic categories in proximate analysis

Category	*Includes*
Water	Water
Crude protein	Essential amino acids, non-essential amino acids, amines, nucleic acids
Total lipid (ether extract)	Triglycerides, phospholipids, sterols, miscellaneous lipids such as waxes, etc.
Crude fibre	Cellulose, hemicellulose, chitin (all insoluble polysaccharides)
Nitrogen-free extract	Mono-, oligo- and soluble saccharides, water-soluble vitamins
Ash	Essential minerals, non-essential minerals, toxic elements

12.6 CHEMICAL EVALUATION

The term proximate composition was used rather extensively throughout this chapter. In the strictest sense, proximate analysis is the partitioning of compounds in a feed into six basic categories, based on their detailed chemical composition. These categories and these constituents are given in Table 12.3. This analysis was first developed in Germany in the 1870s and has been in use since then as a means of evaluation of feedstuffs, diets and produce.

Apart from the preceding chemical analyses, there are others which are used in diet evaluation. However, the details of these fall beyond the scope of this book. Such analyses involve the testing of protein and lipid quality. Protein quality evaluation is performed by pepsin digestibility, the protein dispersibility index, urease activity and available lysine. Methods of measuring lipid quality include oxidative rancidity by its peroxidase value, anisidine value or Kries' test.

12.7 ORGANOLEPTIC PROPERTIES (FLAVOUR)

In contrast to other foods, a concerted effort has not been made to control organoleptic factors in aquaculture. However, the factors that contribute to organoleptic properties, are numerous and complex. The major factors relate to the diet and the culture environment. Flavour is determined by a trained panel, much as wine or tea tasting is done. No explicit methods have been developed to test flavour.

References

Akiyama, D. M. (1988) Soybean Meal Utilization by Marine Shrimp, American Soybean Meal Association, Singapore.

Ako, H., Kraul, S. and Tamaru, C. (1991) Pattern of fatty acid losses in several warmwater fish species during early development, in *Larvi '91* (eds P. Lavens, P. Sorgeloos, E. Jaspers and F. Ollevier), Special Publication No. 15, European Aquaculture Society, Ghent, pp. 23–25.

Alanärä, A. (1992) The effect of time-restricted demand feeding on feeding activity, growth and feed conversion in rainbow trout (*Oncorhynchus mykiss*). *Aquaculture*, **108**, 357–68.

Albrecht, M. L. and Breitsprecher, B. (1969) Untersuchungen ueber die chemische Zusammensetzung von Fischenahrtieren und Fischfutter-mittein. *Z-fish N. F.*, **17**, 143–63.

Allen, J. R. M. and Wootten, R. J. (1982) The effect of ration and temperature on the growth of the three-spined stickleback, *Gasterosteus aculeatus. Journal of Fish Biology*, **20**, 409–22.

Alsted, N. S. (1991) Studies on the reduction of discharges from fish farms by modification of the diet, in *Nutritional Strategies and Aquaculture Waste* (eds C. B. Cowey and C. Y. Cho), University of Guelph, Ontario, pp. 77–89.

Anderson, T. A. (1986) Histological and cytological study of the gastro-intestinal tract of the luderick, Girella tricuspidata (Pisces, Kyphosidae), in relation to diet. Journal of Morphology, **190**, 109–19.

Anderson, T. A. (1987) Utilization of algal cell fractions by the marine herbivore, the luderick, *Girella tricuspidata* (Quoy and Gaimond). *Journal of Fish Biology*, **31**, 221–28.

Anderson, T. A. (1989) Mechanism of digestion in the marine herbivore, the luderick, *Girella tricuspidata* (Quoy and Gaimond). *Journal of Fish Biology*, **39**, 535–47.

Anderson, T. A. and Braley, H. (1993) Appearance of nutrients in the blood of the golden perch *Macquaria ambigua* following feeding. *Comparative Biochemistry and Physiology*, **104A**, 349–56.

Anderson, T. A., Bennett, L. R., Conlon, M. A. and Owens, P. C. (1993) Immunoreactive and receptor-active insulin-like growth factor-I (IGF-1) and IGF-binding protein in blood plasma from the freshwater fish *Macquaria ambigua* (golden perch). *Journal of Endocrinology*, **136**, 191–98.

Atkinson, J. L., Hilton, J. W. and Slinger, S. J. (1984) Evaluation of acid-insoluble ash as an indicator of feed digestibility in rainbow trout (*Salmo gairdneri*). *Canadian Journal of Fisheries and Aquatic Sciences*, **41**, 1384–86.

Austreng E. (1981) Digestibility determinations in fish using chromic oxide marker and analysis of contents from different segments of the gastrointestinal tract. *Aquaculture*, **13**, 265–72.

Balon, E. K. (1975) Terminology of intervals in fish development. *Journal of the Fisheries Resource Board of Canada*, **32**, 1663–70.

Baragi, V. and Lovell, R. T. (1986) Digestive enzyme activities in striped bass from first feeding through larval development. *Transactions of the American Fisheries Society*, **115**, 478–84.

Bardach, J. E., Ryther, J. H. and McLarney, W. O. (1972) *Aquaculture: the Farming and Husbandry of Freshwater and Marine Organisms*, John Wiley, USA.

Barrington, E. J. W. (1957) The alimentary canal and digestion, in *The Physiology of Fishes*, Vol. 1 (ed. M. E. Brown), Academic Press, New York, pp. 105–61.

Bauermeister, A. E. M., Pirie, B. T. S. and Sargent, J. R. (1979) An electron microscope study of lipid absorption in the pyloric caeca of rainbow trout (*Salmo gairdneri*) for wax ester-rich zooplankton. *Cell Tissue Research*, **200**, 475–86.

Beamish, F. W. H. and Medland, T. E. (1986) Protein sparing effects in large rainbow trout, *Salmo gairdneri. Aquaculture*. **55**, 35–42.

Beamish, F. W. H. and Thomas, E. (1984) Effects of dietary protein and lipid on nitrogen losses in rainbow trout. *Salmo gairdneri. Aquaculture*, **41**, 359–71.

Bergheim, A, Aabel, J. P. and Seymour, E. A. (1991) Past and present approaches to aquaculture waste management in Norwegian net pen culture operations, in *Nutritional Strategies and Aquaculture Waste* (eds C. B. Cowey and C. Y. Cho), University of Guelph, Ontario, pp. 117–36.

Bergot, F. and Breque, J. (1983) Digestibility of starch by rainbow trout: effects of physical state of starch and of the intake level. *Aquaculture*, **34**, 203–12.

Bever, K., Chenoweth, M., and Dunn, A. (1977) Glucose turnover in kelp bass (*Paralabrax* sp.): *in vivo* studies with [6-3H, 6-14C]-glucose. *American Journal of Physiology*, **232**, R66–72.

Bimbo, A. and Crowther, J. B. (1991) Fishoils: processing beyond crude oil. INFOFISH International, **6**, 20–25.

Bjornsson, B. T., Ogasawara, T., Hirano, T. *et al.* (1988) Elevated growth hormone levels in stunted Atlantic *salmon. Salmo salar. Aquaculture*, **73**, 275–81.

Blyth, P. and Purser, J. (1992) Feeding strategies for sea-cage salmonids. *Austasia Aquaculture*, **6(2)**, 38–40.

Boonyaratapalin, M. (1991) Nutritional studies on seabass (*Lates calcarifer carifer*), in Fish Nutrition In Asia (ed. S. S. De Silva), Special Publication 5, Asian Fisheries Society, Manila, pp. 33–41.

Boonyaratpalin, M. and Akiyama, D. M. (1990) The aquaculture industry in Southeast Asia, in *The Current Status of Fish Nutrition in Aquaculture* (eds M. Takeda and T. Watanabe), Japan Translation Centre, Tokyo, pp. 41–56.

Bowen, S. H. (1978) Chromic oxide in assimilation studies – a caution. *Transactions of the American Fisheries Society*, **107**, 755–56

Bowen, S. H. (1981) Digestion and assimilation of periphytic detrital aggregate by *Tilapia mossambica. Transactions of the American Fisheries Society*, **110**, 239–45.

Bowen, S. H. (1987) Dietary protein requirements of fishes – a reassessment. *Canadian Journal of Fisheries and Aquatic Sciences*, **44**, 1995–2001.

Braley, H. and Anderson, T. A. (1992) Changes in blood metabolite concentrations in response to repeated capture, anaesthesia and blood sampling in the golden perch, *Macquaria ambigua. Comparative Biochemistry and Physiology*, **103A**, 445–50.

Brett, J. R. and Glass, N. R. (1973) Metabolic rates and critical swimming speeds of sockeye salmon (*Oncorhynchus nerka*) in relation to size and temperature. *Journal of the Fisheries Research Board of Canada*, **30**, 379–87.

Brett, J. R. and Groves T. D. D. (1979) Physiological energetics in *Fish Physiology*, Vol. 8 (eds W. S. Hoar, D. J. Randall and J. R. Brett), Academic Press, New York, pp. 279–352.

Brett, J. R. and Higgs, D. A. (1970) Effects of temperature on the rate of gastric digestion in fingerling sockeye salmon *Oncorhynchus nerka. Journal of the Fisheries Research Board of Canada*, **27**, 1967–79.

Brett, J. R., Shelbourne, J. E. and Shoop, C. T. (1969) Growth rate and body composition of fingerling sockeye salmon, *Oncorhynchus nerka*, in relation to temperature and ration size. *Journal of Fisheries Research Board of Canada*, **26**, 2363–94.

Bromage, N. R. and Cumaranatunga, P. R. T. (1988) Egg production in the rainbow trout, *in Recent Advances in Aquaculture*, Vol. 3 (eds J. F. Muir and R. J. Roberts), Croom Helm, London, pp. 63–138.

Brown, M E. (1946) The growth of brown trout (*Salmo trutta* Linn). I. Factors influencing the growth of trout fry. Journal of Experimental Biology, **22**, 118–29.

Buddington, R. K. (1979) Digestion of an aquatic macrophyte by *Tilapia zillii. Journal of Fish Biology*, **15**, 449–56.

Buddington, R. K. (1980) Hydrolysis resistant organic matter as a reference for measurement for fish digestive efficiency. *Transactions of the American Fisheries Society*, **109**, 653–56.

Buddington, R. K. and Diamond, J. M. (1986) Aristotle revisited: the function of pyloric caeca in fish. *Proceedings of the National Academy of Sciences, USA*, **83**, 8012–14.

Cao, Q., Duguay, S. J., Plisetskaya, E. *et al.* (1989) Nucleotide sequence and growth hormone-regulated expression of salmon insulin-like growth factor 1 mRNA. *Molecular Endocrinology*, **3**, 2005–10.

Chiu, Y. N. (1989) Considerations for feeding experiments to quantify dietary requirements of essential nutrients in fish, in *Finfish Nutrition in Asia*, (ed. S. S. De Silva), Special Publication No. 4, Asian Fisheries Society, Manila, pp. 46–57.

Cho, C. Y. (1982) Effect of dietary protein and lipid levels on energy metabolism of rainbow trout (*Salmo gairdneri*). *Proceedings of the 9th Symposium on Energy Metabolism of Farm Animals, Norway*. Publication No. 29, European Association for Animal Production, Copenhagen, pp. 176–79.

Cho, C. Y. and Cowey, C. (1991) Rainbow trout, *Oncorhynchus mykiss*, in *Handbook of Nutrient Requirements of Finfish*, (ed. R. P. Wilson), CRC Press, Boca Raton, FL, pp. 131–44,

Cho, C. Y. and Kaushik, S. J. (1990) Nutrition energetics in fish: energy and protein utilization in rainbow trout (*Salmo gairdneri*). World Review, Nutrition and Diets, **61**, 132–72.

Cho, C. Y., Bayley, H. S. and Slinger, S. J. (1974) Partial replacement of herring meal with soybean meal and other changes in a diet for rainbow trout (*Salmo gairdneri*). *Journal of the Fisheries Research Board of Canada*, **31**, 1523–28.

Cho, C. Y., Bayley, H. S. and Slinger, S. J. (1976) Influence of level and type of dietary protein, and level of feeding on feed utilization by rainbow trout. *Journal of Nutrition*, **106**, 1547–56.

Cho, C. Y., Slinger, S. J. and Bayley, H. S. (1982) Bioenergetics of salmonid fishes: energy intake, expenditure and productivity. *Comparative Biochemistry and Physiology*, **73A**, 239–47.

Cho, C. Y., Cowey, C. B. and Watanabe, T. (1985) Finfish Nutrition in Asia – Methodological Approaches to Research and Development, IDRC, Ontario.

Cho, C. Y., Castledine, A. J. and Lall, S. P. (1990) The status of Canadian aquaculture with emphasis on formulation, quality and production of fish feeds, in *The Current Status of Fish Nutrition in Aquaculture* (eds M. Takeda and T. Watanabe), Japan Translation Centre, Tokyo, pp. 67–82.

Cho, C. Y., Hynes, J. D., Wood, K. R. and Yoshida, H. K. (1991) Quantitation of fish culture waste by biological (nutritional) and chemical (limnological) methods: the development of high nutrient dense (HND) diets, in *Nutritional Strategies and Aquaculture Waste* (eds C. B. Cowey and C. Y. Cho), University of Guelph, Ontario, pp. 37–50.

Choubert, G., De La Noüe, J. and Luquet, P. (1982) Digestibility in fish: improved device for the automatic collection of faeces. *Aquaculture*, **20**, 185–89.

Chow, K. W. (1980) Storage problems of feedstuffs. *Fish Feed Technology*, ADCP/REP/80/11,UNDP/FAO, Rome, pp. 355–61.

Christiansen, J. S., Ringø, E. and Jobling, M. (1989) Effects of sustained exercise on growth and body composition of first-feeding fry of Arctic charr. *Salvelinus alpinus. Aquaculture*, **79**, 329–35.

Chua, T. E. and Teng, S. K. (1978) Effect of feeding frequency on the growth of young estuary grouper, *Epinephalus tauvina* (Forskal), cultured in floating net-cages. *Aquaculture*, **14**, 31–47.

Cui, Y. and Liu, J. (1990) Comparison of energy budget among six teleosts. III. Growth rate and energy budget. *Comparative Biochemistry and Physiology*, **97A**, 381–84.

Cockerell, I., Francis, B. and Halliday, D. (1971) Changes in nutritive value of concentrate feedstuffs during storage. Proceedings of the Conference on Development of Feed Resources and Improvement of Animal Feeding Methods in the CENTO Region Countries, Tropical Products Institute, London, pp. 181–92.

Cowey, C. B. (1990) The present status and problems of world aquaculture with special reference to fish feeds. Aquaculture in the UK, in *The Current Status of Fish Nutrition in Aquaculture* (eds M. Takeda and T. Watanabe), Japan Translation Centre, Tokyo, pp. 13–26.

Cowey, C. B. and Luquet, P. (1983) Physiological basis of protein requirements of fishes. Critical analysis of allowances, in *Protein Metabolism and Nutrition*, Vol. 1 (eds R. Pion, M. Arnal and D. Bonin), **1**, INRA, Paris, pp. 365–84.

Cowey, C. B. and Sargent, J. R. (1979), Nutrition, in *Fish Physiology*, Vol. 8 (eds W. S. Hoar, D. J. Randall and J. R. Brett), Academic Press, New York. pp. 1–69.

Crampton, E. W. and Maynard, L. A. (1938) The relation of cellulose and lignin content to the nutritive value of animal feeds. *Journal of Nutrition*, **15**, 383–95.

Cravedi, J. P., De La Noüe, J. and Delaus, G. (1987) Digestibility of chloramphenicol oxolinic acid and oxytetracycline in rainbow trout and influence of these antibiotics on lipid digestibility. *Aquaculture*, **60**, 133–41

Csavas, I. (1990) Aquaculture development and environmental issues in the developing countries of Asia, in Environment and Aquaculture in Developing Countries (eds R. S. V Pullin, H. Rosenthal and J. L. Maclean), ICLARM Conference Proceedings 31, ICLARM, Manila.

Cumaranatunga, P. R. T. and Thabrew, H. (1989) Effects of legume (*Vigna catiang*) substituted diets on the ovarian development of *Oreochromis niloticus*. *Proceedings of the Third International Symposium on Feeding and Nutrition in Fish*, 28 August–1 September. Toba, Japan Translation Centre, Tokyo, pp. 333–44.

Dabrowski, K. (1984) The feeding of fish larvae: present state of the art and perspectives. *Reproduction, Nutrition and Development*, **24**, 807–33.

Dabrowski, K. R. and Kaushik, S. J. (1985) Rearing of coregonid (*Coregonus schinzi palea* Cuv et Val) larvae using dry and live food. III. Growth of fish and developmental characteristics related to nutrition. *Aquaculture*, **48**, 123–35.

Dabrowski, K., Kaushik, S. J. and Fauconneau, B. (1987) Rearing of sturgeon (*Acipenser baeri* Brandt) larvae. III. Nitrogen and energy metabolism and amino acid absorption. *Aquaculture*, **65**, 31–41.

Dave, G., Johanssen-Sjobeck, M. L., Larson, A., Lewander, K., and Lidman, U. (1975) Metabolic and hematological effects of starvation in the European eel, *Anguilla anguilla* L.-1. Carbohydrate, lipid, protein and inorganic ion metabolism. *Comparative Biochemistry and Physiology*, **52A**, 423–30.

Degani, G. and Viola, A. (1987) The protein sparing effects of carbohydrates in the diets of eels (Anguilla anguilla). *Aquaculture*, **64**, 283–91.

De La Noüe, J. and Choubert, G. (1986) Digestibility in rainbow trout: comparison of the direct and indirect methods of measurement. *Progressive Fish-Culturist*, **48**, 190–5

De La Noüe, J., Choubert, G. and Pouliot, T. (1989) Digestibility in the rainbow trout as affected by the number of pyloric caeca. Dry matter, amino acid, carbohydrate and energy, in *Aquaculture – A Biotechnology in Progress*, Vol. 2

(eds N. de Pauw, E. Jaspers, H. Ackefors and N. Wilkins), European Aquaculture Society, Belgium, pp. 829–37.

De Silva, S. S. (1985a) Evaluation of the use of internal and external markers in digestibility studies, in *Finfish Nutrition in Asia: Methodological Approaches* (eds C. Y. Cho, C. B. Cowey and T. Watanabe), International Development Research Centre, Ottawa, pp. 96–102.

De Silva, S. S. (1985b) Performance of *Oreochromis niloticus* (L.) fry maintained on mixed feeding schedules of differing protein content. *Aquaculture and Fisheries Management*. **16**, 331–40.

De Silva, S. S. (1989a) Digestibility estimations of natural and artificial diets, in *Fish Nutrition Research in Asia* (ed. S. S De Silva), Asian Fisheries Society, Manila, pp. 36–45.

De Silva, S. S. (1989b) Reducing feed costs in semi-intensive aquaculture systems in the tropics. *NAGA ICLARM Quarterly*, **12**, 6–7.

De Silva, S. S. and Davy, F. B. (1992) Strategies for finfish nutrition in research for semi-intensive aquaculture in Asia. *Asian Fisheries Science*, **5**, 129–44.

De Silva, S. S. and Gunasekera, R. M. (1989). Effect of dietary protein level and amount of plant ingredient (*Phaseolus aureus*) incorporated into diets on consumption, growth performance and carcass composition in *Oreochromis niloticus* L. fry. *Aquaculture*, **80**, 121–33.

De Silva, S. S. and Gunasekera, R. M. (1991) An evaluation of the growth of Indian and Chinese major carps in relation to the dietary protein content. *Aquaculture*, **92**, 237–41.

De Silva, S. S. and Owoyemi, A. A. (1983) Effect of dietary quality on the gastric evacuation and intestinal passage in *Sarotherodon mossambicus* (Peters) fry. *Journal of Fish Biology*, **23**, 347–55.

De Silva, S. S. and Perera, M. K. (1983) Digestibility of an aquatic macrophyte by the cichlid Etroplus suratensis with observations on the relative merits of three indigenous components as markers and daily changes in protein digestibity. Journal of Fish Biology, 23, 675–84.

De Silva, S. S. and Perera, M. K. (1984) Digestibility in *Sarotherodon niloticus* fry: effect of dietary protein level and salinity with further observations on daily variability in digestibility. *Aquaculture*, **38**, 293–306.

De Silva, S. S. and Radampola, K. (1990) Effect of dietary protein level on the reproductive performance of *Oreochromis niloticus*, in *The Second Asian Fisheries Forum* (eds R. Hirano and I. Hanyu), Asian Fisheries Society, Manila, Philippines, pp. 559–63.

De Silva, S. S., Gunasekera, R. M. and Atapattu, D. (1989) The dietary protein requirements of young tilapia and an evaluation of the least cost dietary protein levels. *Aquaculture*, **80**, 271–84.

De Silva, S. S., Maitipe, P. and Cumaranatunga, P. R. T. (1984a) Aspects of the biology of the euaryhaline Asian cichlid, *Etroplus suratensis. Environmental Biology of Fishes*, **10**, 77–82.

De Silva, S. S., Perera, M. K. and Maitipe, P. (1984b) The composition, nutritional status and digestibility of the diets of *Sarotherodon mossambicus* from nine man-made lakes in Sri Lanka. *Environmental Biology of Fishes*, **11**, 205–19.

De Silva, S. S., Shim, K. F. and Ong, A. K. (1990) An evaluation of the methods used in digestibility estimations of a dietary ingredient and comparisons on external and internal markers and time of faeces collection in digestibility studies in the fish *Oreochromis aureus* (Steindachner). *Reproduction, Nutrition, Development*, **30**, 215–26.

De Silva, S. S., Gunasekera, R. M. and Shim, K. F. (1991) Interactions of varying dietary protein and lipid levels in young red tilapia: evidence of protein sparing. *Aquaculture*, **95**, 305–18.

Devendra, C. (1985) Non-conventional Feed Resources in Asia and the Pacific, 2nd edn. FAO/APHCA Publication No. 6, FAO, Bangkok.

Dominy, W. G. and Lim, C. (1991) Performance of binders in pelleted shrimp diets, in Proceedings of the Peoples' Republic of China Aquaculture and Feed Workshop, September, 1989 (eds. D. M. Akiyama and K. H. Tan), American Soybean Association, Singapore, pp. 320–26.

Dos Santos, J. and Jobling, M. (1988) Gastric emptying in cod, Gadus morhua: effects of food particle size and dietary energy content. *Journal of Fish Biology*, **33**, 511–16.

Duray, M. and Kohno, H. (1988) Effects of continuous lighting on growth and survival of first feeding larval rabbitfish, *Siganus guttatus*. *Aquaculture*, **72**, 73–79.

El-Sayed, A. F. and Teshima, S. I. (1992) Protein and energy requirements of Nile tilapia, *Oreochromis nilotica*, fry. *Aquaculture*, **103**, 55–63.

Elliot, J. M. (1972) Rates of gastric evacuation in brown trout, *Salmo trutta* L. *Freshwater Biology*, **2**, 1–18.

Elliott, J. M. (1976) Energy losses in the waste products of brown trout (Salmo trutta). *Journal of Animal Ecology*, **45**, 561–80.

Elliott, J. M. and Davison, W. (1975) Energy equivalents of oxygen consumption in animal energetics. *Oecologia*, **19**, 195–201.

Embody G. C. and Gordon, M. (1924) A comparative study of natural and artificial foods of brook trout. *Transactions of the American Fisheries Society*, **54**, 185–200.

Fange, R. and Grove, D. (1979) Digestion, in Fish Physiology, Vol. VIII (eds. W. S. Hoar, D. J. Randall and J. R. Brett), Academic Press, New York, pp. 161–260.

FAO (1990) Aquaculture Production (1985–1988), FAO Fisheries Circular No. 815, revision 2, *FAO*, Rome.

FAO (1991) Aquaculture Production (1986–1990), FAO Fisheries Circular No. 815, revision 3, *FAO*, Rome.

FAO (1992) Aquaculture production (1984–1990), FAO Fisheries Circular No. 815, revision 4, *FAO*, Rome.

Farbridge, K. J. and Leatherland, J. F. (1987) Lunar cycles of coho salmon, *Oncorhynchus kisutch*. *Journal of Experimental Biology*, **129**, 165–78.

Feist, G. and Schreck, C. B. (1990) Hormonal content of commercial fish diets and of young salmon (*Oncorhynchus kisutch*) fed these diets. *Aquaculture*, **86**, 63–75.

Foltz, J. W. (1982) A feeding guide for a single cropped catfish (*Ictalurus punctatus*). Proceedings of the World Mariculture Society, **13**, 274–81.

Fostier, A., Jalabert, B., Billard, R. *et al.* (1983) The gonadal steroids, in Fish Physiology, Vol. IX (eds W. S. Hoar, D. J. Randall and E. M. Donaldson), Academic Press, London, pp. 277–372.

Foy, R. M. and Rosell, R. (1991) Loading of nitrogen and phosphorus from a Northern Ireland fish farm. *Aquaculture*, **96**, 17–30.

French, C. J., Hochachka, P. W., and Mommsen, T. P. (1983) Metabolic organisation of liver during spawning migration of sockeye salmon. *American Journal of Physiology*, **245**, 827–30.

Furuichi, M. and Yone, Y. (1980) Effect of dietary dextrin levels on the growth and feed efficiency, chemical composition of liver and also muscle and the absorption of dietary protein and dextrin in fishes. *Bulletin of the Japanese Society of Scientific Fisheries*, **46**, 225–29.

Furuichi, M. and Yone, Y. (1981) Change of blood sugar and plasma insulin levels of fishes in glucose tolerance test. *Bulletin of the Japanese Society of Scientific Fisheries*, **47**, 761–64.

Furukawa, A. and Tsukahara, H. (1966) On the acid digestion method for the determination of chromic oxide as an index substance in the study of digestibility in fish feed. *Bulletin of the Japanese Society of Scientific Fisheries*, **32**, 502–506.

Gannam, A. L. and Lovell, R. T. (1991) Effects of feeding 17α-methyltestosterone, 11 ketotestosterone, 17β-estradiol, and 3,5,3-triidothyronine to channel catfish, *Ictalurus punctatus*. *Aquaculture*, **92**, 377–88.

Gardner, M. L. G. (1985) Production of pharmacologically active peptides from foods in the gut, in Food and the Gut (eds J. O. Hunter and V. A. Jones), Bailliere Tindall, London, pp. 121–34.

Gatlin III, D. M., Bai, S. C. and Erickson, M. C. (1992) Effects of vitamin E and synthetic antioxidants on composition and storage quality of channel catfish, *Ictalarus punctatus*. *Aquaculture*, **106**, 323–39.

Goddard, J. S. (1988) Food and feeding, in *Freshwater Crayfish: Biology, Management and Exploitation*. (eds D. M. Moldich and R. S. Lowery), pp. 145–166, London, Croom-Helm.

Goh, Y. and Tamura, T. (1980) Olfactory and gustatory responses to amino acids in two marine teleosts – red sea bream and mullet. *Comparative Biochemistry and Physiology*, **66C**, 217–24.

Gohl, B. (1981) Tropical feeds. *Animal Production and Health, Series* No. 12, FAO/UNDP, Rome.

Gordon, M. S., Bartholomew, C. A., Grinnell, A. D., Jorgensen, C. B. and White, F. N. (1977) *Animal Physiology: Principles and Adaptations*, 3rd edn, New York, Macmillan.

Gowen, R. J. and Bradbury, N. B. (1987) The ecological impact of salmon farming in coastal waters: a review. *Oceanography and Marine Biology Annual Review*, **25**, 563–75.

Guillaume, J., Coustans, M., Metailler, R., Person-Le Ruyet, J. and Robin, J. (1991) Flatfish, turbot, sole and plaice, in *Handbook of Nutrient Requirements of Finfish* (ed. R. P. Wilson), CRC Press, Boca Raton, FL, pp. 77–82.

Gwiazda, S., Noguchi, A., Kitamura, S. and Saio, K. (1983) *Agricultural Biological Chemistry*, **47**, 623–25.

Hajen, W. E., Beames, R. M., Higgs, D. A. and Dosanjh, B. S. (1993) Digestibility of various feedstuffs by post-juvenile Chinook salmon (*Oncorhynchus tshawytscha*) in sea water. 1. Validation of technique. *Aquaculture*, **112**, 321–32.

Hakanson, L. (1986) Environmental impact of fish cage farms, NITO Conferences, June 1986, Norway.

Hall, D. N. F. (1962) Observations on the taxonomy and biology of some Indo-west Pacific penaeids (Crustacea: *Decapoda*). *Fisheries Publication*, **17**, 1–225.

Halver, J. E. (ed.) (1989) *Fish Nutrition*, 2nd edn, Academic Press, San Diego, CA.

Hanley, F. (1987) The digestibility of foodstuffs and the effects of feeding selectivity on digestibility determinations in tilapia, *Oreochromis niloticus*. *Aquaculture*, **66**, 163–67.

Hanley, F. (1991) Effects of feeding supplementary diets containing varying levels of lipid on growth, food conversion, and body composition of Nile tilapia, *Oreochromis niloticus* (L.). *Aquaculture*, **93**, 323–34.

Hasan, M. R. and Macintosh, D. J. (1992) Optimum food particle size in relation to body size of common carp, *Cyprinus carpio* L., fry. *Aquaculture and Fisheries Management*, **23**, 315–25.

Hashimoto, Y., Konosu, S., Fusetami, N. and Nose, T. (1968) Attractants for eels in the extracts of short-necked clam – I. Survey of constituents eliciting feeding behaviour by the omission test. *Bulletin of Japanese Society of Scientific Fisheries*, **34**, 78–83.

Hastings, W. H. (1969) Nutritional scores, in *Fish in Research* (eds. O. W. Neuhaus and J. C. Halver), New York, Academic Press, pp. 263–93.

Hastings, W. H. and Higgs, D. (1980) Feed milling process, in *Fish Feed Technology*, ADCP/REP/ 80/11, UNDP/FAO, Rome, pp. 293–313.

Hazel, J. R. and Prosser, C. L. (1974) Molecular mechanisms of temperature compensation in poikilotherms. *Physiological Reviews*, **54**, 620–77.

Hempel, E. (1993) Constraints and possibilities for developing aquaculture. *Aquaculture International*, **1**, 2–19.

Hendricks J. D. and Bailey G. S. (1989) Adventitious toxins, in *Fish Nutrition*, 2nd edn (ed. J. Halver), Academic Press, San Diego, pp. 605–51.

Henken, A. M., Kleingeld, D. W. and Tijssen, P. A. T. (1985) The effects of feeding level on apparent digestibility of dietary dry matter, crude protein and gross energy in the African catfish *Clarias gariepinus* (Burchell 1822). *Aquaculture*, **51**, 1–11.

Hepher, B. (1975) Supplementary feeding in fish culture. *Proceedings of the International Congress on Nutrition*, **9** (3), 183–98.

Hepher, B. (1978) Ecological aspects of warmwater fish pond management, in *Ecology of Freshwater Fish Production* (ed. S. D. Gerking), Blackwell Scientific Publications, Oxford, pp. 447–65.

Hepher, B. (1988a) Nutrition of Pond Fishes, Cambridge University Press, Cambridge.

Hepher, B. (1988b) Principles of fish nutrition, in *Fish Culture in Warmwater Systems: Problems and Trends* (eds. M. Shilo and S. Sarig), CRC Press, New York, pp. 121–42.

Hepher, B., Chervinski, J. and Tugari, H. (1971) Studies on carp nutrition. III. Experiments on the effect on fish yields of dietary protein source and concentration. *Bamidgeh*, **23**, 11–37.

Hibaya, T. (1982) *An Atlas of Fish Histology: Normal and Pathological Features*, New York, Gustav Fischer Verlag.

Hickling, C. F. (1962) *Fish Culture*, Faber & Faber, London.

Hickling, C. F. (1966) On the feeding process in the white amur, *Cternopharyngodon idella*. *Journal of Zoology*, **148**, 408–19.

Hidalgo, F. and Alliot, E. (1988) Influence of water temperature on protein requirement and protein utilization in juvenile sea bass, *Dicentrarchus labrax*. *Aquaculture*, **72**, 115–29.

Hill, C. H. and Matrone, G. (1970) Chemical parameters in the study of in vivo and in vitro interactions of transition elements. Feed Proceedings, **29**, 1474–81.

Hilton, J. W. (1984) Ascorbic acid–mineral interactions in fish, in *Ascorbic Acid in Domestic Animals* (eds I. Wegger, F. J. Tagwerker and J. Moustaard), The Royal Danish Agricultural Society, Copenhagen, pp. 218–24.

Hilton, J. W. (1989) The interaction of vitamins, minerals and diet composition in the diet of fish. *Aquaculture*, **79**, 223–34.

Hilton, J. W., Cho, C. Y. and Slinger, S. J. (1981) Effect of extrusion processing and steam pelleting diets on pellet durability, pellet water absorption, and the physiological response of rainbow trout (*Salmo gairdneri* R.). *Aquaculture*, **25**, 185–94.

Himick, B. A. Higgs, D. A. and Eales, J. G. (1991) The acute effects of alteration in the dietary concentrations of cabohydrate, protein and lipid on plasma T4, T3 and glucose levels in the rainbow trout, Oncorhynchus mykiss. General and Comparitive Endocrinology, 82, 451–58.

Hirao, S., Yamada, J. and Kikuchi, R. (1960) On improving efficiency of feed for fish culture. I. Transit and digestibility of diet in eel and rainbow trout observed by use of P32. *Bulletin of the Tokai Regional Fisheries Laboratory*, **7**, 67–74.

Hofer, R. and Nasiruddin, A. N. (1985) Digestive processes during the development of the roach, *Rutilus rutilus*. *Journal of Fish Biology*, **26**, 683–89.

Hofer, R. and Schiemer, F. (1983) Feeding ecology, assimilation efficiencies and energetics of two herbivorous fish: *Sarotherodon mossambicus* (Peters), *Puntius filamentosus* (Cuv et Val), in Studies in Hydrobiology, Vol. 12 (ed. F. Schiemer), Dr W. Junk Publishing, Holland, pp. 153–64.

Holmgren, S., Grove, D. J. and Fletcher, D. J. (1983) Digestion and the control of gastrointestinal mobility, in *Control Processes in Fish Physiology* (eds J. C. Rankin, T. J. Pitcher and R. Duggan), Croom Helm, London, pp. 23–40.

Horn, M. H. (1989) Biology of marine herbivorous fishes. *Oceanography and Marine Biology. Annual Reviews*, **27**, 167–272.

Howell, B. R. and Tzoumas, T. S. (1991) The nutritional value of *Artemia* nauplii for larval sole, *Solea solea* (L.),with respect to their (n-3) HUFA content, in *Larvi '91* (eds P. Lavens, P. Sorgeloos, E. Jaspers, and F. Ollevier), Special Publication No. 15, European Aquaculture Society, Ghent, pp. 63-65.

Huang, H. J. (1989) Aquaculture feed binders, in *Proceedings of a Workshop on Aquaculture and Feeds* (ed. D. M. Akiyama), American Soybean Association, Singapore, pp. 316–19.

Hung, S. (1991) Sturgeon, *Acipenser* spp., in *Handbook of Nutrient Requirements of Finfish* (ed. R. P. Wilson), CRC Press, Boca Raton, FL, pp. 153–60.

Hunt, B. P. (1960) Digestion rate and food consumption of Florida gar, Warmouth and Largemouth bass. *Transactions of the American Fisheries Society*, **89**, 206–10.

Hunter, G. A. and Donaldson, E. M. (1983) Hormonal sex control and its application to fish culture, in Fish Physiology, Vol. IX b (eds W. S. Hoar, J. Randall and E. M. Donaldson), Academic Press, London, pp. 223–304.

van Ihering, R. (1937) A method for inducing fish to spawn. *Progressive Fish Culturist*, **34**, 15–16.

Iversen, E. S. (1968) *Farming the Edge of the Sea*, Fishing News, London.

Iwama, G. K. (1991) Interaction between aquaculture and the environment. *Critical Reviews in Environmental Control*, **21**, 177–216.

Jana, B. B. and Chakrabarti, R. (1990) Exogenous introduction of live plankton – a better approach to carp growth than direct-manure system. *Progressive Fish-Culturist*, **52**, 252–60.

Jantrarotai, W. (1993) On-farm feed preparation and feeding strategies for catfish and snakehead, in Farm-Made Aquafeeds (eds M. B. New, A. G. J. Tacon and I. Csavas), FAO-RAPA/AADCP, Bangkok, pp. 101–19.

Jobling, M. (1981a) The influence of feeding on the metabolic rate of fishes: a short review. *Journal of Fish Biology*, **18**, 385–400.

Jobling, M. (1981b) Mathematical models of gastric emptying and the estimation of daily rates of food consumption for fish. *Journal of Fish Biology*, **19**, 245–57.

Jobling, M. (1982) Some observations on the effects of feeding frequency on the food intake and growth of plaice, *Pleuronectes platessa*. *Journal of Fish Biology*, **20**, 431–44.

Jobling, M. (1988) A review of the physiological and nutritional energetics of cod, Gadus morhua, with particular reference to growth under farmed conditions. *Aquaculture*, **70**, 1–19.

Jobling, M. and Wandsvik, A. (1983) An investigation of factors controlling food intake in Arctic charr (*Salvelinus alpinus*). *Journal of Fish Biology*, **23**, 397–404.

Johnsen, P. B. and Dupree, H. K. (1991) Influence of feed ingredients on the flavour quality of farm-raised fish. *Aquaculture*, **96**, 139–50.

Jones, F. (1987) Controlling mould growth in feeds. *Feed International*, **8**, 20–29.

Jones, K. A. (1990) Chemical requirements of feeding in rainbow trout, *Oncorhynchus mykiss* (Walbaum); palatability studies on amino acids, amides, amines, alcohols and aldehydes, saccharides, and other compounds. *Journal of Fish Biology*, **37**, 413–23.

Jones, R. (1974) The rates of elimination of food from the stomach of haddock *Melanogrammus aeglefinus, cod Gadus morhua,* and whiting *Merlangius merlangus. Journal du Conseil,* **35**, 225–43.

Jonsen, P. B. and Dupree, H. K. (1991) Influence of feed ingredients on the flavour quality of farm-raised catfish. *Aquaculture,* **96**, 139–50.

Josupeit, H. (1992) Recent developments in aquaculture and its impact on trade. Paper presented at CIHEAM, Institute Agronomico Mediterraneo de Zaragoza, March, Zaragoza, Spain.

Juell, J. E., Furevik, D. M. and Bjordal, Å. (1993) Demand feeding in salmon farming by hydroacoustic food detection. *Aquacultural Engineering,* **12**, 159–67.

Kamstra, A. and Heinsbroek, L. T. N. (1991) Effects of attractants on start of feeding of glass eel, *Anguilla anguilla L. Aquaculture and Fisheries Management,* **22**, 47–56.

Kanazawa, A. (1988) Broodstock nutrition, in *Fish Nutrition and Mariculture* (ed. T. Watanabe), Kangawa International Fisheries Training Centre, Japan International Cooperation Agency, pp. 147–59.

Kanazawa, A. (1991a) Puffer fish, *Fugu rubripes,* in *Handbook of Nutrient Requirements of Finfish* (ed. R. P. Wilson), CRC Press, Boca Raton, FL, pp. 123–30.

Kanazawa, A. (1991b) Ayu, *Plecoglossus altivelis,* in *Handbook of Nutrient Requirements of Finfish* (ed. R. P. Wilson), CRC Press, Boca Raton, pp. 23–29,.

Kapoor, B. G., Smit, H. and Verighina, I. A. (1975) The alimentary canal and digestion in fish, in *Advances in Marine Biology,* Vol. 13, (eds. C. M. Young and F. S. Russell), Academic Press, New York, pp. 109–213.

Karas, P. (1990) Seasonal changes in growth and standard metabolic rate of juvenile perch, *Perca fluviatilis. Journal of Fish Biology,* **37**, 913–20.

Kaushik, S. (1979) Application of a biochemical method for the estimation of amino acid needs in fish: quantitative arginine requirements of rainbow trout in different salinities. Proceedings of the World Symposium on Finfish Nutrition Technology, Vol. 1, 20–23 June. Hamburg,.

Kaushik, S. J. (1990) Status of European aquaculture and fish nutrition, in *The Current Status of Fish Nutrition in Aquaculture* (eds M. Takeda and T. Watanabe), Japan Translation Centre, Tokyo, pp. 3–12.

Kaushik, S. J. and Fauconneau, B. (1984) Effects of lysine administration on plasma arginine and on some nitrogenous catabolites in rainbow trout. *Comparative Biochemistry and Physiology,* **79A**, 459–62.

Kawai, S. and Ikeda, S. (1972) Effect of dietary change on the activities of digestive enzymes in carp intestine. *Bulletin of Japanese Society of Scientific Fisheries,* **38**, 265–70.

Kawai, S. and Ikeda, S. (1973) Studies on digestive enzymes of fishes. III. Development of the digestive enzymes of rainbow trout after hatching and the effect of dietary change on the activities of digestive enzymes in the juvenile stage. *Bulletin of Japanese Society of Scientific Fisheries,* **39**, 819–23.

Kay, D. E. (1979) Food Legumes, Tropical Products Institute Crop and Product Digest No. 3. Tropical Products Institute, London.

Keembiyahetty, C. N. and Gatlin III, D. M., (1992) Dietary lysine requirement of juvenile hybrid striped bass (*Morone chrysops x M. saxatilis*). *Aquaculture,* **104**, 271–77.

Ketola, H. G. (1983) Requirement for dietary lysine and arginine by fry of rainbow trout. *Journal of Animal Science,* **56**, 101–107.

Ketola, H. G. (1985) Mineral nutrition: effects of phosphorus in trout and salmon feeds on water pollution, in *Nutrition and Feeding in Fish* (eds C. B. Cowey, A. M. Mackie and J.G. Bell), Academic Press, London, pp. 465–73.

Kiaerskou, J. (1991) Production and economics of 'low pollution and diets' for the aquaculture industry, in *Nutritional Strategies and Aquaculture Waste* (eds C. B. Cowey and C. Y. Cho), University of Guelph, Ontario, pp. 65–76.

Kissil, G. W. (1991) Gilthead sea bream, *Sparus aurata*, in *Handbook of Nutrient Requirements of Finfish*, (ed. R. P. Wilson), CRC Press, Boca Raton, FL, pp. 83–88.

Kim, K. I., Kayes, T. B. and Amundson, C. H. (1991) Purified diet development and re-evaluation of the dietary protein requirement of fingerling rainbow trout (*Oncorhynchus mykiss*). *Aquaculture*, **96**, 57–67.

Knights, B. (1983) Food particle size preferences and feeding behaviour in warmwater aquaculture of European eel, *Anguilla anguilla* (L.). *Aquaculture*, **30**, 173–90.

Knights, B (1985) Feeding behaviour and fish in culture, in *Nutrition and Feeding in Fish* (eds C. B.Cowey, A. M. Mackie and J. G. Bell), Academic Press, London, pp. 223–41.

Knox, D. Cowey, C. B. and Adron, J. W. (1981) Studies on the nutrition of salmonid fish. The magnesium requirement of rainbow trout (*Salmo gairdneri*). *British Journal of Nutrition*, **45**, 137–48.

Koch, F. and Weiser, W. (1983) Partitioning of energy in fish: can reduction of swimming activity compensate for the cost of production? *J. Exp. Biol.*, **107**, 141–46.

Kryvi, H. (1989) Pollution loading and recipient effects from fish farms. *Vann*, **24**, 285–92.

Lall S. (1991) Concepts in the formulation and preparation of a complete diet, in *Finfish Nutrition in Asia*, (ed. S. S. De Silva), Asian Fisheries Society, Manila, pp. 1–12.

Lauff, M. G. and Hofer, R. (1984) Proteolytic enzymes in fish development and the importance of dietary enzymes. *Aquaculture*, **37**, 335–46.

Lavens, P., Sorgeloos, P., Jaspers, E. and Ollevier, F. (1991) *Larvi '91*, Special Publication No. 15, European Aquaculture Society Ghent, Belgium.

Leatherland, J. R., Cho, C. and Hilton, J. W. (1984) Effect of diet on serum thyroid hormone levels in rainbow trout (*Salmo gairdneri* Richardson). *Comparative Biochemistry and Physiology*, **78A**, 601–605.

Leavitt, D. F. (1985) An evaluation of gravimetric and inert marker techniques to measure digestibility in the American lobster. *Aquaculture*, **47**, 131–42.

Leech, A. R., Goldstein, L., Cha, C. and Goldstein, J. M. (1979) Alanine biosynthesis during starvation in skeletal muscle of the spiny dogfish, *Squalus acanthias*. *Journal of Experimental Zoology*, **207**, 73–80.

Leger, C., Gatesoupe, F., Metailler, R., Luquet, P. and Fremont, L. (1979) Effect of dietary fatty acids differing by chain lengths and *w* series on the growth and lipid composition of turbot *Scophthalmus maximus* L. *Comparative Biochemistry and Physiology*, **64B**, 345–50.

Leger, P., Bengston, D. A., Simpson, K. L. and Sorgeloos, P. (1986) The use and nutritional value of *Artemia* as a food source. *Oceanography and Marine Biology Annual Review*, **24**, 521–623.

Lie, O., Lied, E. and Lambertsen, G. (1988) Feed optimization in Atlantic Cod (*Gadus morhua*): fat versus protein content in the feed. *Aquaculture*, **69**, 333–41.

Lied, E., Julshamn, K. and Braekkan, O. R. (1979) Determination of protein digestibility in Atlantic cod (*Gadus morhua*) with internal and external indicators. *Canadian Journal of Fisheries and Aquatic Sciences*, **39**, 854–58.

Liener, I. E. (1980) Toxic Constituents of Plant Foodstuffs, Academic Press, Sydney.

Lim, C. and Akiyama, D. M. (1992) Full-fat soybean meal utilization by fish. *Asian Fisheries Science*, **5**, 181–97.

Lin, H., Romsos, D. R., Tack, P. I. and Leveille, G. A. (1977a) Influence of dietary lipid on lipogenic enzyme activities in coho salmon, *Oncorhynchus kisutch* (Walbaum). *Journal of Nutrition*, **107**, 846–54.

Lin, H., Romsos, D. R., Tack, P. I. and Leveille, G. A. (1977b) Effects of fasting and feeding various diets on hepatic lipogenic enzyme activities in coho salmon (*Oncorhynchus kisutch* (Walbaum). *Journal of Nutrition*, **107**, 1477–83.

Lin, H., Romsos, D. R., Tack, P. I. and Leveille, G. A. (1978) Determination of glucose utilization in Coho salmon [*Oncorhynchus kisutch*] (Walbaum) with (6-^{3}H)- and (U-^{14}C)-glucose. *Comparative Biochemistry and Physiology*, **59A**, 189–91.

Lindsay, G. J. H. and Gooday, G. W. (1985) Chitinolytic enzymes and the bacterial microflora in the digestive tract of cod, *Gadus morhua*. *Journal of Fish Biology*, **26**, 255–65.

Ling, S. W. (1967) Feeds and feeding of warmwater fishes in ponds, in Proceedings of the World Symposium on Warmwater Pond Fish Culture (ed. T. V. R. Pillay), FAO, Rome, pp. 251–309.

Lovell, R. T. (1980) Feeding tilapia. Aquaculture Magazine, 7, 42–43.

Lovell, T. (1989a) *Nutrition and Feeding of Fish*, van Nostrand Reinhold, New York.

Lovell, T. (1989b) Energy requirements for growth; fish versus farm animals. *Aquaculture Magazine*, July/August, pp. 65–66.

Lukton, A. (1958) Effect of diet on imidazole compounds and creatine in chinook salmon. *Nature*, **182**, 1019.

Luo, D. and McKeown, B. A. (1989) Immunological evidence of growth hormone releasing factor-like substances in salmon (*Oncorhynchus Kisutch* and *O. Keta*). *Comparative Biochemistry and Physiology*, **93B**, 615–20.

Luquet, P. (1989) Practical considerations on the protein nutrition and feeding of tilapia. *Aquatic Living Resources*, **2**, 99–104.

Luquet, P. (1991) Tilapia, *Oreochromis* spp., in *Handbook of Nutrient Requirements of Finfish* (ed. R. P. Wilson), CRC Press, Florida, USA, 169–180.

Machiels, M. A. M. and Henken, A. M. (1985) Growth rate, feed utilization and energy metabolism of the African catfish, *Clarias gariepinus* (Burchell 1822), as affected by the dietary protein and energy content. *Aquaculture*, **44**, 271–84.

Mackie, A. M. and Mitchell, A. I. (1985) Identification of gustatory feeding stimulant for fish – application in aquaculture, in *Nutrition and Feeding in Fish* (eds C. B. Cowey, A. M. Mackie and J. G. Bell), Academic Press, London, pp. 177–89.

MacLatchy, D. L. and Eales, J. G. (1990) Growth hormone stimulates hepatic thyroxine 5′-monodeiodinase activity and 3,5,3′-triiodothyronine levels in rainbow trout, *Salmo gairdneri*. *General and Comparative Endocrinology*, **78**, 164–72.

Magnuson, J. J. (1969) Digestion and food consumption by skipjack tuna, *Katsuwonus pelamis*. *Transactions of the American Fisheries Society*, **98**, 379–92.

Mahr, K., Grabner, M., Hofer, R. and Moser, H. (1983) Histological and physiological development of the stomach in *Coregonus spp. Archives für Hydrobiology*, **98**, 344–53.

Marte, C. L. (1980) The food and feeding habits of Penaeus monodon Fabricus collected from Makato River, Aklan, Philippines (Decapoda, Natantia). Crustaceana, **38**, 225–36.

Matty, A. J., Chaudhry, M. A. and Lone, K. P. (1982) The effect of thyroid hormones and temperature on protein and nucleic acid contents of liver and muscle of *Sarotherodon mossambica*. *General and Comparative Endocrinology*, **47**, 497–507.

Medda, A. K. and Ray, A. K. (1979) Effect of thyroxine and analogues on protein and nucleic acid contents of liver and muscle of lata fish (*Ophiocephalus punctatus*). *General and Comparative Endocrinology*, **37**, 74–80.

Medland, T. E. and Beamish, F. W. H. (1985) The influence of diet and fish density on apparent heat increment in rainbow trout, *Salmo gairdneri*. *Aquaculture*, **47**, 1–10.

Merican, Z. O. and Phillips, M. J. (1985) Solid waste production from rainbow trout *Salmo gairdneri* Richardson, cage culture. *Aquaculture and Fisheries Management*, **16**, 55–69.

Mironova, N. V. (1978) Energy expenditure on egg production in young *Tilapia mossambica* and the influence of maintenance conditions on their reproductive intensity. *Journal of Ichthyology*, **17**, 627–33.

Mommsen, T. P., French, C. J. and Hochachka, P. W. (1980) Sites and patterns of protein and amino acid utilization during the spawning migration of salmon. *Canadian Journal of Zoology*, **58**, 1785–99.

Moore, B. J., Hung, S. S. O. and Medrano, J. F. (1988) Protein requirement of hatchery-produced juvenile white sturgeon (*Acipenser transmontanus*). *Aquaculture*, **71**, 235–45.

Mourente, G. and Odriozola, J. M. (1990) Effect of broodstock diets on lipid classes and their fatty acid composition of larvae of gilthead sea bream (Sparus aurata) during yolksac stage. *Fish Physiology and Biochemistry*, **8**, 103–10.

Murat, J. C., Castilla, C. and Paris, H. (1978) Inhibition of gluconeogenesis and glucagon-induced hypoglycemia in carp (*Cyprinus carpio* L.). *General and Comparative Endocrinology*, **34**, 243–50.

Nagai, M. and Ikeda, S. (1971) Carbohydrate metabolism in fish. I. Effects of starvation and dietary composition on the blood glucose level and the hepatopancreatic glycogen and lipid contents in carp. *Bulletin of Japanese Society of Scientific Fisheries*, **37**, 404–409.

Nagayama, F. and Saito, Y. (1968) Distribution of several hydrolytic enzymes in fish. FAO/FI/EIFAC 68/56 11-8, FAO, Rome, pp. 1–12.

Nandeesha, M. C. N. (1993) Aquafeeds and feeding strategies in India, in Farm-Made Aquafeeds (eds. M. B. New, A. G. J. Tacon and I. Csavas), FAO-RAPA/AADCP, Bangkok, pp. 213-54.

Nandeesha, M. C., Srikanth, G. K., Keshvanath, P. *et al.* (1990). Effects of non-defatted silkworm-pupae in diets on the growth of common carp, *Cyprinus carpio*. *Biological Wastes*, **33**, 17–23.

Nandeesha, M. C., De Silva, S. S., Krishna Murthy, D. and Dathatri, K. (1994a) Use of mixed feeding schedules in fish culture. Field trials on catla, *Catla catla* (Hamilton-Buchanan), rohu, *Labeo rohita* (Hamilton), and common carp, *Cyprinus carpio* L. *Aquaculture and Fisheries Management*, **25**, 659–670.

Nandeesha, M. C., De Silva, S. S. and Krishna Murthy, D. (1994b) Use of mixed feeding schedules in fish culture. II. Performance of common carp on plant and animal based diets. *Aquaculture and Fisheries Management* (in press).

Nellen, W. (1986) Live animal food for larval rearing in aquaculture: non-*Artemia* organisms, in *Realism in Aquaculture: Achievements, Constraints, Perspectives* (eds Bilio, M., Rosenthal, H. and Sindermann, C. J.), European Aquaculture Society, Bredene, Belgium, pp. 215–49.

New, M. B. (1976) A review of dietary studies with shrimps and prawns. *Aquaculture*, **9**, 101–44.

New M B (1987) Feed and Feeding of Fish and Shrimp, *ADCP*/REP/87/26, FAO/UNDP, Rome.

New, M. B. (1991). Compound feeds – a world view. *Fish Farmer*, March/April, 39–44.

New, M. B. and Csavas, I (1993) Aquafeed in Asia – a regional overview, in Farm-Made Aquafeeds (eds M. B. New, A. G. J. Tacon and I. Csavas), FAO-RAPA/AADCP, Bangkok, pp. 1–23.

New, M. B. and Wijkstrom, U. N. (1990) Feed for thought. *World Aquaculture*, **21 (1)**, 17–23.

New, M. B., Csavas, I. and Tacon, A. G. J. (1993) *Farm-Made Aquafeeds*, Proceedings of the FAO/AADCP Regional Expert Consultation on Farm-Made Aquafeeds, 14–18 December 1992, Bangkok, FAO-RAPA/AADCP, Bangkok.

Newman, M. W., Huezo, H. E, and Hugues, D. G. (1979) The response of all male tilapia hybrids to four levels of protein in isocaloric diets. Proceedings of the World Mariculture Society, **10**, 788–92.

Ng, W. K. and Wee, K. L. (1989) The nutritive value of cassava leafmeal in pelleted feed for Nile tilapia. *Aquaculture*, **83**, 43–58.

Nikolski, G. V. (1965) Theory of Population Dynamics as the Background for Rational Exploitation and Management of Fishery Resources, Nauka, Moscow (translation by J. E. S. Bradley, 1968, Oliver & Boyd, Edinburgh).

Nikolskii, G. V. (1969) *The Ecology of Fishes*, Academic Press, London.

Nose, T. and Halver, J. E. (1981) Bioenergetics and Nutrition of Fish. Symposia from XII International Congress of Nutrition, Alan R. Liss, New York, pp. 939–43.

Nose, T. and Toyama, K. (1966) Protein digestibility of brown fishmeal in rainbow trout. *Bulletin of Freshwater Fisheries Research Laboratory*, **15**, 213–24.

NRC (National Research Council) (1981) Nutrient Requirements of Coldwater Fishes, National Academy of Sciences, Washington, DC.

NRC (National Research Council) (1983) Nutrient Requirements of Warmwater Fishes and Shellfishes, National Academy of Sciences, Washington, DC.

Odum, W. (1970) Utilization of the direct grazing and plant detritus food chain by *Mugil cephalus*, in *Marine Food Chains* (ed. J. H. Steele), Oliver & Boyd, London, pp. 222–40.

Ogino, C. (1980) Requirements of carp and rainbow trout for essential amino acids. *Bulletin of the Japanese Society of Scientific Fisheries*, **46**, 171–74.

O'Keefe, T. and Grant, B. F. (1991) Stable forms of vitamin C: essentiality, stability and bioavailibility. *American Soybean Association Technical Bulletin*, **AQ 29**, 1–12.

Omar, E. A. and Günther, K. D. (1987) Studies on feeding of mirror carp (*Cyprinus carpio* L.) in intensive aquaculture. *Journal of Animal Physiology and Animal Nutrition*, **57**, 172–80.

Palmer, T. N. and Ryman, B. E. (1972) Studies on oral glucose intolerance in fish. *Journal of Fish Biology*, **4**, 311–19.

Pandey, H. S. and Singh, R. P. (1980) Protein digestibility by Khosti fish *Colisa fasciatus (Pisces, Anabantidae)* under the influence of certain factors. *Acta Hydrochimica Hydrobiologia*, **8**, 583–85.

Pandian, T. J. (1967) Intake, digestion, absorption and conversion of food in the fishes *Megalops cyprinoides and Ophiocephalus striatus. Marine Biology*, **1**, 16–32.

Pandian, T. J. and Marian, M. P. (1985) Nitrogen content of food as an index of absorption efficiency. *Marine Biology*, **85**, 301–11.

Pantastico, J. B. (1988) Non-conventional feed resources in aquaculture: an overview of work done in the Philippines, in *Finfish Nutrition Research in Asia* (ed. S. S. De Silva), Heinemann Publishers Asia, Singapore, pp. 71–87.

Parazo, M. M. (1990) Effect of dietary protein and energy level on growth, protein utilization and carcass composition of rabbitfish, *Siganus guttatus. Aquaculture*, **86**, 41–49.

Parker, N. C. (1987) Feed conversion indices: controversy or convention. *Progressive Fish Culturist*, **49**, 161–66.

Pelissero, C. and Sumpter, J. P. (1992) Steroid and 'steroid-like' substances in fish diets. *Aquaculture*, **107**, 283–301.

Persson, G. (1991) Eutrophication resulting from salmonid fish culture in fresh and salt waters: Scandinavian experiences, in *Nutritional Strategies and Aquaculture Waste* (eds C. B. Cowey and C. Y. Cho), University of Guelph, Ontario, pp. 163–86.

Pfeffer, E., Beckmann-Toussaint, J., Henrichfriese, B. and Jensen, H. D. (1991) Effect of extrusion on efficiency of utilization of maize starch by rainbow trout, *Oncorhynchus mykiss*. *Aquaculture*, **96**, 293–303.

Phillips, A. M., Tunison, A. V. and Brockway, D. R. (1948) The utilization of carbohydrate by trout, *Fisheries Research Bulletin*, No. 11, Conservation Department Bureau of Fish Culture, New York.

Phillips, M. J., Clarke, R. and Moroat, A. (1993) Phosphorous leaching from Atlantic salmon diets. *Aquacultural Engineering*, **12**, 47–54.

Piedrahita, R. and Tchobanoglous, G. (1987) The use of human wastes and sewage in aquaculture, in *Detrital and Microbial Ecology in Aquaculture* (eds D. J. W. Moriarty and R. S. V. Pullin), ICLARM Conference Proceedings 14, Manila, pp. 336–52.

Pierce, R. J. and Wissing, T. E. (1974) Energy cost of food utilization in the bluegill (*Lepomis macrochirus*). *Transactions of the American Fisheries Society*, **103**, 38–45.

Piggott, G. M. and Tucker, B. W. (1985) Special feeds, in *Fish Nutrition* (ed. J. E. Halver), pp. 653–79, Academic Press, San Diego, USA.

Pillay, T. V. R. (1953) Studies on the food, feeding habits and alimentary tract of the grey mullet, *Mugil tade* Forskal. *Proceedings of National Institute of Science, India*, **19**, 777–827.

Pillay, T. V. R. (1979) The state of aquaculture, in *Advances in Aquaculture* (eds. T. V. R. Pillay and W. A. Dill), Fishing News, Farnham, pp. 1–10.

Pillay, T. V. R. (1990) *Aquaculture: Principles and Practices*, Fishing News Books, Oxford.

Pillay, T. V. R. (1992) Aquaculture and the Environment. *Fishing News Books*, Oxford.

Ping, W. (1993) Aquafeeds and feeding strategies in China, in Farm-Made Feeds (eds M. B. New, A. G. J. Tacon and I. Csavas), FAO-RAPA/AADCP, Bangkok, pp. 201–12.

Piper, R. G. *et al.* (1982) Fish Hatchery Management, US Department of Interior, Fish and Wildlife Service, Washington.

Plisetskaya, E. M. (1989) Physiology of fish endocrine pancreas. *Fish Physiology and Biochemistry*, **7**, 39–48.

Preffer, E., Beckmann-Toussaint, J., Henrichfreise, B. and Jansen, H. D. (1991) Effect of extrusion on efficiency of utilization of maize starch by rainbow trout (*Oncorhynchus mykiss*). *Aquaculture*, **96**, 293–303.

Prejs, A. and Blaszczyk (1977) Relationship between food and cellulase activity in freshwater fishes. *Journal of Fish Biology*, **11**, 447–52.

Pridmore, R. D. and Rutherford, J. C. (1992) Modelling phytoplankton abundance in a small enclosed bay used for salmon farming. *Aquaculture and Fisheries Management*, **23**, 525–42.

Prus, T. (1970) Calorific value of animals as an element of bioenergetical investigations. *Polish Archives Hydrobiologie*, **17**, 183–99.

Pullin, R. S. V. (1988) Tilapia Genetic Resources for Aquaculture. Proceedings of the workshop on Tilapia Genetic Resources for Aquaculture, March 1987, Bangkok. *ICLARM*, Manila.

Pullin, R. S. V and Lowe-McConnell, R. H. (1982). The Biology and Culture of Tilapias, ICLARM Conference Proceedings 7, ICLARM, Manila.

Reis, L. M., Reutebuch, E. M. and Lovell, R. T. (1989) Protein-to-energy ratios in production diets and growth, feed conversion and body composition of channel catfish, *Ictalurus punctatus*. *Aquaculture*, **77**, 21–27.

Reznick, D. (1983) The structure of guppy life histories: the trade-off between growth and reproduction. *Ecology*, **64**, 862–73.

Rimmer, D. W. and Wiebe, W. J. (1987) Fermentative microbial digestion in herbivorous fishes. *Journal of Fish Biology*, **31**, 229–36.

Rottman, R. W., Shrieman, J. V. and Lincoln, E. P. (1991) Comparisons of live foods and two dry diets for intensive culture of grass carp and bighead carp larvae. *Aquaculture*, **96**, 269–80.

Russell, J. (1992) Profile of an intelligent feeder. *Austasia Aquaculture*, **6 (2)**, 41.

Rutledge, W. P. and Rimmer, M. A. (1991) Culture of larval sea bass, *Lates calcarifer* (Bloch), in saltwater rearing ponds in Queensland, Australia. *Asian Fisheries Science*, **4**, 345–55.

Rychly, J. and Spannhof, L. (1979) Nitrogen balance in trout. I. Digestibility of diets containing varying levels of protein and carbohydrates. *Aquaculture*, **16**, 39–46.

Santiago, C. B and Lovell, R. T. (1988) Amino acid requirement for growth of Nile tilapia. Journal of Nutrition, 118, 1540–46.

Santiago, C. B. and Reyes, O. S. (1991) Optimum dietary protein level for growth of bighead carp (*Aristichthys nobilis*) fry in a static water system. *Aquaculture*, **93**, 155–65.

Sargent, J., Henderson, R. J. and Tocher, D. R. (1989) The lipids, in *Fish Nutrition* , 2nd edn (ed. J. E. Halver), Academic Press, London, pp. 154–219.

Sbikin, Y. N. (1974) Age related changes in the role of vision in the feeding of various fishes. *Journal of Ichthyology*, **14**, 133–39.

Scott, D. P. (1962) Effect of food quantity on fecundity of rainbow trout, *Salmo gairdneri*. *Journal of the Fisheries Research Board of Canada*, **19**, 715–31.

Seymour, E. A. (1985) Appetite in eels: a puzzle to unravel. *Fish Farmer*, **8**, 37–38.

Seymour, E. A. and Bergheim, A. (1991) Towards a reduction of pollution from intensive aquaculture with reference to the farming of salmonids in Norway. *Aquaculture Engineering*, **10**, 73–89.

Smith L. S. (1980) Digestion in teleost fishes, in Fish Feed Technology, lectures presented at the FAO/UNDP Training Course in Fish Feed Technology, College of Fisheries, University of Washington, Seattle, 9 October–15 December 1978. FAO, Rome, pp. 3–18.

Smith, R. R. (1981) Energy Metabolism in Fishes. Symposia from XII International Congress of Nutrition. Alan R. Liss, New York, pp. 945–53.

Smith R. R. (1989) Nutritional energetics, in *Fish Nutrition*, 2nd edn (ed. J. E. Halver), Academic Press, San Diego, CA.

Soofiani, N. M. and Hawkins, A. D. (1982) Energetic costs of different levels of feeding in juvenile cod, *Gadus morhua* L. *Journal of Fish Biology*, **21**, 577–92.

Spataru, P., Hepher, B. and Halevy, A. (1980) The effect of the method of supplementary feed application on the feeding habits of carp (*Cyprinus carpio* L.) with regard to natural food. *Hydrobiologia*, **72**, 171–78.

Springate, J. R. C. and Bromage, N. R. (1985) Effect of egg size on early growth and survival in rainbow trout (*Salmo gairdneri* Richardson), *Aquaculture*, **47**, 163–72.

Springate, J. R. C., Bromage, N. R. and Cumaranatunga, P. R. T. (1985) The effect of different rations on fecundity and egg quality in the rainbow trout (*Salmo gairdneri*), in *Nutrition and Feeding in Fish* (eds C. B. Cowey, A. M. Mackie and J. G. Bell), Academic Press, London, pp. 371–93.

Steffansson, S. O., Nortvedt, R., Hansen, J. T. and Taranger, G. L. (1990) First feeding of Atlantic salmon, *Salmo salar* L. under different photoperiods and light intensities. *Aquaculture and Fisheries Management*, **21**, 435–41.

Steffens, W. (1989) *Principles of Fish Nutrition*, Ellis Horwood, London.

Stickney, R. R. and Andrews, J. W. (1972) Effects of dietary lipids on growth, food conversion, lipid and fatty acid composition of channel catfish. *Journal of Nutrition*, **102**, 249–58.

Stickney, R. R. and Shumway, S. E. (1974) Occurrence of cellulase activity in the stomach of fishes. *Journal of Fish Biology*, **6**, 779–90.

Stirling, H. P. and Dey, T. (1990) Impact of intensive cage fish farming on the phytoplankton and periphyton of a Scottish freshwater lake. *Hydrobiologia*, **190**, 193–214.

Stivers, T. E. (1970) Feed manufacturing, in Report of the 1970 Workshop on Fish Feed Technology and Nutrition (ed. J. L. Gudet), FAO/EIFAC and USDI/BSFW, US Government Printing Office, Washington DC, pp. 14–42.

Stob, M. (1983) Naturally occurring food toxicants: estrogens, in Handbook of Naturally Occurring Food Toxicants (ed. M. Rechcigl Jr), CRC Press, Boca Raton, FL, pp. 80–100.

Stroband, H. W. J. and Dabrowski, K. (1981) Morphological and physiological aspects of the digestive system and feeding in freshwater fish larvae, in *Nutrition des Poissons* (ed. M. Fontaine), CNRS, Paris, pp. 355–78.

Sumagaysay, N. S. (1993) Growth, daily ration, and gastric evacuation rates of milkfish (*Chanos chanos*) fed supplemental diet and natural food. *Journal of Applied Ichthyology*, **9**, 65–73.

Sumagaysay, N. S., Chiu-Chern, Y. N., Estillo, V. J. and Sastrillo, M. A. S. (1990) Increasing milkfish (*Chanos chanos*) yields in brackishwater ponds through increased stocking rates and supplementary feeding. *Asian Fisheries Science*, **3**, 251–56.

Sumagaysay, N. S., Marquez, F. E. and Chiu-Chern, Y. N. (1991) Evaluation of different supplemental feeds for milkfish (*Chanos chanos*) reared in brackishwater ponds. *Aquaculture*, **93**, 177–89.

Sumpter, J. P., Le Bail, P. Y., Pickering, A. D., *et al.* (1991) The effect of starvation on growth and plasma growth hormone concentrations of rainbow trout, *Oncorhynchus mykiss. General and Comparative Endocrinology*, **83**, 94–102.

Sweeting, R. M. and McKeown, B. A. (1986) Somatostatin reduces plasma growth hormone concentrations of rainbow trout, *Oncorhynchus mykiss. General and Comparative Endocrinology*, **64**, 2062–63.

Tacon A. G. J. (1987a) The Nutrition and Feeding of Farmed Fish and Shrimp – A Training Manual, Vol. 1. The Essential Nutrients, GCP/075/ITA, FAO/UNDP, Rome.

Tacon, A. G. J. (1987b) The Nutrition and Feeding of Farmed Fish and Shrimp – A Training Manual, Vol. 2. Nutrient Sources and Composition, GCP/RLA/075/ITA, FAO/UNDP, Rome.

Tacon, A. G. J. (1988) The Nutrition and Feeding of Farmed Fish and Shrimp – A Training Manual, Vol. 3, Feeding Methods, GCP/PLA/075/ITA, Field Document 7, FAO, Rome.

Tacon A. G. J. and De Silva S. S. (1983) Mineral composition of some commercial feeds available in Europe, *Aquaculture*, **31**, 11–20.

Tacon, A. G. J. and Ferns, P. N. (1979) Activated sewage sludge – a potential animal foodstuff. I. Proximate and mineral content: seasonal variation. *Agriculture and Environment*, **4**, 257–69.

Tacon, A. G. J. and Jackson, A. (1985) Utilization of conventional and non-conventional protein sources in practical feeds, in *Nutrition and Feeding in Fish* (eds C. B. Cowey, A. M. Mackie and J. G. Bell), Academic Press, London, pp. 119–45.

Tacon, A. G. J. and Rodriguez, A. M. P. (1984) Comparison of chromic oxide, crude fibre, polythylene and acid-insoluble ash as dietary markers for the estimation of apparent digestibility coefficient in rainbow trout. *Aquaculture*, **43**, 391–99.

Tacon, A. G. J., Haaster, J. V., Featherstone, P. B. *et al.* (1983a) Studies on the utilization of full-fat soybean and solvent extracted soybean meal in a complete diet for rainbow trout. *Bulletin of the Japanese Society for Scientific Fisheries*, **49**, 1437–43.

Tacon, A. G. J., Stafford, E. A. and Edwards, C. A. (1983b) A preliminary investigation of the nutritive value of three terrestrial worms for rainbow trout. *Aquaculture*, **35**, 187–99.

Takeuchi, T., Watanabe, T. and Ogino, C. (1978a) Supplementary effect of lipids in a high protein diet of rainbow trout. *Bulletin of Japanese Society of Scientific Fisheries*, **44**, 677–81.

Takeuchi, T., Watanabe, T. and Ogino, C. (1978b) Optimum ratio of protein to lipid in diets of rainbow trout. *Bulletin of Japanese Society of Scientific Fisheries*, **44**, 683–88.

Takeuchi, T., Watanabe, T. and Ogino, C. (1979) Availability of carbohydrate and lipid as dietary energy source for carp. *Bulletin of Japanese Society of Scientific Fisheries*, **45**, 977–82.

Takeuchi, T., Watanabe, T., Ogino, C. *et al.* (1981) Effects of low protein–high calorie diets and deletion of trace elements from a fish meal diet on reproduction of rainbow trout. *Bulletin of the Japanese Society of Scientific Fisheries*, **47**, 645–54.

Thorpe, J. E. and Wankowski, J. W. J. (1979) Feed presentation and food particle size for juvenile Atlantic salmon, *Salmo salar* L, in *Finfish Nutrition and Fish Feed Technology*, Vol. 1 (ed. J. E. Halver and K. Tiews), H. Heenermann, Berlin, pp. 501–13.

Tidwell, J. H., Webster, D. D. and Knaub, R. S. (1991) Seasonal production of rainbow trout, *Oncorhynchus mykiss*, in ponds using different feeding practices. *Aquaculture and Fisheries Management*, **22**, 335–41.

Urban, E. R. and Pruder, G. D. (1991) A method of economic comparisons for aquaculture diet development. *Aquaculture*, **99**, 127–42.

Van Dyke, J. M. and Sutton, D. L. (1977) Digestion of duckweek (*Lemna sp.*) by grass carp (*Ctenopharyngodon idella*). *Journal of Fish Biology*, **11**, 273–78.

Van Ihering, R. (1937) A method for inducing fish to spawn. *Progressive Fish Culturist*, **34**, 15–16.

Van Keulen, J. and Young, B. A. (1977) Evaluation of acid-insoluble ash as a natural marker in ruminant digestibility studies. *Journal of Animal Science*, **44**, 282–87.

Van Marrewijk, W. J. A. and Zandee, D. I. (1975) Amino acid metabolism of Astacus leptodactylus (Esch). II: biosynthesis of the non-essential amino acids. *Comparative Biochemistry and Physiology*, **50B**, 449–55.

Verrina, S. S., Nandeesha, M. C. and De Silva, S. S. (a, in prep). Analysis of input–output interrelationship in the Indian major carp farming system in Andhra Pradesh, India.

Verrina, S. S., Nandeesha, M. C., Rao Gopal, K. and De Silva, S. S. (b, in prep). Status and Technology of Indian Major Carp Farming in Andhra Pradesh, India. Asian Fisheries Society, Indian Branch, Special Publication 9, Mangalore.

Viola, S., Lahav, E. and Angeoni, H. (1992) Reduction of feed protein levels and of nitrogenous N-excretions by lysine supplementation in intensive carp culture. *Aquatic Living Resources*, **5**, 277–85.

Voet, D. and Voet, J. G. (1990) *Biochemistry*, Wiley, New York.

Von Gongnet et al (1987) The influence of different protein/energy rations and increasing feeding level on nitrogen excretion in growing mirror carp (*Cyprinus carpio L.*). *Journal of Animal Physiology and Animal Nutrition*, **58**, 173–88.

Walford, J. and Lam, T. J. (1993) Development of digestive tract and proteolytic enzyme activity in seabass (*Lates calcarifer*) larvae and juveniles. *Aquaculture*, **109**, 187–205.

Walton, M. J., Cowey, C. B. and Adron J. W. (1982) Methionine metabolism in rainbow trout fed diets of differing methionine and cystine contents. *Journal of Nutrition*, **112**, 1525–35.

Wannigama, N. D., Weerakoon, D. E. M. and Muthukumarana, G. (1985) Cage culture of S. *niloticus* in Sri Lanka: effect of stocking density and dietary crude protein levels on growth, in *Finfish Nutrition in Asia. Methodological Approaches to Research and Development* (eds. C. Y. Cho, C. B. Cowey and T. Watanabe), International Development Research Centre (Canada), Ontario, pp. 113–17.

Watanabe, T. (1991) Past and present approaches to aquaculture waste management in Japan, in *Nutritional Strategies of Aquaculture Waste* (eds C. B. Cowey and C. Y. Cho), University of Guelph, Ontario, pp. 137–54.

Watanabe, T., Kitajima, C. and Fugita, S. (1983) Nutritional value of live organisms used in Japan for mass propagation of fish: a review. *Aquaculture*, **34**, 115–43.

Watanabe, T., Itoh, A., Kitajima, C. and Fujita, S. (1984a) Effect of dietary protein levels on reproduction of red sea bream. *Bulletin of the Japanese Society of Scientific Fisheries*, **50**, 1015–22.

Watanabe, T., Arakawa, T., Kitajima, C. and Fujita, S. (1984b) Effect of nutritional quality of broodstock diets on reproduction of red sea bream. *Bulletin of the Japanese Society of Scientific Fisheries*, **50**, 495–501.

Watanabe, T., Ohashi, S., Itoh, A., Kitajima, C. and Fujita, S. (1984c) Effect of nutritional composition of diets on chemical components of red sea bream broodstock and eggs produced. *Bulletin of the Japanese Society of Scientific Fisheries*, **50**, 503–15.

Watanabe, T., Takeuchi, T., Saito, M. and Nishimura, K. (1984d) Effect of low protein–high calorie or essential fatty acid deficiency diet on reproduction of rainbow trout. *Bulletin of the Japanese Society of Scientific Fisheries*, **50**, 1207–15.

Watanabe, T., Itoh, A., Murakami, A., Tsukashima, Y., Kitajima, C. and Fujita, S. (1984e) Effect of nutritional quality of diets given to broodstock on the verge of spawning on reproduction of red sea bream. *Bulletin of the Japanese Society of Scientific Fisheries*, **50**, 1023–28.

Watanabe, T., Itoh, A., Satoh, S., Kitajima, C. and Fujita, S. (1985a) Effect of dietary protein levels and feeding period before spawning on chemical component of eggs produced by red sea bream broodstock. *Bulletin of the Japanese Society of Scientific Fisheries*, **51**, 1501–509.

Watanabe, T., Koizumi, T., Suzuki, H., Satoh, S., Takeuchi, T., Yoshida, N., Kitada, T. and Tsukashima, Y. (1985b) Improvement of quality of red sea bream eggs by feeding broodstock on a diet containing cuttlefish meal or on raw krill shortly before spawning. *Bulletin of the Japanese Society of Scientific Fisheries*, **51**, 1511–21.

Watanabe, T., Satoh, S. and Takeuchi, T. (1988) Availability of minerals in fishmeal to fish. *Asian Fisheries Science*, **1**, 175–95.

Watanabe, T., Davy, F. B. and Nose, T. (1990) Aquaculture in Japan, in *The Current Status of Fish Nutrition in Aquaculture* (eds M. Takeda and T. Watanabe), Japan Translation, Tokyo, pp. 115–30.

Watanabe, W. O. (1975) Identification of the essential amino acids of the freshwater prawn, *Macrobrachium rosenbergii*. University of Hawaii, MSc thesis.

Watanabe, Y. (1984) Morphological and functional changes in rectal epithelial cells of pond smelt during post-embryonic development. *Bulletin Japanese Society of Scientific Fisheries*, **50**, 805–14.

Weatherly, A. H. (1963) Notions of niche and competition among animals with special reference to freshwater fish. *Nature*, **197**, 14–17.

Wee, K. L. (1989) Pond experimental methodology, in Fish Nutrition Research in Asia (ed. S. S. De Silva), Asian Fisheries Society Special Publication No. 4, Asian Fisheries Society, Manila, pp. 68–79.

Wee, K. L. (1991) Use of non-conventional feed stuff of plant origin as fish feeds – is it practical and economically feasible? in *Fish Nutrition Research in Asia* (ed. S. S. De Silva), Asian Fisheries Society Special Publication No. 5, Asian Fisheries Society, Manila.

Wee, K. L. and Ngamsnae, P. (1987) Dietary protein requirement of fingerlings of the herbivorous carp tawes, *Puntius gonionotus* (Bleeker). *Aquaculture and Fisheries Management*, **18**, 121–29.

Wee, K. L. and Wang, S. S. (1987) Nutritive value of *Lucaena* leafmeal in pelleted feed for Nile filapia. *Aquaculture*, **62**, 97–108.

Werner, E. E. and Hall, D. J. (1976) Niche shifts in sunfishes, experimental evidence and significance. *Science*, **191**, 404–406.

White, F. (1991) Use of mechanical feeders in aquaculture. *Austasia Aquaculture*, **5**, 22–24.

Wijkstrom, U. N. and New, M. B. (1989) Fish for feed: a help or a hindrance to aquaculture 2000? *INFOFISH International*, **6** , 48–52.

Wilson, R. P. (1989) Amino acid and proteins, in Fish Nutrition, 2nd edn (ed. J. E. Halver), Academic Press, San Diego, pp. 111-51.

Wilson, R. P. (ed.) (1991) *Handbook of Nutrient Requirements of Finfish*, CRC Press, Boca Raton, FL.

Wilson, R. P. and Halver, J. E. (1986) Protein and amino acid requirements of fishes. Annual Review of Nutrition, 6, 225–44.

Wilson, R. P., Poe, W. E. and Robinson, E. H. (1980) Leucine, isoleucine, valine and histidine requirements of fingerling channel catfish. *Journal of Nutrition*, **110**, 627–33.

Wilson, R. P., Poe, W. E., Nemetz, T. G. and MacMillan, J. R. (1988) Effect of recombinant bovine growth hormone administration on growth and body composition of channel catfish. *Aquaculture*, **73**, 229–36.

Windell, J. T. (1967) Rate of digestion in fishes, in *The Biological Basis of Freshwater Fish Production* (ed. S. D. Gerking), Blackwell Scientific Publications, Oxford, pp. 151–73.

Windell, J. T., Foltz, J. W. and Sarokon, J. A. (1978) Effect of fish size, temperature and amount fed on nutrient digestibility of a pelleted diet by rainbow trout, *Salmo gairdneri. Transactions of the American Fisheries Society*, **107**, 613–616.

Windsor, M. and Barlow, S. (1981) Introduction to Fishery By-products, Fishing News Books, Farnham.

Wohlfarth, G. W. and Schroeder, G. L. (1979) Use of manure in fish farming – review. Agricultural Wastes , 1, 279–99.

Wohlfarth, G. W. and Hulata, G. (1987) Use of manures in aquaculture, in *Detritus and Microbial Ecology in Aquaculture* (eds D. J. W. Moriarty and R. S. V. Pullin), ICLARM Conference Proceedings, Vol. 14, August 1985, Bellagio, Italy. ICLARM, Manila, pp. 353–67.

Xie, X. J. and Sun, R. (1993) Pattern of energy allocation in the southern catfish (*Silurus meridionalis*). *Journal of Fish Biology*, **42**, 197–207.

Yang, H. Z. (1985) The Major Chinese Integrated Fish Farming Systems and their Effects, Freshwater Fisheries Research Centre, Wuxi, China.

Yone, Y. (1976) Nutrition studies in red sea bream, in Proceedings of the First International Conference on Aquculture Nutrition, October 1975, Lewes/Rehoboth, DE.

Zanuy, S. and Carrillo, M. (1985) Annual cycles of growth, feeding rate, gross conversion efficiency and hematocrit levels of sea bass (*Dicentrarchus labrax L.*) adapted to two different osmotic media. *Aquaculture*, **44**, 11–25.

Zeitoun, I. H., Tack, P. I., Halver, J. E. and Ullrey, D. E. (1973) Influence of salinity on protein requirements of rainbow trout (*Salmo gairdneri*) fingerlings. *Journal of the Fisheries Research Board of Canada*, **30**, 1867–73.

Zeitoun, I. H., Ullrey, D. E. and Magee, W. T. (1976) Quantifying nutrient requirements of fish. *Journal of the Fisheries Research Board of Canada*, **33**, 167–72.

Zuercher, P. (1987) Training personnel to cope in hot climates. *Feed International*, **8**, 38–42.

Index